Marketing and Selling Construction Services

Hedley Smyth BA (Hons),

Director, Centre for Construction [illegible]ing in the Centre for Construction Management, Oxford Brookes University

Blackwell
Science

Editorial Offices:
Osney Mead, Oxford OX2 0EL
25 John Street, London WC1N 2BL
23 Ainslie Place, Edinburgh EH3 6AJ
350 Main Street, Malden
MA 02148 5018, USA
54 University Street, Carlton
Victoria 3053, Australia
10, rue Casimir Delavigne
75006 Paris, France

Other Editorial Offices:

Blackwell Wissenschafts-Verlag GmbH
Kurfürstendamm 57
10707 Berlin, Germany

Blackwell Science KK
MG Kodenmacho Building
7-10 Kodenmacho Nihombashi
Chuo-ku, Tokyo 104, Japan

First published 2000

Set in Palatino and produced by Gray Publishing, Tunbridge Wells, Kent
Printed and bound in Great Britain
by MPG Books Ltd, Bodmin, Cornwall

The Blackwell Science logo is a trade mark of Blackwell Science Ltd, registered at the United Kingdom Trade Marks Registry

DISTRIBUTORS

Marston Book Services Ltd
PO Box 269
Abingdon
Oxon OX14 4YN
(*Orders*: Tel: 01235 465500
Fax: 01235 465555)

USA
Blackwell Science, Inc.
Commerce Place
350 Main Street
Malden, MA 02148-5018
(*Orders*: Tel: 800 759 6102
781 388 8250
Fax: 781 388 8255)

Canada
Login Brothers Book Company
324 Saulteaux Crescent
Winnipeg, Manitoba R3J 3T2
(*Orders*: Tel: 204 837-2987
Fax: 204 837-3116)

Australia
Blackwell Science Pty Ltd
54 University Street
Carlton, Victoria 3053
(*Orders*: Tel: 03 9347 0300
Fax: 03 9347 5001)

A catalogue record for this title is available from the British Library

ISBN 0-632-04987-1

Library of Congress Cataloging-in-Publication Data:
Hedley, Smyth.
Marketing and selling construction services/
Hedley Smyth
p. cm.
Includes bibliographical references (p.) and index.
ISBN 0-632-04987-1
1. Construction industry–Great Britain–Marketing. 2. Construction industry–Europe–Marketing case studies. 3. Construction industry–Customer services–Great Britain–Marketing. 4. Construction industry–Customer services–Europe–Marketing case studies. I. Title
HD9715.G72H3 1999
624′.068–dc21 99-37765
CIP

For further information on
Blackwell Science, visit our website:
www.blackwell-science.com

Contents

I dedicate this book to
Nicola, Claire and Phoebe with love

Preface

This book is about *marketing*. It is also a book about *selling*. Yet, more than anything, it is a book that *integrates* the two. Selling is handled from the standpoint of how it is informed by marketing. Marketing is handled in terms of how selling has a great deal to contribute to the development of effective strategy.

It is a book for senior management, those undertaking MBAs and other Master's programmes and those in the academic community, as well as those aspiring to become senior managers and students studying for their first degree. It is also a book for senior members of project teams. And, of course, it is for all those working in marketing and sales.

The seeds for this book have been sown over the past 12 years. Time and time again my co-marketers in contracting and consultants would say that there were few marketing degree and training courses that were really relevant to the construction sector. After all, we all knew that construction was 'special and different'. We have clients, not customers; our business is long-term and lumpy, with every project being a prototype for the production line, which never comes! I also remember a particular example: a marketing conversation with a director of a major contractor who asked me where I went to obtain information and what inspired me regarding construction marketing.

Fortunately things are somewhat better today. Frankly, contractors and consultants have become more professional and adept at marketing, or at least some have done. We are more educated in marketing than we once were. Marketers, business development managers or sales people have begun to learn lessons from other sectors, which share some similar characteristics: defence contractors, the hotel trade, aircraft and ship assembly, the legal and accountancy professions for example. Indeed, one international contractor recently introduced a policy called 'Customer First' and a national UK contractor has been a pioneer of long-term client auditing. Research in the area has been growing internationally and training has improved substantially, for example from the universities such as Leeds, Strathclyde and the Centre for Construction Marketing at Oxford Brookes University in the UK.

This development of sales and marketing activity in the construction industry is still in its early stages and there is still a long way to go. Many people wish to take the next steps, finding that the main constraints are within their own organisations. Attitudes take a while to change – a generation in some organisations. Investment is always hard to lever out of a cash-generating sector that is dependent on keeping overheads low, especially when surviving recessions.

This book is about taking marketing to the next stage both strategically and practically. It is visionary, being *primarily* for senior managers and those in practice:

- ❑ Managing directors
- ❑ Divisional heads
- ❑ Senior members of the project team
- ❑ Marketing personnel
- ❑ Sales personnel.

I use the words 'sales' and 'selling' where this is accurate and make no attempt to dress them up into marketing or 'business development', for sales is a credible and important function in its own right.

The *secondary* market is the academic community and consultants. There are important conceptual and practical issues addressed of relevance to consultants in construction, who share many of the same experiences and concerns as contractors. It is not about how consultants can obtain work from contractors, although there are implied lessons to be derived from reading the book. It is highly relevant in the consultant–client relationship. Academics, especially Master's students and lecturers, will find a great deal of useful information that is pioneering for construction. Researchers, whose area embraces procurement, client satisfaction and quality, as well as marketing, should also find this an important text.

The book draws on 45 case studies in the text and uses examples that range from the mediocre through to 'best practice'. The case studies are almost exclusively taken from the construction sector and include:

- ❑ Over 45 organisations, half of which are international in operation
- ❑ Organisations with their origins from over 10 different countries

- ❑ Case studies, which are presented anonymously for several reasons, including confidentiality; however, the cases are draw from many well-known organisations.

Marketing and Selling Construction Services is really only the springboard for what comes next. If the book manages to turn a few heads, challenge and help to change a number of practices, then it has done its job. If it is used by senior management and board members to steer their future policies, organisational structures and procedures, so much the better. Construction surely needs a continuing injection of imagination to successfully develop into the first decade of 2000. With that in mind, it is with some sadness that the recession of the early 1990s was not used as an opportunity to restructure into *client-orientated* organisations. It was only as the savage cuts had made their deepest scars that contractors and consultants again began to address what form of organisations might best suit serving the changing nature of their clients.

Perhaps the construction sector is not quite so special and different as we once thought. Surely we need to create imaginative, coherent marketing policies as much as any other sector? We need a consistent sales effort that does not follow the same old well-trodden path, but strikes out with a spirit of excitement and adventure. I hope that this book will form part of that adventure, and that it will push back the boundaries of our current experience. The policies and practices suggested may act as the stimulus for some of the templates for future marketing and selling of construction activities.

The construction industry became aware of the need for more professional marketing and sales efforts in line with the other industries during the 1980s. The recession of the 1990s has added further impetus in construction as organisations tried to climb towards greater prosperity once again. Earlier books have helped to generate interest and inform practice. The rate of change continues to accelerate. Management must keep pace with the major shifts in organisational theory and thinking. Business and sales strategy must keep pace with changes in the market. Management can grasp the market dynamics, help to influence the way in which the market unfolds and begin to create their own clients as they form their marketing policies in the image of those they serve.

As such, *Marketing and Selling Construction Services* builds on the existing, if still somewhat slender, literature on marketing and sales of construction and consultant services. It is a springboard for the future. Coupled with the envisioning, there is also a practical edge to the book. Achieving a balance between strategic oversight and the practical outworking of marketing involves providing opportunities to reflect on our day-to-day activities, learning from our mistakes and shortcomings. More than this, it is necessary to implement dynamic and evolving sales efforts, sales systems which respond to client needs, add real service value for the client and deliver an actual *competitive advantage* for the consultant or contractor. Thus, the book does not provide a single blueprint for marketing and selling. It offers distinctive options, from which the reader may make informed choices. Linked with the pioneering and strategic content, there is plenty of scope for contractor and consultant differentiation. Contractors and consultants cannot simply continue to follow the same sales and marketing practices – that is a zero-sum game. Everyone has faced the competition at its most intensive, driving down prices and enhancing the adversarial environment. Clients are diversifying in organisational structure and managerial styles. The market will force the changes on the construction industry. The winners will be those who embrace change, encourage further change and carve out the territory on their own terms.

I have tried to combine the traditional approach to writing a construction text with those that are more familiar on the business management bookshelves. This provides variety, allowing the reader to dip into the material as well as taking an in-depth, comprehensive look at the content. The material is presented in a traditionally written format, interspersed with highlighted information:

- ❑ *Themes* – setting out the main chapter themes and keywords
- ❑ ... *On the case* – presents case study material, mainly from within the construction and consultant business
- ❑ *Summary* – setting out the points from the chapter.

These will help the reader to identify at a glance some of the main issues and examples of marketing and selling construction services.

Hedley Smyth
Director of the Centre for Construction Marketing
in the Centre for Construction Management
Oxford Brookes University

Acknowledgements

Most of the case study material has been collected over a long period from a variety of sources. I would particularly like to thank all those organisations that have provided additional case study material for this book. For reasons of market competition and confidentiality, not every organisation could be named. Actually, it is not always helpful to do so as we can become distracted by their situation, rather than relating the issues and lessons to our own. To all those who have contributed I wish to say 'thank you', and I list below those who have given their consent individually or by company:

- Daryl Atkinson
- Murray Bean
- Miranda Bellord
- Keith Carey
- Tom Carroll
- Chris Cole
- Martin Fusco
- Patrick Hammill
- Tony Harris
- Geoff Reynolds
- Tom Smith
- Michael Warr
- Gordon Wright

- John R Harris Architects
- Geoffrey Reid Associates
- Laing Partnership Housing
- Levitt Bernstein Associates
- Ridge & Partners
- Tarmac Building
- WSP Group plc.

I would also like to thank those who have, no doubt unwittingly, contributed a fundamental idea or element to this book, by causing me to prick up my ears and take note, set me on a train of thought or taken me down a practical avenue in commercial life. They include the late Alan Adams, formerly of Pearce Construction; many of my colleagues in the School of Architecture at Oxford Brookes University; Tom Davies, School of Policy Studies, Bristol University; Stephen Drewer, University of the West of England; Graham Ive and Graham Winch from the Bartlett Graduate School, University College London; former colleagues at Geoffrey Reid Associates and Fred Wellings from the time when he was with Laing & Cruickshank prior to the 'Big Bang'.

Purpose of the Book

Themes

1 The *aims* of this book are to:
- ❑ Pioneer the further development of marketing and sales for management and marketing personnel
- ❑ Show what can be achieved in construction marketing and sales.

2 The *objectives* of this book are to:
- ❑ Demonstrate dynamic marketing strategies, enhancing job satisfaction, client satisfaction, and hence turnover and profit
- ❑ Develop flexibility in relation to the market
- ❑ Ensure that all the competing options for marketing and selling are explored and systematically implemented
- ❑ Creatively manage clients
- ❑ Identify and address marketing and sales paradoxes.

3 The *outcomes* of this book are to demonstrate that creating competitive advantage recognises that:
- ❑ Management approach is the main means of differentiating a service
- ❑ People are the main asset when working to a marketing strategy and within a marketing management system.

Keywords

Corporate culture, Management, Marketing mix, Relationship marketing and implementation

Introduction

Marketing and Selling Construction Services is written to help you to pioneer the development of professional marketing and sales practice. The book examines marketing and selling as practised in construction. Contractors are the prime focus, yet much is of relevance to the professions. Its overall aim is to ask the question, 'Where do we go from here?' The objective is to sketch out new ways forward; identify opportunities and equip, whether at board level, in managing marketing or at the coal face of sales; explore new ways of being competitive and creating profits.

Where do board members learn their strategic marketing skills? Where do marketing and sales personnel learn the tools of the trade? Mainstream marketing has largely ignored construction. Construction may not have the glamour and status of other sectors, yet it constitutes around 10% of gross national product in times of growth throughout most economies. Contracting and the consultant professions have ignored marketing until quite recently. Even now, there is a strong argument to say that the emphasis is on implementation rather than strategy – on sales rather than enhancing sales with strategic marketing.

Marketing and Selling Construction Services is a practical text. This book addresses the gaps, tailoring mainstream marketing into ideas and applications for contractors and consultants to use. The book develops current practice in far-reaching ways, invests in the future, and dynamically creates opportunities for competitive advantage.

The marketing tools developed to serve mass consumer markets in the USA 30–40 years ago are difficult to translate into business-to-business contexts. This *marketing mix* approach – a recipe of four ingredients blended into the offer made to customers – has limited application, especially in construction. The more recent *relationship marketing* approach – building close relationships over a long period, in particular between businesses – is not easily transferable either. How can marketing be created to suit the diverse and unique combination of characteristics of the construction sector? I hope that this book will show you the way. Indeed, there are many ways in which to create a unique place in the market. This book is designed to be pioneering and therefore part of your inspiration. I say:

> *Copycats and zero-sum games are out; differentiation for competitive advantage is in! That is the only road to survival in an increasingly competitive environment.*

It is time to push back the boundaries. Success can be the result of daring to be different and committing ourselves to that. It is risky, yet staying where we are is no longer a low-risk option.

Poor image

Only the oil industry is viewed less favourably than construction. A recent survey showed that 58% of 15–17-year-olds think working in construction means being a 'builder', many believing that this is about trades such as bricklaying or being an electrician.[1] This is what managers portray to the public and to those who could be potential employees of the industry, key to the future health of the industry. Construction undersells itself. Managing the complexity of the multi-billion Hong Kong dollar infrastructure job at Chek Lap Kok airport in China's Special Administrative Region of Hong Kong is no mean feat in terms of size.[2] Appreciating the complexity of the British–Chinese–Japanese joint venture for constructing the passenger terminal, the largest airport contract worth over HK$10bn (approximately US$1.3bn and £0.85bn), is only one dimension. The consortium, comprising the China State Construction Engineering Corporation, Amec, Balfour Beatty, Kumagai Gumi and the Maeda Corporation, had not worked together before. And then there were the design teams, works package contractors and subcontractors. There is no management textbook on how to organise this. Companies in other sectors have been advocating returning to core business and outsourcing the remainder, while construction has been subcontracting for years. Major corporations are inventing project teams, yet contractors and consultants have been working on projects in 'virtual organisations' throughout the century. Construction may have low status, yet a good proportion of construction work has been ground breaking. There is a vast range of management experience between implementing large construction projects and the local builder.

Many companies and practices will acknowledge that they do not use the full palette of marketing and sales tools. The purpose of this book is to try and help you to change the image and the substance of what you deliver in the eyes of the client. Previous texts have concentrated upon marketing the small- to medium-sized company,[3] or they have focused upon practical guidance lifted from elsewhere and the employment of a mechanistic approach to exploring the concepts of marketing.[4] This was foundational stuff. There are also numerous standard textbooks on mainstream marketing. This book draws in threads from marketing theory, weaving them into methods that are tailored to the specific characteristics of construction. It is time to embrace marketing and to generate a level of sophistication that matches the complexity of the work and other areas of management. It is time to scale the walls of institutionalised approaches to marketing, the traditional thinking on selling construction and consultant services – to be pioneers and investors in marketing.

Background

The management aim

As you meet a potential client for the first time, what are you selling to that organisation? Is it track record, the size of the company, its reputation, financial stability, quality of service, all of these and more? In essence you are selling *confidence*. On what is your confidence founded? An increasing trend over recent years has been for expert commentators to say that management is the main source of competitive advantage today. Marketing has been at the sharp end of that. That may be true for a standard product, where managing the delivery and after-sales service is the only variable. It is also true for construction because the design and engineering is not usually under the control of the contractor, unless it is a design and build or BOOT contract. The quality of executing the design is the only physical aspect of the construction product over which a contractor has control. In fact, consultants spend more time managing the development process than designing their product, the structure.

A survey of construction, defence, IT, consultant and utility contractors managing large projects found that the most critical activity in winning a major project is developing person-to-person relationships with potential clients.[5] The recognition that management starts with selling means that a consistent approach is necessary in order to be competitive, not only in tendering and project management, but also from the outset of the sales process.

What should the manager at board level, partner of a consultant organisation and personnel in marketing and sales be aiming to do? The aim is to *manage marketing*. Does business work like that in practice? The truth is usually not. We can divide what people do into familiar categories of:

- ❑ Important things to do
- ❑ Urgent things to do.

We recognise that there is a tension between them. The telephone always seemed urgent until voice mail came along, even if the call was not very important. Important long-term decisions can be put off until tomorrow. Selling is always urgent, but many of the enquiries and leads turn out to be futile, which would have become abundantly clear had we stood back and collected more information. Strategic marketing is something for improving in the future, but not today if we are honest (Figure 0.1 illustrates this).

Ideally sales and marketing should be seamless. They are part of the same picture, linked through management. This is shown in Figure 0.2, where marketing management bridges the gap between sales and marketing, bonding the urgency and importance through that role.

Managing marketing is a theme that will run throughout the book. There are two dimensions to this theme:

- ❑ Develop consistent management at a corporate level
- ❑ Manage marketing to create differentiated services for selling.

These are not in contradiction. Overall consistency is needed in order to coherently mark out and create a service offering that is both strong and distinctive in comparison to competitors. In the words of Theodore Levitt:[6]

... the purpose of a business is to create and keep a customer.

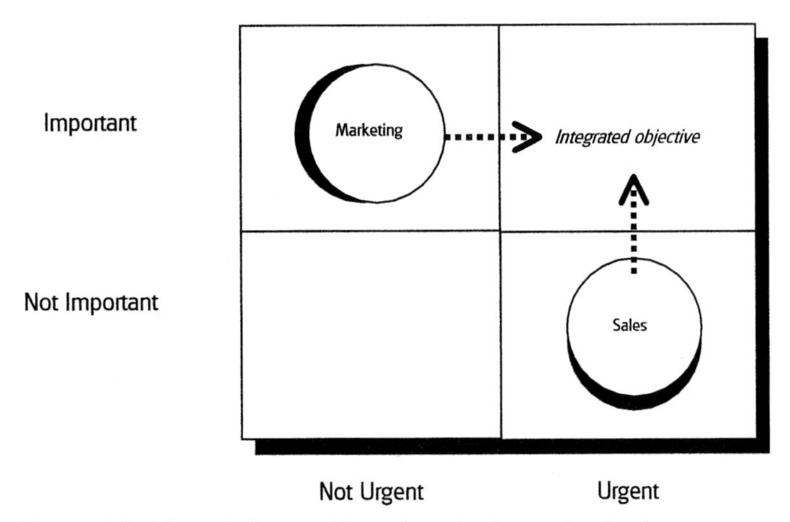

Figure 0.1 The relative position of marketing and sales in corporate actions.

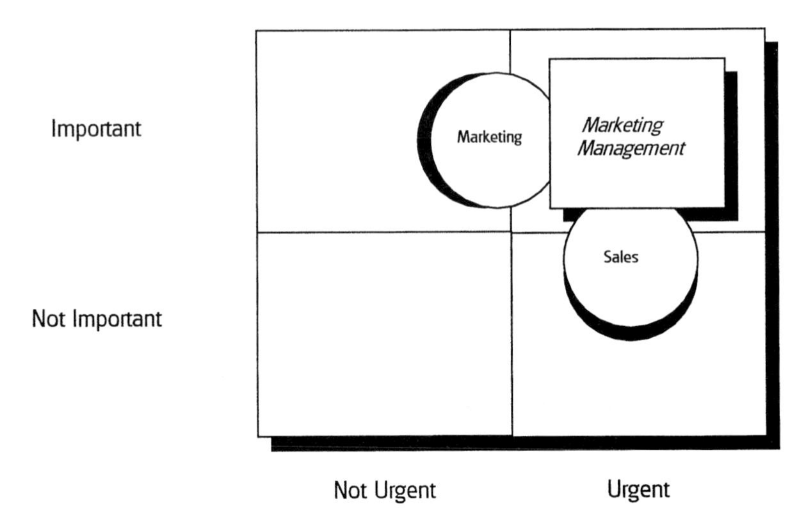

Figure 0.2 The ideal attitude shift for marketing and sales in corporate actions.

Definitions of marketing are multitudinous, Levitt's being the one I will use for this book as the definition focuses upon the desired outcome.

Management in practical terms is all about people. People make the difference. People are the differentiating variable in construction and consultant services. Other traditional marketing variables – place of delivery and procurement route being determined by the client and 'product design' being undertaken by the design team – are largely outside the control of the contractor. How are the skills, creativity and energies of people harnessed for the best endeavours of marketing and for selling? This is the key question. The corporate management of people has not been an area of excellence, in construction in particular, nor has it been among consultants. During training courses I have frequently alluded to the construction sector as 'the most conservative industry run by anarchists'. There may be good reasons why industry management attitudes have not fundamentally changed very much,[7] but these are not overwhelming. This book gives good reasons for change – to create a pleasant place of work, a client-orientated service, improved quality and relevance of what the client receives, high client satisfaction and higher levels of repeat business. Ultimately, this improves profitability.

It is people who add value and who create competitive prices. It is your people who *serve* those in the client organisation. From the client perspective, this is not just a matter of clients making choices as to which consultant or contractor gets the job; it has consequences for the *size of the market*. There are trade-offs in capital investment as to which are the more profitable or efficient, which will yield least hassle, which will increase the political standing of those making key budget decisions within their organisations. Construction probably does itself few favours in this respect. It does little to help *grow* the market.

Management changes

If management changes have been few, structures have changed dramatically in recent years. There have been a series of management changes that have affected all sectors and construction has been forced to respond. Two factors have been dominant:

- ❑ Mechanisation of office work through the introduction of information technology
- ❑ Recession of the first half of the 1990s.

The way in which organisations change have been fivefold:

- ❑ *Decentralisation of management* and financial accountability
- ❑ Increased overall *centralisation of policy and financial control*
- ❑ *Delayering* of the management hierarchy
- ❑ Encouragement and promotion of *generalists* within the organisation
- ❑ *Empowerment* of staff to accept delegation with responsibility for decision-making and central support.

Organisations that were mechanistic or autocratic have found it more difficult to adjust and to embrace the changing business environment. Many consultants and contractors fall into this category. Successful adjustment demands a more flexible approach. Ironically, it requires greater levels of team working, frequently across functional departments. One example would be where the sales effort to a potential client during the courting and prequalification stages has led to offers of a high quality specification design-and-build project, a fact that is not conveyed to the estimating department during pricing of the negotiated contract. Another example concerns sales promises about team working and harmonious working, which is expressed in practice in a defensively negotiated construction contract by the legal department or legal representative. Such cross-functional co-operation is not merely a marketing question. It is an issue for total quality management, value management, training and human resource development. It is the ability to work effectively in this co-ordinated way that can yield the competitive edge in terms of both profitability and client satisfaction. According to the electronic banking pioneer, First Direct:[8]

> *The only real competitive advantage a business can achieve is the ability to effect more change more quickly than its rivals, and an ability to get more out of its people.*

This is a significant comment from an organisation working at arm's length from its customers. It underscores its importance in construction where face-to-face contact is high over long periods. This should be set alongside the typical problems identified in construction. Construction project teams have been mechanistic in nature, failing to respond both flexibly and adequately.[9] Table 0.1 is taken from a survey in North America. It shows that effective management is inextricably linked to communication and that communication is identified as a major problem for most people in the workplace: management and communication issues represent 75% of all problems from this sample of 2855 respondents.

The problem of communication is an international one, occurring both within contracting organisations and between the organisation and the client, because failure to address issues adequately internally must filter through to affect levels of service delivery.

These broad management changes in commerce have led to specific changes in contracting and consultant practices. Responses to recession have decimated some marketing and sales functions. In line with many sectors, the role of separate marketing and sales functions from other functions has been questioned. In some cases, departments, in some cases, have given way to everyone having sales responsibilities, guided by board policy. Contractors have begun to embrace this thinking too, but is it appropriate

Table 0.1 Survey of problems

Issue	Number of mentions	Percentage of sample
Job management	1146	40
Communicating with others	984	35

for construction? This has been the common model for the professions across Europe, with senior management performing the function, although not in North America, where even quite small design practices have had a separate sales function.[11] Small- to medium-sized contractors have followed the European consultant practice and large contractors having favoured the sales department. Could everyone adopt the marketing function as a normal or preferred option in large concerns? This is an important question, which this book tries to answer in the context of the organisational approaches put forward for securing competitive advantage through marketing. A right or wrong answer cannot be provided. Indeed, advantage comes from how organisations apply the guidance provided here. It becomes one piece in the competition jig-saw. An appreciation of the overall marketing options is a necessary starting point.

Marketing mix and relationship marketing paradigms

Marketing has been dominated by one approach until quite recently. The *marketing mix* approach was developed 40 or so years ago in North America to serve the growing consumer markets derived from the wide availability of a growing range of cheaply manufactured products. The primary ingredients are:

- Product
- Place
- Promotion
- Price.

Blending these variables gives rise to the offer in the marketplace. Product and place, however, are outside the hands of the contractor. To some degree this is also the case for architects and design engineers, because consultants only design the complete 'product' once a contract is secured. Price is a primary focus. However, in terms of the understanding of the marketing mix approach, any out-turn price – that is the final account and percentage professional fees – has little to do with the estimated or original tender price in most cases. Yet price has been *the* powerful tool of selection among clients. That has been the consequence of undifferentiated services.

Promotion has been low key. With low levels of differentiation there is slender justification for substantial budgets for advertising and other promotional activities.[12] The company name is often the real difference between companies, and consultants for most companies are reasonably competent service providers. The low cost of entry into the construction market also means that any new service is relatively easy to copy. Recently, for example, many contractors have jumped upon the 'partnering bandwagon'. A way of constructing the service is needed so that it becomes hard to copy, either in terms of cost or rapidity of response. Another approach or paradigm may be helpful in this context.

The *relationship marketing* approach has emerged over the last 10 years or so. In many ways it is still in its infancy. Developed by the so-called Nordic School, it has arisen from the service sector, centred upon business-to-business sales. Some of the key characteristics of marketing under this approach are:

- Understanding the client business
- Understanding client needs
- Developing close relations
- Inducing trust
- Delivering promises and giving satisfaction
- Fostering loyalty
- Seeking repeat business.

Relationship marketing recognises that a great deal of selling involves people. Relationship marketing therefore recognises their creativity, feelings, subjectivity, as well as the cultural context within which they operate. This approach is more difficult to think through and certainly more difficult to formulate and implement into a strategy. This may be a weakness, but within it also lies the primary strength – different paths and patterns of investment to pursue. A coherent approach takes investment, time to adopt and refine. It is therefore difficult to copy easily, if at all. Thus a systematic approach, a more comprehensive one, is needed in order to differentiate the strategy and to manage the marketing and sales process of relationship marketing.

The most obvious contrast between relationship marketing and the marketing mix is about flexibility. The marketing mix is relatively *static*, subject to periodic review and amendment. The relationship marketing approach is more concerned with process and hence tends to be more *dynamic* and ideally more *holistic*. Figure 0.3 sets out in more detail some of the defining qualities.

In practice, many contractors and consultants are combining elements of both relationship marketing and the marketing mix. Yet

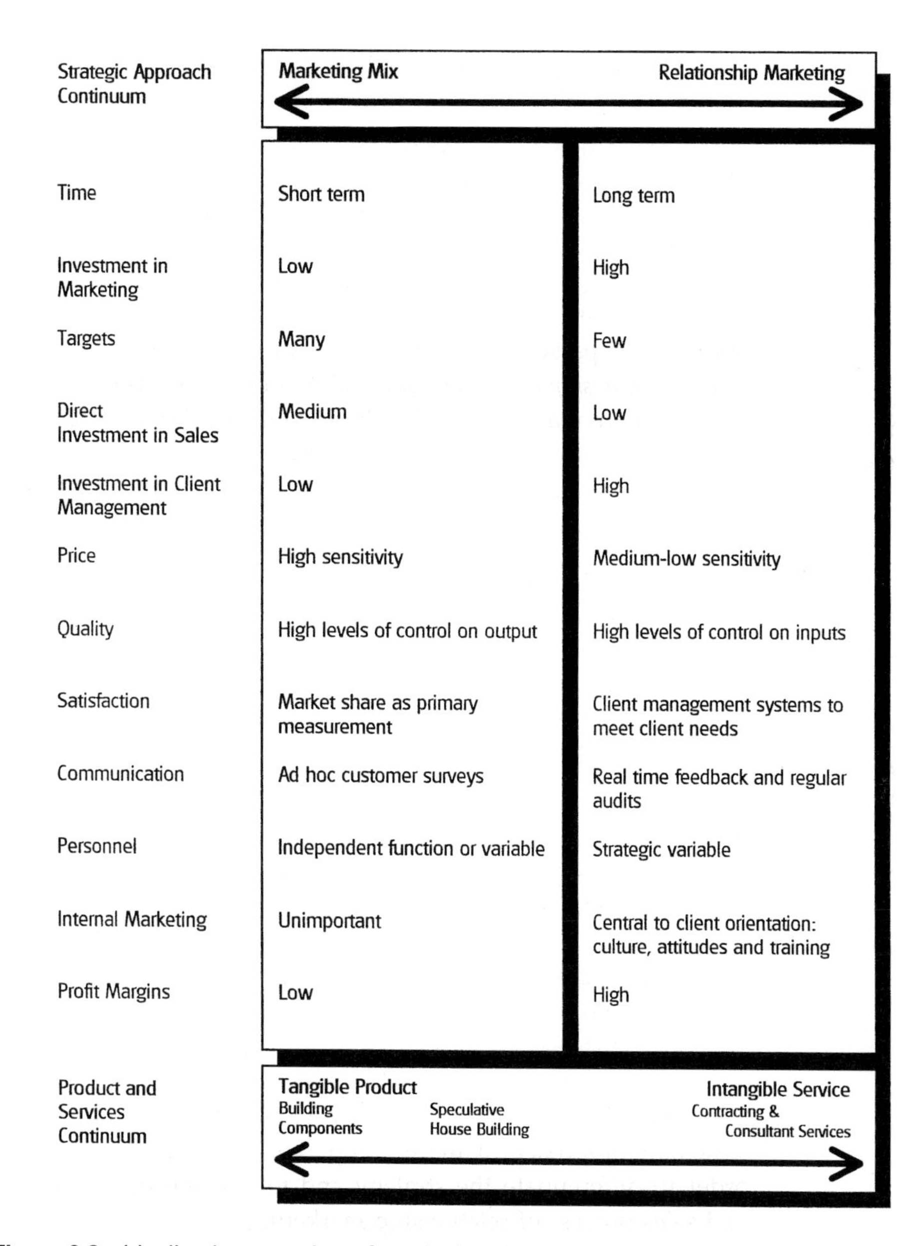

Figure 0.3 Idealised conception of marketing theory.

there is scope for improvement in the marketing and sales efforts of all contractors, whatever the balance between the two approaches. As the book progresses, there will be a switch in emphasis from the marketing mix to relationship marketing, particularly from Chapter 7, as relationships become the focus.

Street-fighting man

Those working in the building environment often perceive themselves and their companies as somehow 'special' and 'different'. This is reinforced by the poor perception that contracting has in the public mind. The sector fails to match up with the clean, flow-line production systems and the just-in-time management techniques of modern industry. It is therefore economically *backward*! There is little or no evidence to support this notion.[13] If it were backward, it would not be able to compete to attract stock market investment, as there would always be higher rates and more stable returns to be yielded in other sectors and therefore a wholesale exit from construction stocks. Yet it is different, the combination of features being unique, although individual characteristics are replicated elsewhere, such as:

- *Assembly* similar to shipbuilding or aircraft construction
- *Contracting*, as in the defence sector
- *Weather dependency*, as in tourism
- *Place-specific delivery*, as in hotel and catering service provision
- *Customised or prototypical design*, for example IT networking, and so on.

The sector is somewhat special, but this should not be overdone. Contractors and consultants do plead for a better understanding from others, yet where a lack of understanding exists the onus is on the industry to educate others. We are to blame.

Why has this not been adequately addressed? Are others not listening? I believe there has been something of an indulgence in being special and different. It provides an excuse when things do not go so well. This operates at a sector level. It also operates very powerfully within corporate hierarchies and for individuals. It is used on the one hand to stifle creativity and to avoid change: 'We've always done it like that'. It is used by individuals to avoid taking responsibility – a thin veil to hide behind. It is a type of posturing resonant of the street culture rather than corporate culture.

Gang-leaders posture at the street-corners vying for a fight. The bully-boys always try to get others to do their dirty work for them. Verbal punches hit hard, physical ones are sometimes let loose, but cowardice lurks in the shadows. The management style of many in contracting companies is based on the street-fighting man. Banter and joking, usually at the expense of others, is used for point scoring when things are on a reasonably even keel. If that fails or the pressure is great, verbal abuse and shouting are the weapons to instil fear and maintain power in the office corridor – a blame culture. This is the street-corner expression of being special and different. It is an excuse for failing to address the insecurities felt by the individual for failing to address the shortcomings of the corporate body. Passing the buck and ruling out of fear is ultimately divisive, both for the employer and for the individual, who is torn apart in managing the conflicts and contradictions of their feelings. The street-fighting man knows that eventually he is to be replaced by someone throwing a harder verbal punch.

The strong man rises above this. Overcoming the internal conflicts enables them to help others, rather than focusing on themselves. This paradox of focusing outside oneself to resolve internal conflict works because the individual *is* taking responsibility. This is actually central for successful marketing. As a focus outside oneself encourages helping others, who are the groups to be helped? The client is the first group. A client orientation moves more to the fore. Posturing has yielded to delivering the goods. The staff is the second group. This helps them to overcome their street-fighting mentality both through direct help, such as training, and indirectly through demonstration, so that they too become more eager to serve. Contracting and consultant practice is, or should be, about working *with* people, not against them. This is true of marketing too, especially relationship marketing, which has relationship building rather than destruction at its core.

Therefore marketing is not only about strategic plans, but also about sales. It embraces processes and organisational structures. Marketing encompasses people's attitudes and how this affects their day-to-day behaviour.

Addressing these issues implies changes in practice. Change is frequently seen as an inconvenience; in fact, it is the norm. One of the primary management skills has become the ability to manage change: fostering internal change and responding to externally generated changes. They remain poorly developed skills in most organisations and for most individuals. One of the main blocks to the individual development of these skills is the fear of change. Senior management frequently impose change without addressing these issues, and staff resist it. A note of caution is needed. Proposing or inducing change requires careful thought and is neither to be undertaken lightly nor with an expectation of immediate response, as one construction commentator noted:

> *Enforced culture change invariably means a culture shock. Change is unsettling, creates fear of the consequences in those lacking confidence in themselves or in their ability to cope with what will be required of them in the future. Thus, change engenders resistance in people, and with or without their full and enthusiastic support the improved systems, methods, machines and materials will never achieve their full potential.*[14]

Pointers and signposts are provided throughout the book to indicate how changes in marketing and sales can begin to be implemented. Exactly how this is done is dependent upon each situation and the success of implementation is of course part of the instigation, development and maintenance of competitive advantage.

Marketing aims

This book has a number of aims, which will be looked at in terms of their levels of operation – sector, company, functional or departmental and individual. The aims can be summarised as:

- Developing dynamic strategies for guiding companies in their marketing and sales, which transform ritualistic lip service into an integrated approach to business development generally and corporate business planning
- Critically exploring competing approaches to marketing theory and attendant sales practice
- Demonstrating how different approaches affect the organisational structures and processes
- Looking at the predominant *top-down* marketing management approach and introducing the *bottom-up* approach, which has been slow to be addressed in mainstream marketing thinking generally and which has a specific relevance to construction
- Examining how the top-down and bottom-up management approaches can be integrated through sales practice
- Developing sales management processes and selling techniques in relation to competing marketing theories, showing the advantages and disadvantages for the construction industry

- Showing how the processes and techniques relate to other areas within the construction business and demonstrating the ideas with practical tools that can be immediately introduced into management systems
- Providing illustrations of company case studies and better practice, so as to identify the exemplary and to show areas for improvement in an applied context
- Introducing selling techniques for implementation and evaluation, tailored to the needs of construction
- Providing understanding about how sales and marketing are integrated activities, and especially how selling can be harnessed for market research and incremental development of strategic marketing.

Objectives

Improvement is the main objective. Who is in a position to facilitate change for improving marketing and sales? This is a book for:

1. *Chief executives, presidents, senior partners*
2. *Senior management* at board level
3. *Other levels of senior management.* They have an advisory responsibility for developing their organisation.
4. *Senior members of the project team.* Marketing or sales personnel do not always hold all the keys to unlocking the marketing and sales potential of their organisation. Sir Peter Davis, Group Chief Executive of the Prudential Corporation said at a memorial lecture:

   > *Marketing is first and foremost a value system ... It must be owned by the chief executive and the board as a whole.*[15]

   He went on to say that marketing must be transparent, rigorous, involving and imaginative in order to fully involve others. Who are they?
5. *Marketing people.* The book will provide valuable ways in which marketing personnel can differentiate services and develop competitive strengths in simple ways. It will also challenge them to work in greater co-operation with senior management and with others across the hierarchy. If the

marketing and sales functions are separated, then approaches for closer management and integration will be set out too. Some marketing people are really sales people with a grand title. Some think that they are marketers and may discover a cool reality that they are really only selling. In both cases, the book will help to equip you to develop your role.

6. *Sales personnel.* It will not provide 20 steps for closing the sale, but will assist you in being more effective. The skills that you will be encouraged to develop are those of facilitating the sales process rather than owning the contacts. Selling is moving away from who you know, towards identifying all the service elements and mobilising the people and resources to more fittingly meet client needs.
7. *Heads of other departments.* They may well be increasingly called upon in the sales process and therefore understanding the marketing and sales thrust is increasingly important in these positions of responsibility. Professionals, especially those trained in the consultant professions, frequently feel it is their role to do the job professionally. This is insufficient in today's evolving markets of intensified competition.
8. *Others in direct client contact.*

Thus the *operational orientation* is to be complemented with a *client orientation*. In other words, the perspective of the profession relies heavily on learnt expertise, imparted through training, whereas a *client orientation* relies on acquired experience derived from seeing the service through their eyes and therefore their needs.

Contractor and consultant

The focus is construction. There is a prime emphasis upon the interface with the client, although the design team and contractor as sources of work, respectively, are referred to and alluded to at various points. There is also an emphasis upon the contractor, although the case study material covers both contractors and consultants. In terms of organisational size, it is the medium and larger organisations, which are the primary audience, who are undertaking large projects and whose clients tend to be large organisations.

Implementation

It has been suggested that effective marketing and sales require a systematic approach. This must take account of two dimensions: *vertically* up and down the hierarchy and *horizontally* across divisional, departmental and functional activities. The first is easier today than at any time since World War II. Through *delayering*, the recent recession has flattened management structures and so the lines of communication have shortened. The approach in contracting and consultants has been largely top-down. There is a bottom-up approach to management too. *Empowerment* is one important aspect. It is not to be confused with task delegation, but is more to do with delegated authority, where initiative is encouraged at all levels and top management support the responsibilities adopted by those below with resources and back up; empowerment is not an excuse for 'passing the buck'.

Systematic marketing from the business plan through to closing a sale and securing repeat business therefore finds a more favourable context than in recent times. This is not to say that the issue has been addressed. Compared to many other sectors, the construction sector in its widest sense has a very poor record and thus there is considerable time spent upon the ways in which this can be improved. Systematic marketing here embraces internal marketing and communications in order to establish the need. This covers attitude, process and structure. It goes further than that. Establishing ownership of the process and embedding this into the everyday working of an organisation is crucial to success and is largely absent in most organisations. The closest most organisations get to beginning to establish systems is having a few procedures, which many staff fail to adhere to in any case.[16]

Horizontal marketing shares many of the features cited above. It does, however, pose greater problems. The recent return to core businesses is of some assistance. The breadth of services and products may have been slimmed down, but the complexity of functional roles of experts remains, indeed grows, at a steady pace. Financial acumen in putting BOOT and development projects together is one example. Whether the expertise is sourced internally or from outside may be a moot point, yet in either case working with the marketing policies and in tandem with the sales process is necessary. Indeed, could it be more helpful to view the financial expert, the accountancy expert, project estimator, procurement, legal and other roles as part of the 'sales team', responsible for the creation and delivery of a seamless offer to the client?

The importance of people and managing them in marketing and selling at the outset has been indicated. People are a source of competitive advantage. Implementing horizontal marketing secures competitive advantage. It is also a major challenge to people's operational or professional focus. A client orientation requires an attitude shift before it begins to impact upon behaviour, regardless of the processes and structures that senior management may put in place. Indeed, one of the paradoxes of marketing is that the closer one looks at specific actions geared to help the client, the more it tends to end up with an internal focus of shifting policy, process, structure and attitudes that others hold onto as their security.

Addressing these issues is a richer source of competitive advantage than the conventional *zero-sum game*. How often has a company said, 'The growth area next year is going to be such and such', and then sales efforts are geared to take a slice of the action. The trouble is that every other company has seen the same signs and makes the same decision. The result is that every contractor or consultant piles into the same market. Instead of being ahead of the game, the company finds itself up against the same old competition, vying for work on price alone.

How can this book provide a range of service and management options for securing competitive advantage and higher margins? It is structured to offer alternative paradigms with differing pathways through each. The summary of coverage is provided below, in order to provide a glimpse of the possibilities and potential.

Summary of content

Overall, the book starts with an emphasis on marketing and develops this into an emphasis on sales; however, it is not as simple as this. The two are closely intertwined and feed off each other in a dynamic environment.

In general terms the book proceeds from strategy, through functions, towards practical implementation. Yet there are feedback loops to create a virtuous circle of development and refinement throughout. It also moves from attitudes to behaviour, from structures to processes as the book unfolds, again creating feedback loops to provide an integrated approach.

Section I looks at the nature and potential of *marketing* from the background of sales and marketing in the 1930s. The gradual emergence of marketing is traced, from the growth of sales and marketing competencies gained through exporting services in developing countries, especially in the Middle Eastern markets from the late 1960s, through to the cutbacks in state-funded work and intensified competition from the mid-1970s into the 1990s. This is the basis for a more sophisticated sales approach.

Chapter 1 deals with marketing within the *business planning* process. Approaches to developing strategic direction within the marketplace are raised. There are different views of the market. The strategy of adopting an integrated approach, yet differentiating services is covered. Establishing a market position using a marketing matrix is developed as one starting point in the planning process.

Commitment is raised as an important yet neglected area in plan implementation. Market risk and market share are addressed. The comparative advantages of top-down and bottom-up approaches to marketing are also introduced.

Chapter 2 moves the analysis from *strategic market positioning* to *selling*. The management implications of adopted market positions are set out, not just from a marketing perspective, but viewing the broader operational implications too. These are related to the way in which the sales offers to clients differ. Decisions upon means of market segmentation and identifying specific client targets flow from the market position occupied by a company or for a service. Key stages – segmentation, niche market and target selection – are identified, coupled with the prioritisation of potential clients. Alternative approaches are introduced; thus, this chapter introduces levels for differentiating services.

Selling is introduced as the implementation stage. It is a continuous process from the moment of first courting a target through to securing repeat business, especially for clients who are regular procurers of construction and consultant services.

Chapter 3 translates the implications of the previous chapter into *sales systems*. Systems are dependent upon the corporate culture and business approach from the top down. It is also dependent upon selling in target markets. The two are, of course, conceptually related. At a practical level, they are related through the sales system.

Attitudes of all staff to selling are addressed, in particular how staff can learn from experiences for future sales opportunities in the context of the learning organisation. Sales staff attitudes are addressed, especially moving away from individually 'owned' sales and contact information to corporately owned information. The changing sales skills are increasingly required for the changing marketplace.

The greater integration of marketing and sales is developed through the practical building blocks of a comprehensive system, both in terms of recognising attitudes and behaviour and the necessary procedures. Key measurements are introduced, which are picked up again towards the end of the book.

Chapter 4 has *market vehicles* as its focus. Developing on the systematic approach from the previous chapter, the emphasis shifts to market penetration through geographical diversification. The international market context is considered in particular. Ways of setting up operations in distant markets are addressed from the perspective of ensuring that the right clients are targeted within the overall marketing plan. Consideration is also given to the support role from the headquarters in the domestic market.

Section II looks at the nature and potential of *selling* in detail. Having considered the marketing–sales interface in parts of the previous section, this section examines the extent to which it is a matter of gifting, training or a professional component of any job.

Chapter 5 addresses the *marketing mix* in relation to *sales promotion*. The choice and mix over marketing approaches affects the sales process. This explores the 4Ps – product, place, promotion and price – in relation to the range of promotional tools available. The chapter evaluates these tools presented in mainstream marketing literature in the construction context. The thrust is to counsel caution on their application in construction and consultant practice, yet it also offers insights and new ways to use shunned techniques in the construction context that offers scope for competitive advantage.

Chapter 6 addresses *relationships* and their implications for *sales promotion*. The chapter develops the concept of relationship marketing. The dimensions of relationship marketing are presented in terms of ways of getting close to, being committed to and hence investing in service relationships. The approach adopted has consequences for the means of client management, models of client handling being explored within the chapter.

Chapter 7 distinguishes between the sale of *selling the service and product*. The steps from targeting through to closing a sale are considered in terms of practical selling. The emphasis is upon the marketing mix approach in this chapter. Techniques for successful selling at each stage are set out, with case studies illustrating points. Account is taken of the different types of procurement option available.

Chapter 8 has *selling through relationships* as its focus, with emphasis on relationship marketing. Methods for mapping the client body and profiling the decision-making unit are explored. These are related back to the method of client management selected in Chapter 6. There are implications for effective client handling, which in turn is affected by the contract procurement route. Trust, an important theme for maintaining successful relationships, underlies much of the chapter content.

Chapter 9 picks up the different contract options in terms of the way in which *added value* can be brought to bear in the effort to serve the client and secure competitive advantage. Applying added value down the supply chain is considered, as well as its relationship to quality assurance and total quality management issues.

Selling added value in the service is the focus of Chapter 10. Client service is the main thrust. Building client relationships prior to and during projects is the context for examining in great depth how to understand and respond to client needs. Engendering and maintaining trust is one of the key foci. Building relations that can be maintained after project completion and that will enhance the potential for securing future work is another focus. Client loyalty through the creation and maintenance of trust is vital for adding value to a service, which will yield business from the same client and in referral markets.

The focus of Chapter 11 is the *construction project team*. The team is a temporary amalgam of people, a virtual organisation, some from outside the contracting or consultant organisations. The question as to how this organisation can best be harnessed to maximise both the service delivery and create opportunities for future is the primary issue. This is related back to client management, sales systems and management support from senior levels.

Chapter 12 centres upon the *client perspective*, particularly *repeat business*, as one of the main aims of creating satisfied clients. The economic drive in the market militates against satisfaction leading to repeat business. Investment in a range of relationship marketing techniques can overcome this drive. The opportunities of and problems for doing this are explored.

Alliances and partnering procurement options are examined in this context.

Section III explores *selling and evaluation techniques*. The type and characteristic of different types of data are analysed in order to evaluate their usefulness and the ways of using and interpreting data are set out. Contractors and consultants feed data and other types of information to their receptive audiences and an evaluation of their purpose and impact is important, especially by those undertaking face-to-face selling.

Chapter 13 has *sales messages* at its core. Distinction is made between messages of fact and myth, between features and benefits in a service. Messages are related back to the market position that the organisation and service occupy. The image and communications of the organisation need to be in harmony from a public relations perspective and specifically in all the sales messages.

Chapter 14 concerns *sales monitoring*. Establishing the balance of activities picks up on themes from Chapter 3. The main purpose here is to look in considerable detail at the mechanisms used within the overall system. Monitoring sales progress and staff performance are the two objectives. This is more complex than is frequently appreciated and a more thorough means is put forward than is used in most construction-related organisations. It is considered an important issue, for if this is not structured in an appropriate way for the markets in which the organisation is operating, the motivation for the sales efforts is frequently diverted into inappropriate ways of selling.

Integrating marketing and selling is the main theme of the next chapter. *Market research* is a neglected area in construction. There are good reasons for this, yet there are means to overcome the obstacles. Chapter 15 considers this through adopting a bottom-up approach to marketing, whereby selling becomes a primary source of market research. Drawing on some of the material presented in Chapters 1, 3, 7 and 10, a systematic and integrated way of selling is set out, which facilitates effective market research with the use of several techniques.

This chapter thus threads together marketing and selling into a holistic and dynamic way of selling construction services. In this way it leads into the *Conclusion* of the book, which has pioneered the need and means for improvement in marketing and sales. It urges commitment to taking this seriously. It is time for immediate change and for preparation for long-term change. It is no longer viable for the long term in construction to be defined as, 'that which never comes'.

Summary

1. The *purpose* has been to:
   - Stimulate a pioneering approach to marketing and sales
   - Indicate that there are sound practical ways to become a pioneer
   - Indicate management as a main means for differentiating services
   - Claim people to be the main asset when working to a marketing strategy and within a marketing management system
   - Show the need to integrate marketing and selling construction services with corporate culture, structure, processes and attitudes.
2. The chapter has:
   - Identified the target reader
   - Outlined the content of the book.

References and notes

1 *Building* (1997) Image building, 19 September.

2 Hong Kong Airport Authority (1997) *Hong Kong's New Airport Development in Brief*, Hong Kong Airport Authority.

3 Honess, S. (1997) *Marketing for Construction Firms*, Thomas Telford, London.
Booth, W.D. (1992) *Design/Build Marketing: Strategies and Procedures for the Small and Mid-size Contractor*, van Nostrand Reinhold, New York.

4 Fisher, N. (1986) *Marketing for the Construction Industry: A Practical Handbook for Consultants and Other Professionals*, Longman, Harlow.
Pearce, P. (1992) *Construction Marketing: A Professional Approach*, Thomas Telford, London.

5 Policy Publications (1997) *Managing Major Bid Projects*, University of Luton and Marketing Business, Bedford.

6 Levitt, T. (1983) *The Marketing Imagination*, The Free Press, Macmillan, New York, p. 7.

7 See for example Smyth, H.J. (1985) *Property Companies and the Construction Industry in Britain*, Cambridge University Press, Cambridge.

8 Lindsay, K. and Hawkes, A. (1997) UK financial services – short-changed on ideas? *Admap*, May.

9 Ahmad, I.U. and Sein, M.K. (1997) Construction project teams for TQM: a factor-element impact model, *Construction Management and Economics*, **15**, 457–467.

10 Stephenson, R.J. (1996) *Project Partnering for the Design and Construction Industry*, John Wiley, New York.

11 Coxe, W. (1983) *Marketing Architectural and Engineering Services*, second edition, van Nostrand Reinhold, New York.
12 Smyth, H.J. and Thompson, N. (1997) *Developing Loyal Clients*, Centre for Construction Marketing, Oxford Brookes University, Oxford.
13 Smyth, H.J. (1985) *Property Companies and the Construction Industry in Britain*, Cambridge University Press, Cambridge. See especially Chapter 3.
14 Baden Hellard, H. (1995) *Project Partnering: Principle and Practice*, Thomas Telford, London.
15 Davis, Sir Peter (1997) Marketing – the undersold business tool, extract from the Sir Kenneth Cork Memorial Lecture, published in *Professional Manager*, July.
16 See for different aspects of this shortcoming:
Branch, R.F., McIlveen, A. and Smyth, H. J. (1995) *Developing Unique Selling Propositions in Fragile Construction Markets*, Centre for Construction Marketing, Oxford Brookes University.
Building (1997) Image building, 19 September.
Buttle, F. (1996) Service quality in the construction industry, *Proceedings of the 1st National Construction Marketing Conference*, 4 July, The Centre for Construction Marketing in association with CIMCIG, Oxford Brookes University, Oxford.
Preece, C.N., Putsman, A. and Walker, K. (1996) Satisfying the client through a more effective marketing approach in contracting – a case study, *Proceedings of the 1st National Construction Marketing Conference*, 4 July, The Centre for Construction Marketing in association with CIMCIG, Oxford Brookes University, Oxford.
Smyth, H.J. (1996) *Effective Selling of Design and Build: Architecture as a Sales Strategy*, paper presented at COBRA '96, RICS Construction and Building Research Conference, 19–20 September, University of the West of England, Bristol.
Smyth, H.J. (1998) The internal market monitoring systems in contracting and consulting firms, *Proceedings of the 3rd National Construction Marketing Conference*, 9 July, The Centre for Construction Marketing in association with CIMCIG, Oxford Brookes University, Oxford.
Stockerl, K.C. (1997) The importance of strategic marketing planning for the UK construction industry in a changing European business environment, *Proceedings of the 2nd National Construction Marketing Conference*, 3 July, The Centre for Construction Marketing in association with CIMCIG, Oxford Brookes University, Oxford.

Section

I What is Marketing?

This section contains four chapters. The emphasis for the section is *marketing* and therefore focuses upon the formulation of *strategic business policy making*. It also covers distinct options and thus provides opportunities for differentiation of marketing policy, which will then be translated into and enhanced by implementation in the sales process. Selling is the emphasis for the second section.

Marketing and sales are frequently confused in people's minds. There are two main reasons. First, a common definition for marketing has yet to be found. In one sense this shows strength, because the different perspectives and approaches adopted by people and their companies are highlighted. So it is not proposed to debate this. One good working definition, which is open-ended, sufficiently loose, will satisfy those that like definitions and will be used in this book, is, 'the purpose of a business is to create and keep a customer'.[1] As a purpose of marketing is, through offering a different service in order to meet the particular needs of their clients, to avoid head-on competition as much as possible, generating differences is the first step, and this should start with the concept of marketing itself.

Second, the fight to keep salary bills down and to give staff status leads to sales people being called marketers, business development managers, in fact anything but sales people.

The focus of marketing within this first section is to move towards an outcome of taking marketing seriously, using creativity and imagination, coupled with some hard graft in practice. The section is designed to help develop a strategic direction appropriate to your organisation. Therefore this section focuses upon marketing as policy *formulation*, while the next section has *implementation* as its focus, through the sales process and the delivery of the sales promises.

It becomes clear that integration between the functions is necessary. Again there are two reasons. Marketing and sales need to be seamless. Conceptually and in practice they are distinctive, yet must be zipped together to be effective in practice. An *implementation gap* occurs where the policies fail to be acted upon or where those responsible for implementation interpret them in different, perhaps inappropriate, ways to those formulating them. The second reason is that management culture and practice demands an increasing feedback and refinement so that a virtuous circle is formed; marketing determining sales and sales feeding back to nudge and change marketing policy in an interactive way.

The subjects of this section cover development of the business plan in the first chapter. Marketing is clearly one of the cornerstones of a plan and sets the context for subsequent decisions. Choice also implies the need to recognise and manage the risks inherent with each option. The choice of approach is therefore of critical importance to contractor and consultant.

Chapter 2 moves the analysis towards selling. This is concerned with initiating seamless contact with sales, and ensuring that the two remain conjoined. There are strategic aspects to and implications for selling, particularly the way in which services are differentiated during the stages and in the levels of the sales process. The view adopted here goes beyond closing the sale; in other words, the successful post-tender negotiations, the project and post-project stages of sales are included.

A systematic approach is required to embrace the content of the second chapter and this third one focuses upon sales systems. The emphasis is upon establishing corporately owned information systems, thus moving away from a culture that places great emphasis upon the individual as the collector and retainer of information to an integrated approach to sales and marketing.

Chapter 4 looks at targeting and how this can be managed strategically. Particular attention is paid to international markets. Different approaches are suggested for markets close to the domestic market and distant markets, the role of the headquarters changing according to geographical location.

It will be useful to put marketing in a brief historical context in terms of construction and consultant practice. This is interesting in its own right, yet the prime purpose is to understand how marketing has evolved in the sector. This provides a valuable insight into its present form. In other words, it becomes clearer what the strengths are, what the weaknesses are and where there are particular 'blind spots'. A phrase that is frequently heard across the industry corridors is, 'We've always done it like that'. In many cases, there are for good economic reasons. Prudence is one thing, yet the phrase is frequently heard when people still have to perceive a possibility for change. This is a 'blind spot'. Add to this blind spot the fear many people have of change, then to move forward in the marketplace becomes very hard work. It is easier to make progress when it can be clearly seen where there are few good reasons for staying put, or indeed it may prove detrimental to maintain the same old ways of doing things.

Why are we where we are today?

Where we have come from

It has been said that there have been four eras of construction:[2]

- Railways in the 1800s
- Urban infrastructures of the 1930s
- Electricity generation era
- Petrochemicals in the post-World War II era.

These civil engineering projects led the way in domestic and overseas markets in the economy and therefore in contracting. These types of projects were the staple diet for the dominant contractors, such as Peto and Brassey of the UK in the 1800s, Morrison-Knudsen and Utah Construction of the USA in the 1930s, the pre-eminent Bechtel in processing facilities during the 1940s. What is the staple diet of today? European and Japanese contractors have spent the 1970s and 1980s trying to match the growth of the North American companies, particularly in international markets.[2] Yet a new era of civil engineering projects, which drive economic growth, has failed to materialise. Micro- and small-scale technologies and the service sector have dominated recent economic development. Competition has therefore intensified. That is the overview. Where does marketing and selling fit into this?

Modern advertising had its advent with Toulouse-Lautrec's poster for the Moulin-Rouge in 1891 and for the Ambassadeurs in the following year. The strategic side of business development for contractors began in the 1930s. Recession was the main driver. Contractors adopted different strategies for that period. Many entered the speculative housing market serving the growing middle classes, while others grew to become substantial firms from that origin. At the other end of the spectrum, there were large infrastructure projects (see ... *On the case 1*).[3,4]

Innovation in procurement was introduced, for example the Bovis fee system with retailer, Marks and Spencer, in the 1930s and the target cost contract used in the later years of World War II in Britain. Both these measures were client led. Indeed, governments dominated the construction market to such an extent in the

... On the case 1

A large British engineering and general contractor decided to accept only large-scale projects, such as power stations during the 1930s, either because they yielded premium profits or because they were prestigious. They turned down opportunities, which did not fully meet these criteria.

This strategy arose from the paternal-style family control, whereby decision making was concentrated in a few hands and a large number of shareholders did not need to be financially rewarded.

This proved successful. Indeed, many competitors failed to survive as major contractors. Rearmament later led to opportunities for diversifying the workload. A number of speculative house builders, along with this large British engineering and general contractor, proved to be in the strongest position to become the next generation of major contractors.

post-war years that marketing and selling remained undeveloped. The reconstruction period in Europe, large oil-related projects and the development of the mass consumer market in North America kept the contractors in business, with healthy workloads. Ironically, the growth in other sectors helped to contribute to lack of consideration of marketing in contracting and consulting.

That continued until there was a slowdown in the world economy. Prompted by the Middle Eastern oil crises and the indebtedness of developing countries in the wake of Keynesian demand-led economic strategies, the governments of many countries began to rein back expenditure, especially capital expenditure, from the mid 1970s. Intensified competition resulted, as has been commented upon in the British context,[4] especially among large contractors:

> *... seeking over 50% of their building turnover in the public sector and frequently over 50% of their turnover from civil engineering work. The situation was exacerbated by the already intense competition for public sector work between the large and medium contractors which emerged during the prosperous 1960s (p. 194).*

Marketing strategies and selling practices were in place to the extent that contractors could positively respond to the situation in order to plug the gap, creating some successful companies while others receded or failed. Success was largely based upon the ability to manage contracts in a high inflation environment.

The oil crises had a 'silver lining'. The Middle Eastern countries had been expanding their construction programmes. Funded out of oil revenues, programmes accelerated in the wake of the oil crisis. The large North American contractors, such as Bechtel and Fluor Daniel, were prime beneficiaries, followed by European contractors.

This situation prevailed from the late 1960s, growing throughout the 1970s, easing in the 1980s. Contractors from the Far East, such as Korea and the Indian sub-continent, were able to undercut contractors, due to low labour costs and ability to construct many of the less demanding projects, as well as a few of the sophisticated ones.

The decision to exploit the market was the key strategic issue. Marketing was *not* an issue. Contractors from the developed countries did take on board selling skills. Networking through government officials, local representatives, with aid agencies and financial institutions was important, frequently over-riding competitive bidding as the means to secure work. This was the beginning of *having to develop* a sales capacity, often at senior management levels. In the domestic markets, this sales networking was to be influential in the 1980s, but in the 1970s the market in the Middle East boosted balance sheets and relieved the intensity of competition at home.

Selling became an important component of North American contracting and design services in the 1970s and 1980s. Any reasonably sized architectural practice had its own sales person or marketing co-ordinator. This was partly a reflection of the levels of activity in the marketplace and partly a reflection of the strong territorial and regional divisions in those markets. In Europe, many marketing and sales activities were creeping more and more into the activities of boards, partnerships and senior management. Strategic decisions were being taken to pursue new markets.

One example was the rise of design and build in the British market. The path pursued by contractors was a familiar one – set up a separate division for the selling of those services. While this has been very successful, it does highlight the key strategic approach, which is to embed the marketing into a *structure*. This demonstrates a lack of confidence in the ability of personnel to handle different client needs in different ways. Creating a divisional structure ensured some control through standardisation.

Standardisation as the solution had its roots in the thinking of the mass markets of the 1960s. Yet the construction market was fragmenting. The government programmes were shrinking and the break up of centralised procurement in some countries was making it more difficult to track the market and project leads early enough and in a pro-active way. There was greater reliance on commercial work as the mainstay during the 1980s. A fragmented market is far more volatile from the viewpoint of any contractor or consultant and one reliant upon the private sector renders the sector more exposed to the booms and slumps of the economic cycle. Government policies no longer offered some counter-cyclical respite.

The need to develop a sales capability became essential for every company as a means of gaining market share and capturing the best of the growth in the upswing of the boom market. Selling became one key for survival in times of slump. The large contractors and consultants had acquired expertise and competencies through networking in international markets of the late 1960s onwards. This was transferred into domestic markets and copied by others. Entrepreneurial selling in North American geographical markets was also copied on both sides of the Atlantic, until a critical mass of selling skills and expertise had been acquired.

The expertise was founded upon selling. Marketing had been avoided in favour of strategic business decisions, which were enshrined in *structural* responses of organisations to the market. This was not only the case for contractors. It was also the case for consultants.[5] Marketing was, and remains in many cases, piecemeal among both contractors and consultants.

The recession of the late 1980s and the first half of the 1990s had a drastic effect on contractors and consultants alike. It proved to be a major lost opportunity. The severity demanded stringent cuts in personnel to the extent that most organisations lost highly valued core staff. It demanded restructuring of management to form flatter structures, *delayering* by necessity rather than design. The opportunity to restructure so that a client-orientated focus emerged, was overlooked. Staff retention, team organisation and management practices could have been reconstituted with this as *the* prime focus. It did not occur.

If only

It is, of course, pure speculation to say what could have been, but in one respect it is useful in order to compare with the current circumstances of the late 1990s and for the twentieth century. A client-orientated restructuring during the early 1990s would have had the aim of gaining short-term *market share*, that is to say, securing contracts at the expense of competitors. The benefit would have accrued to the contractor or consultant with an effect of helping to shake out competition, which would have improved margins in the slow upswing in the latter half of the 1990s.

A client-orientated restructuring during the late 1990s would have had the aim of gaining the 'lion's share' of the growth market. This is essentially *market penetration* in existing or new geographical markets. In either case, marketing is an investment for creating a competitive edge, although in accountancy terms it is always considered an overhead. This is not a plea for more resources. It could be a plea for less in some instances, as will be seen later in the book, but rather for the careful application of resources, so that wherever allocated in the organisational budget headings, the impact is felt across entire organisations. The plea is for people to embrace marketing whatever their functional role in proportion to the positive contribution they can make for their employer.

This is simple to say but difficult to accomplish. Contractors may have repeatedly proved themselves when it has come to setting up a new structure through which to deliver a new service, whether it was design and build, management contracting, or facilities management and maintenance services. Consultants have also proved their ability to develop new services, architects with a landscape division or surveyors with project management services, but a genuinely seamless service remains elusive. The client does not get a 'one-stop-shop'. In reality, the experience is more like a department store, where the client moves from counter to counter. At most a partner acts as the store guide. The solution for the consultant and contractor has been structural. It is for their convenience, not the client's. The convenience is twofold. It is operationally simple. More importantly, it avoids having to address some rather awkward questions. These marketing questions again take us from the structural level right down into the realms of how people behave and the underlying attitudes.

A new era

The need is simple. The fragmented market is going to get tougher unless there are changes, of which marketing and selling are a vital part. Effective marketing and sales are required to beat the competition and to serve clients more efficiently in productivity and value. Achieving results depends upon behaviour and the underlying attitudes.

The issues of behaviour and attitudes can be summed up in one phrase: 'It is time to move from sales to marketing'. What are some of the key characteristics? As we look at these some surprising things emerge. Contractors and consultants are, ironically, better placed to tackle some of the issues now than they have been or may be in the years to come.

This brings us on to a paradox. *Marketing*, according to literature and the majority view in practice, requires analytical skills. It is strategic, the aim being to devise ways of selling and delivering services in a creative and competent way given the context of both the organisation and the marketplace. Strategic analysis is therefore a *cerebral* competence. It is for the type of person who 'lives in their head', who likes time to work things through and does not usually respond well to short-term pressure. The person probably prefers to work at a distance or 'arm's length' from others, and therefore may not be described as a 'people person'. What they do excel at is analysing ways through difficult circumstances in order to come up with ways to beat the competition.

Sales requires a 'seat of the pants' approach. It is *intuitive*, requiring skills to be able to respond quickly. Typically, this type of person is good at 'thinking on their feet'. They will worry later as to whether the rapidly required response fits the strategy or whether the organisation can possibly deliver. Their objective is to maintain the sales momentum, to lead the client towards the next stage of the sales process by overcoming any objections they raise. Interest in long-term issues of strategy is quickly lost, in preference to 'get on with the job'. They embrace the excitement of the immediate pressure.

Intelligence is not the issue here. It is how intelligence is applied within the personality profile of the individual. Both types of intellectual skills and competencies are needed. Of course, the contrasting approaches potentially give rise to conflict. The marketing person will blame the sales staff for failing to implement what they have carefully thought through, which has subsequently been adopted as policy. The sales people will tend to deride marketing ideas as being conceived by those who have no idea what it is like to be in the field and certainly do not appreciate the 'cut and thrust' of selling. Pragmatism rules!

Organisations succeed where these two paradoxical competencies are harnessed in harmony. Laurel and Hardy, Lennon and McCartney, Punch and Judy are famous double acts that contain elements of these apparent opposites. They worked well together in fiction or reality because one was a foil for the other. Tensions are present, which can cause breakdown under some circumstances, but the management job is to avoid this. In the contracting culture it is the *intuitive* sales type that dominates. The flexibility of this 'seat of the pants' approach has ideally suited the contracting context and to some extent the consulting environment too. The downside is the impetuous approach that leads to 'more haste, less speed', missing opportunities by failing to stand back and take a long hard look at where all the frenetic action is leading. At worst, any sign of strategy is ruthlessly weeded out. Any strategy that is left is safe in the knowledge that the long term never becomes today!

Yet in many contracting and professional organisations the same person is responsible for co-ordinating marketing strategy and selling. This can lead to a kind of *schizophrenia*. The double act is incorporated into one person. How can the person cope with this tension? The answer is that most people have not been taught to do so. The best solution is to neglect one. The one to be neglected will be the one that fits least easily with your personality profile. For cultural reasons, in contracting this is likely to result in sales surviving and marketing being neglected. Frequently, this paradox can culminate in the worst combination:

- *Operational bias*, where people become sales staff from the contracting ranks because they get on well with people and are good at 'thinking on their feet'

- *Sales bias*, where these people feel they are often working in a vacuum, that is without strategic guidance and day-to-day support
- *Rigid approach*, where there is reliance on using what has worked rather than creatively looking forward, a matter of 'getting by' or 'that's how we've always done it'
- *Reactive response*, whereby anticipation of market changes fail to be picked up and acted upon ahead of the competition and the zero-sum games flourish.

The final twist to this paradox is a general perception that these two elements do now need to be increasingly integrated. The *schizophrenia* is to be encouraged and managed. Although the restructuring during the early 1990s was not grasped as a way of reconstituting marketing and selling in a coherent way, nonetheless marketing and selling activities are merging for several reasons:

- Flatter management structures
- Faster changing markets
- Internationalisation of markets or merely working over larger geographical markets in ratio to the management and sales resources
- 'Top-down' marketing being complemented by a 'bottom-up' marketing and sales response for competitive advantage
- Relationship marketing, involving everyone with marketing and selling.

The contracting culture is characterised as short-term, flexible, reactive and intuitive. Because the sector has survived it could be argued that there are great strengths to the approach. The same cannot be argued for individual companies over the long term. While there is flux in any sector, there is lineage that can be traced back to the previous century in most cases. The same is not true in contracting. The great railway contractors of the last century cannot be found in the heritage of major contractors of today.[2,6] Indeed, many of those with the longest histories can only trace their origins over the last 50 or 75 years.[2,4] If the culture is reactive, the management style is based upon the street-fighting man. Banter and joking at the expense of others is used for point scoring. Attitudes and the corresponding behaviour can be aggressive. Aggression is used in a negative sense, assertiveness being the positive counterpart. A good analogy is the court jester. The function of the jester is to use wit and banter to highlight the errors in the ruler's ways. It is a strategic role. It combines spontaneity. The jester provokes a repartee in order to engage the ruler and encourages the ruler to see a different viewpoint and therefore to rethink. The court jester potentially embraces the schizophrenia of the cerebral and the intuitive, the strategic marketing and tactical sales functions in a positive and applicable way. Contracting and consultant organisations will benefit from the court jester, who will fit reasonably easily into the current culture. There is even a recent corporate precedent. One of the leading airlines employed a jester to challenge perceived wisdom.

Contractors are ironically in the best position to integrate marketing and sales. Consultants and contractors have selling, it is the marketing that is largely lacking. The seed of marketing has been sown, yet has to germinate and flourish. Managing the schizophrenia will be important for nurturing marketing in its infancy. Success will be evaluated by organisations if they can look back and say they have brought up a balanced personality. Today's schizophrenia is tomorrow's norm. Success for the sector will be measured by the extent to which it has proved to be a role model for others, with enhanced status in the process.

The context mapped out for this section is encapsulated in the matrix below. Underlying attitudes and behaviour are expressed on the vertical scale as *commitment* and on the horizontal scale *investment* reflects resource issues (see Figure I.1). A note of caution concerning interpretation is required here. Investment is not intended to reflect the amount of resources put into marketing and selling, but concerns the way in which existing resource allocation is applied; in other words, to what extent the way in which resources are used involves taking marketing and selling issues seriously. The lost opportunities in restructuring during the early 1990s would be a good example. Training estimating staff and legal consultants to liase and act upon sales promises in project bidding stages would be another. The top right-hand corner of the matrix concerns *strategic support*. Sales expenditure in the sense of a dedicated team could actually be reduced at this stage. If a client orientation becomes instilled into staff thinking in estimating, on site and so on, then marketing and sales support could move away from having as many sales staff securing new clients towards seeking repeat business from existing ones and securing more work through third-party advocates. Strategic support for marketing and sales could also change. Instead of the emphasis on the induction, diffusion, demonstration, training on and

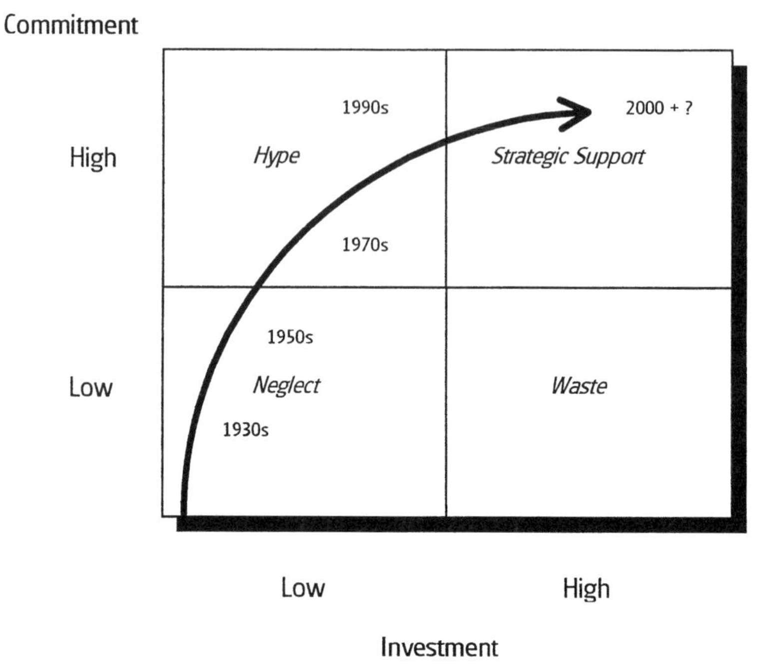

Figure I.1 Changing approach to marketing and sales.

around marketing ideas, marketing and selling would have become infused into everyday practice. In other words, it would have become part of *taken for granted thinking* about the way we work, and hence strategy could direct that effort rather than instigate it.

In order to move along the curve in the matrix, the strategic appreciation of the aspects of marketing at board and senior management levels is needed and that is the subject of the following chapter.

References and notes

1 Levitt, T. (1983) *The Marketing Imagination*, The Free Press, Macmillan, New York.
2 Linder, M. (1994) *Projecting Capitalism: A History of the Internationalization of the Construction Industry*, Greenwood Press, London.
3 McAlpine, Sir Robert and Sons, (1955) *Sir Robert McAlpine and Sons*, Civil Engineering Publications, London.

4 Smyth, H.J. (1985) *Property Companies and the Construction Industry in Britain*, Cambridge University Press, Cambridge.
5 The *Financial Times* and *New Civil Engineer* both reported an absence of marketing among consulting engineers according to French, M. (1996) Information systems to support business planning and marketing for professional services in the construction industry, *Proceedings of the 1st National Construction Marketing Conference, 4 July*, Oxford Brookes University, Oxford.
6 Joby, J.S. (1983) *The Railway Builders: Lives and Works of the Victorian Railway Contractors*, Charles, London.

Chapter 1

The Business Plan and Marketing Strategy

Themes

1. The *aims* of this chapter are to:
   - ❑ Strengthen marketing plans to have long-term, medium-term and short-term objectives
   - ❑ Set the scene for marketing as a means to increase market share and improve financial performance using objectives that are not simply focused upon the bottom line
   - ❑ Strengthen marketing plans as robust yet dynamic working documents
   - ❑ Acknowledge the *top-down* approach to plan formulation requires management commitment.
2. The *objectives* of this chapter are:
   - ❑ Planning for differentiation in the marketplace
   - ❑ Playing to and developing genuine strengths
   - ❑ Selecting approaches to support the implementation
   - ❑ Embracing the four views of the marketplace.
3. The *outcomes* of this chapter are to demonstrate that competitive advantage recognises:
   - ❑ Determining the pecking order as to what really drives the company – operations, finance, personalities or marketing
   - ❑ Robust market planning usually demands changes to company structures and procedures.

Keywords

Business strategy, Marketing plan, Marketing position, Models of growth

The need for a plan

Consider this situation: the top management sit around a table to produce their annual marketing plan. This annual ritual ensures that the old plan is dusted down and a rapid conclusion is reached: 'Well that didn't work!'. And so the whole process is started again, often from scratch. A new plan is proposed, worked over, written up and then put into a filing cabinet for the year.

Another scenario: the senior management, often the managing director or senior partner, says, 'Maintenance work will increase this year – we must get into that in a big way.' This insight has been gained from conversations among clients, professional peers and competitors at the same conferences, client lunches and promotion events. Not surprisingly, every competitor has someone who has had the same thought. The net result is that everyone is piling out of one market in decline and meeting up again in the growth market. This makes intensive competition a certainty. Contractors and consultants end up offering the same services in the same way in the same market. Is it surprising that the industry always competes mainly on price – the classic outcome of this zero-sum game?

These scenarios may be exaggerated, yet we probably all have seen elements of this at one time or another. The way to avoid these scenarios is to develop strategic plans for the business as a whole and for marketing specifically. A decade ago there was 'no tradition of any serious and planned research activity' in construction.[1] The sector was generally short-term in its thinking, although some national industries were thought to be more short-term than their overseas competitors.[2] There has been welcome change, particularly during the second half of the 1990s, when the culture for planning has noticeably improved. These scenarios are therefore not the experience of everyone. Even so, there are some companies, even quite substantial ones, which still do not have a marketing plan and maybe not even an overall business plan. Others have developed sophisticated plans in recent years, yet there is room for improvement, and there are still substantial differences in the way organisations approach strategic business and market planning across cultures and with markedly different results (see Table 1.1). In broad terms, the Japanese tend to pursue a long-term strategic focus, whereas the British and North Americans favour short-term objectives. Quality and market share is preferred to accountancy driven factors. Construction is no different in this respect.

The temptation is to draw the conclusion that British-type companies have somehow got it wrong, that the Japanese style of management is more appropriate. This is to miss the point. The marketplace cannot be occupied by organisations all adopting the same approach. Clearly, some approaches are more suited to particular markets and certain economic circumstances and eras. One objective is to secure a balance between the long and short

Table 1.1 Survey of 90 large companies on attitudes to business

Issue	Japanese companies (%)	US companies (%)	British companies (%)
Objectives			
Objective of short-term profitability	27	80	87
Strategic focus			
Entering emerging market segments	77	50	40
Winning market share	83	73	53
Cost reductions and improved productivity	43	70	83
Competitive advantage			
Quality and reliability	93	77	47
Service and support	70	50	44
Product/Service innovation	68	63	33
Low Prices	60	40	60
Importance of performance indicators			
Sales	53	33	47
Cash flow	7	33	40
Profitability	7	87	60
Management style			
Teamwork	73	63	50
Hierarchical	35	57	83
Variable/*ad hoc* job specifications	73	30	27
Top-down and bottom-up communications	62	30	27

Source: Doyle *et al*. (1992).[3]

terms. That is not to say that there are not lessons to be learnt from the differing approaches. Re-assessment is, of course, vital in the fast changing marketplace.

Certainly, there are important lessons to be learnt in markets that are predominantly price-led in terms of the *marketing mix*. Construction has been price-led for a long time, particularly in North American and European markets. The lessons required for changing a price-dominated market have to be learnt at sector as well as company level. This needs a mutual understanding across organisations, which set the game plan for the sector. Everyone then plays by the rules, even in recession. A cartel is not being advocated here. Apart from being illegal, cartels are to the detriment of the client or customer. The game plan must benefit the client, in addition to or in some way other than lowest price, such as quality of service and added value.

Such an approach starts at company level, with strategic planning. For the majority of construction and consultant companies, strategic planning has still to reach maturity. Therefore, the majority would identify with at least some of the scenarios presented above and would also be able to locate themselves within Table 1.1. There are some key questions that arise from the scenarios:

- Why do we end up competing alongside everyone else?
- Why is so much of the competition largely price driven?
- Why do we all have the same view of the market?
- What is the real purpose of a business and marketing plan?
- How long is our marketing horizon?

The short answer to these questions is to make plans far more robust in the future. A sample of 133 responses spread across 30 construction and consultant organisations, albeit an unrepresentative delegate sample,[4] showed that 58% recognised the need to address marketing issues and one in seven of these identified the necessity for a review of marketing strategy. In order to be robust, a sound starting place is needed. Some strategic issues to address may include:

- Speak to your board of directors, head of marketing and departmental heads and agree to reassess the marketing strategy
- Initiate a programme to obtain adequate information at top-management level from those at the front end of sales and client management
- Undertake market research
- Assess whether competing in sector growth markets is really the best source of work for your organisation
- Agree not to offer the same as everyone else
- Assess the *real* strengths of the organisation.

First, long-term horizons, embracing the next 7 years and over, as well as medium-term and annual horizons, are needed. Second, the operational side of the contractors and consultants absorbs around 80% of the costs. The impact upon the client is around 20%, whereas the softer service issues only account for 20%, yet may account for as much as 80% of the impact upon the client.[5] It is therefore important to address the issues that are most likely to impact the client.

It can be guaranteed that most contracting and consultant organisations, even many of the largest, are inadequately equipped to fully address strategy. Many confidently assume they do it all, when in fact on close questioning they do very little. The source of this is commonly:

- Inadequate *support* from the chief executive
- An absence of a *plan* for market planning
- Low *commitment* levels from top to bottom of the organisation, reinforced by a lack of understanding of purpose, jargon, procedures
- *Hostility* among some, frequently as a consequence of the above.[6]

According to one commentator, managers are reticent to plan, tend:

> *... to opt out from the process by letting vested interests rule. They fail to take ownership of the plans and show commitment to the process. ...*

Research has shown that the following problems are common, including management consultants:

- Poor grasp of the marketing concept and marketing management process
- Little or no marketing analysis undertaken
- Strategy determined in isolation of analysis or implementation programmes
- Action programmes poorly related to recommended strategies
- Narrow view of external trading environment
- Poor and inadequate market intelligence
- Little internal sharing of internal market data
- Inadequate understanding and support from senior management
- Poor internal communication, within the marketing department and functional areas, and through management hierarchies
- Planning personnel run out of steam
- Planning and personnel taken over by events
- Lack of confidence and conviction in marketing plans and planning activity
- Little opportunity during the planning process for lateral thinking and space.[8]

This schedule reinforces the issues about plan formulation. It goes further and addresses, in general terms, the common shortcomings in *implementation*. Many companies develop sound strategies, yet fall down on implementation.[9] There are regional differences

... On the case 2

A highly successful multi-disciplinary surveying practice, employing around 150 staff across four regional offices has a marketing manager responsible for developing its marketing strategy, which ... *principally outlines the current situation with little partner input as to recommendations for future direction and how we get there* (p. 1).

Partners do not take a great deal of notice of the marketing strategy, tending to pursue their own interests. They have considerable scope to do this providing they meet strict turnover targets. The sales volume approach can lead to partners pinching clients or retaining those that might be better suited to another partner.

There is therefore a thin corporate shell, with a need for greater holism below that shell. The marketing manager accepts this situation because the market is highly fragmented and partners are most likely to be successful in areas in which they are most interested. However, a focus is slowly beginning to emerge, the manager encouraging recognition that there will be even greater sales success if market analysis and partner interests can be combined in a co-ordinated and coherent strategy. Increasing market segmentation is seen as one tool for managing this situation internally.

This partnership has succeeded in doubling its turnover during 5 years, so an enhancement of a marketing strategy that is followed in practice could yield great growth opportunities.

in success on this. For example, the North American view of planning is *in order to do* something, whereas the European approach is to plan *because of* some factor or other: the difference between means and ends, between actions and politics.

These implementation problems occur in many contracting and consultant organisations. Business strategies and plans focus upon *what* to do rather than *how* to do it. There is an *implementation gap,* which starts at the planning stage before delegation is begun.

It is easy to sign up to all the objectives set out in ... *On the case 3*, but that is not the issue. There is an absence of information or guidance as to *how* to pursue the objectives. It is legitimate to give

... On the case 3

A civil engineering and building contractor in the south of England states its group vision in its marketing plan as:

- ❑ *What:* Opportunities within the construction sphere, which will yield a profitable return
- ❑ *How:* Expand on existing markets where projected returns are better.

Look at new markets, which are construction-related and offer opportunities for profitable diversification.

Expand on present investment policy in order to produce above average returns by active management, while a secure income stream is maintained to underpin the group activities in less profitable times.

- ❑ *Where:* Primarily in the southern half of England, with expansion into other areas where core clients' requirements enable us to operate profitably
- ❑ *Size:* Turnover will be governed by our ability to manage effectively and profitably at any given level
- ❑ *Review:* There should be an annual review of all activities within the group to ensure that they meet the above criteria, together with action plans to remedy any areas of under-performance.

each group operating division scope to put the flesh on the bones, yet they should be accountable for the results. The marketing document fails to spell this out. One common reason for this sort of shortfall is that marketing and sales personnel are not always involved in calculating rates of return at the final account stage and may not even be present at deciding the tender mark-up. Other senior management may also be unclear as to the whole sales picture across the group, especially at the detailed level of client contact for opening doors and closing sales. In other words, everyone has a partial picture. Implementation does not involve co-ordination of the different market views. Thus, much of it is left to guesswork. All need to know:

- ❑ Preferred new markets, even at a general level
- ❑ Existing average and desired rates of return from a group perspective
- ❑ Critical range of expanding present investment policy in order to calculate the trade-offs between desired returns and resource inputs
- ❑ The kind of work that yields above average cash-flow for investment purposes, in order to improve the reserves for less profitable times
- ❑ Limits of turnover level for effective management
- ❑ Criteria for annual review and under-performance that are not encapsulated in the above.

The sort of criteria that are helpful fall into two categories:

1. *Performance*-based criteria
2. *Results*-based criteria.

In this context, *performance*-based criteria focus upon the marketing process, analysing whether those responsible are implementing the marketing plan comprehensively. These are measurable outputs, such as detailed plans, management actions for steering and monitoring the sales process, and so on. Other performance criteria may be more qualitative, such as evaluating whether sales people are getting close to potential clients. Performance-based criteria are important because they show whether those responsible are doing their jobs – which is not quite the same as securing the work, nor as carrying it out profitably.

Results-based criteria focus upon turnover and profitability, the sales being measured against internal targets, but also against the size of the market. Too often marketing personnel are judged against targets set for them at the start of the year, which are generated by *internally driven decisions* and have nothing or little to do with what is happening in the marketplace. *Market share* is therefore an important statistic for monitoring results, even in a highly fragmented construction and consultant market. Market share may be measured across the sector and in a more targeted way by *market segment*.

Therefore, have a *structure* for developing the marketing plan, which is robust yet dynamic. Be open to the possibility that the marketing plan may require organisational restructuring in order to deliver the content. A *process* is needed, so create a system for:

- Formulation
- Implementation
- Monitoring
- Evaluation and refinement of the details and emphasis within a systematic framework.

Implementation requires understanding and commitment from the top down in two ways – being convinced of the approach and following it through with action every day. Having a robust marketing plan is therefore important.

The changing marketplace

At the end of the 1960s Dexion Ltd, for example, formed an office planning operation in its Materials Handling Division. It entered the market in response to a client opportunity and it grew without any market analysis until the early 1980s, when the operation was closed in the face of recession. The market was for composite wall partitioning; however, this was not a product. The market was for system suppliers putting packages together, yet Dexion supplied through wholesalers that constituted 50% of the whole market. Integrated suppliers catered for 30%, with interior contractors and general builders each occupying 10% of the market. The structural change meant that Dexion were not controlling the point of sale. Their market was entirely at the mercy of the wholesale stockists in terms of how they specified the package to their clients. The consequences were not recognised for some while, the solution being to either sell the packages direct or to invest in selling to the wholesalers. The solution requires addressing whether to sell primarily upon the basis of added value or price.[10]

Why plan when the marketplace is always changing? As pace of change accelerates, punctuated by the occasional event that threatens to overturn everything, it is a good question to ask. It is possible to be entirely pragmatic for a while, as Dexion, or where there is a regular bedrock of work, as happened with public sector work during the first three decades after World War II. It is definitely unsustainable in the long run. Uncertainty needs to be managed.

A business strategy and marketing plan are needed. These may not always anticipate changes, especially the rate and scale of change; however, plans provide a foundation or point of departure for actions. Planned actions include avoiding the paradox of the zero-sum game, creating differentiated services, as well as crisis response. The aim is therefore to develop strategies that promote, allow for and embrace change. Business strategies and marketing plans must therefore operate at different levels and time scales. A framework for this is provided in Figure 1.1.

Marketing is part and parcel of the overall business strategy. In some companies it drives the organisation. In other cases, it is another factor that may follow from other business decisions. In essence it boils down to why the organisation is in business: 'Are you in the architecture business in order to make money or are you making money because you are a good architect?'. Changes in the external environment can also impact on marketing through changes to other key dimensions of the business strategy. The key dimensions are:

- Marketing – for getting the business
- Human resources – for doing the business
- Operations – the business itself
- Finance – oiling the wheels of business.

Long-term aims should be specific and realistic. In other words they go beyond mission statements and good intentions. Kotler[11] says, 'The mission states the philosophy of a company whereas the strategic objectives are measurable goals.' Setting realistic goals may, for example, be to increase market share in the Thai market among overseas contractors from 0.5% to 7% and at least double turnover during a 10-year period. Another example would be to establish an operating base and workload in the facilities management market in the southern half of Finland within the next 7 years. Medium-term objectives may cover matters as those shown in ... *On the case 4*.

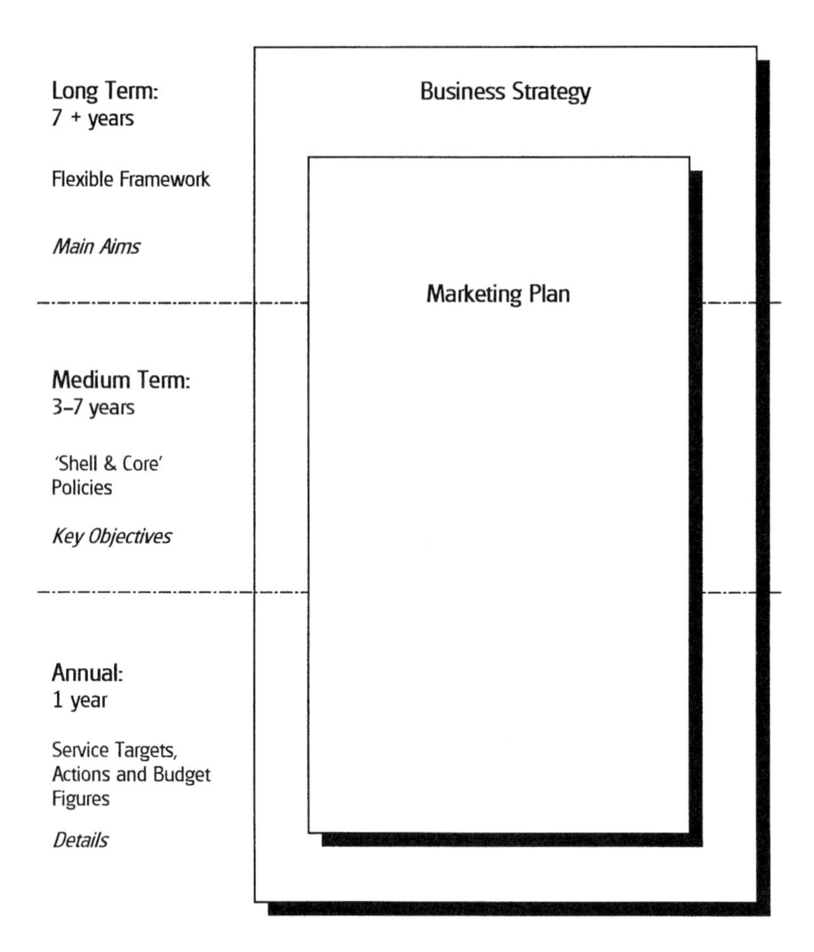

Figure 1.1 Time-scales and levels of strategic planning.

The flexible approach seems to come more naturally to those operating overseas, particularly for service organisations, because of the risks created by distance, as well as political and economic instabilities in some developing nations.[12] This is echoed among contractors and consultants.[13] An overseas presence can help in the home market through increasing credibility. However, the main lesson is having a longer term and more flexible approach, combined with more detailed strategic requirements for marketing.

Starting the plan

Many organisations start the market planning with a SWOT (strengths–weaknesses–opprtunities–threats) analysis. The analysis should use some market research or audit data. Even where this is available, time and again organisations make two fundamental mistakes. They start with the internal factors: the strengths and weaknesses. They focus on these, never fully moving onto the external factors of the opportunities and threats. Why is this?

The first is that the majority of senior management focus primarily on internal issues; the politics and management of chasing,

... On the case 4

The domestic business of a large international contractor set 'sales objectives' as:

- Breaking down internal barriers
- Providing a seamless and flexible service to tailored customer needs
- Increasing repeat orders, negotiated and partnership contracts.

A regional contractor adopted 3-year strategic objectives for the contracting divisions from the group geared to improving and developing:

- Building: staff development and training
- Civils: new procurement methods
- Rail: new markets
- Homes: expansion of existing markets
- Maintenance: continuous improvement of safety and quality
- Central finance: monitoring and improving site efficiency plus staff development
- Historic restoration: investment in future services and techniques
- Specialist subcontractor: maximise return and support from central services
- Property: maintain and improve portfolio.

encouraging, controlling, reprimanding, and so on. The second reason is that these are the factors they know best and feel are within their power to influence. Compared to any amount of externally commissioned information, the internal information is linked to power, while much of the external market information can seem amorphous. On the other hand, the external forces of the bankers and shareholders appear at times more demanding than clients, hence the tendency to fix financial targets driven by these factors rather than what the market will stand. Finally, where senior management have a good perception of the market it is partial and they seldom seek to supplement this with the views from their own sales team, who are often better placed to see the evidence that is anticipating key market changes. While the sales team may not be able to interpret the significance of their observations, senior management can.

In essence *four views* of the market can be gleaned, which senior management should take on board in starting the planning process.

The next issue concerns the analysis of strengths and weaknesses. Senior management tend to be unaware of their real strengths. A frequent pattern of SWOT analysis is to generate a list. One regional contractor of a national group generated a list of 13 strengths at the outset of the marketing plan process. The problem here is that most organisations seldom have more than

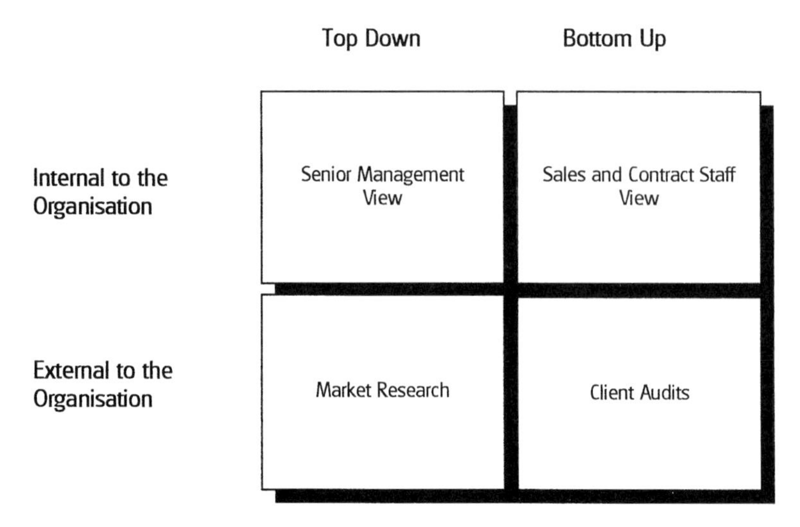

Figure 1.2 Four market views.

three genuine strengths. Longer lists are simply competencies, which they perform, probably as well as many competitors. Strengths have to be seen as being well above the average. Strengths have to be offered in a market where few or no competitors can match the strength. Strengths are the basis for the strategic differentiation of services. McKinsey & Co[14] have organised the sources from which strengths can be developed into the seven Ss:

- Hardware
  - Strategy
  - Structure
  - Systems
- Software
  - Style
  - Staff
  - Skills
  - Shared values.

In sales terms, strengths are to be offered into the market as *unique selling propositions* or *USPs*. The cynic may say there is nothing anyone can offer that is unique, but a combination of factors, including inherent strengths, can form a powerful basis for a USP.

A SWOT analysis is not therefore the best place from which to start the marketing planning process. The established starting point and subsequent stages from marketing literature is presented in Figure 1.3.[15]

The marketing audit, embracing the *four views* of the market is presented as a prior step and one that needs to be taken more seriously than has hitherto been the case. Auditing will be examined on a number of occasions in this book. On each occasion the results can be fed back into this stage of the market planning process, as well as having a more immediate application at the point of sale and project implementation.

However, the classic textbook approach is not the only starting point. There are a number of possible starting points. Merely selecting a different one may result in approaching the market in a different way to competitors, hence beginning to create some competitive advantage. For example, the planning process may start with one of the following:

- Research and development as the driver for service differentiation
- Innovation and value management coupled with market segmentation

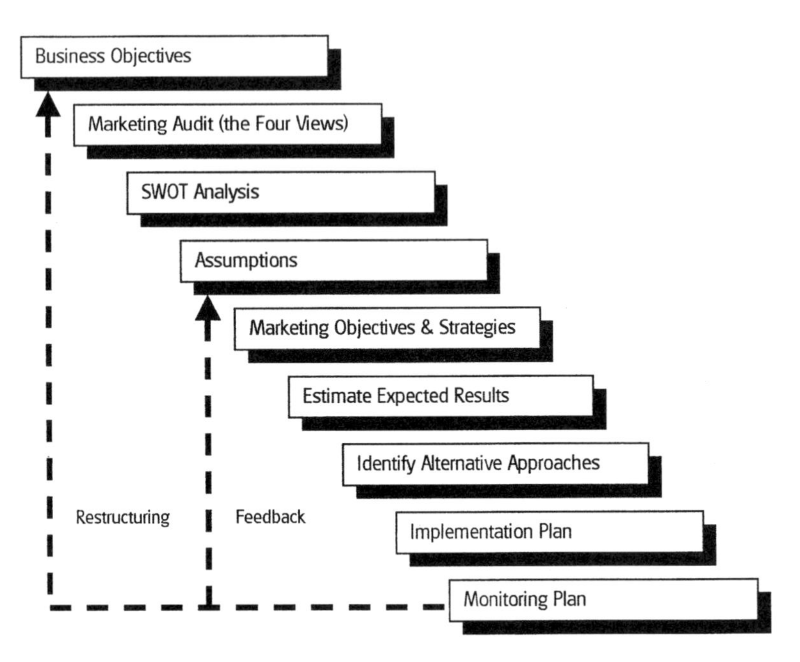

Figure 1.3 The marketing planning process.

- Creation of financial packages to facilitate projects
- Market creation in niche areas.

Each of these approaches is specific and cannot be applied to all organisations. In a sense, that is the aim. However, a different start point from the SWOT analysis or the classic textbook approach will be used here, one that can be applied more generally. It may be that your response is that you and your organisation do not have time for this. This is probably true, but you will not break away from the intensity of this market without differentiating services in order to be in a more manageable position.

The starting point is the *business objectives* identified in Figure 1.3. These flow from the business strategy and will be given a marketing slant. Instead of using the hackneyed phrase, 'What business are we in?', ask 'Where are we in the market?'. To establish this, a marketing matrix can be used that identifies six basic *market positions*. Having established this, attention will then be focused on how organisations can grow in or change their respective positions, with the attendant risks.

Establishing the market position

A *market position* is established around two main variables. On the vertical axis are the technological factors and the methods used by the organisation in its day-to-day work. The technological factors embrace those technologies in the headquarters and on site. The methods concern the way in which the technologies are applied on site and in the office. The methods also cover the general approach that management takes towards successfully completing projects. The methods and the technologies help to determine what is delivered to the client.

Technology and methodology

The technology and methodology can be divided into three categories. The first approach is the *routine* one. This is used by organisations that have developed standard products and services. They used tried and tested solutions. They will adapt these to specific circumstances, without departing from their routine approach. The overall approach to the market is to secure turnover levels. The profit margin will be low. Delivering services at low cost will be the primary consideration of management. Therefore, the overall level of company profitability comes from creating a mass of annual profits, which are achieved from sticking to the routine management practices in head office and on site. The marketing strategy will focus on a few mass markets.

The product and service delivery is standard. Price competition is keen. Selling is undertaken by a dedicated sales team, who are responsible for the identification of leads and following these through until the tender opportunity is secured. They will work independently from the other functions in the organisation and from senior management. Their remuneration will tend to be organised around some sort of incentive package on top of a basic salary.

The classic example of this approach is the volume speculative house-builder in the UK. A range of standard house-types is offered with a few variations on each. Warehouse construction for property developers tends to fall into this category. The standardisation of the service, the product and the routinised management approach permit more of the services being undertaken in-house.

The second approach is the analytical one. In this approach, organisations draw upon their experience and generic skills to solve issues and problems that arise on contracts. It is therefore a problem-solving approach. The contractor or consultant will ideally be seeking an average or above average profit margin. Being responsive to the client and understanding their needs is the key to delivering the required service and creating the profile and reputation in the market.

This approach needs an outward-looking posture by senior management. The service is tailored to the client. Management will become involved in selling, as well as determining the marketing strategy. Indeed, selling will be a co-ordinated effort. Those responsible for selling benefit from having had operational experience in order to interpret and respond to client needs in order to establish credibility up to the pre-qualification stage. They will bring in experts at the pre-qualification stage, if they have not already done so. They will also involve senior management, usually at board level, to provide overall reassurance and confidence that their people can rise to the challenges of the project. Their remuneration should be high, as should their status and ability to command respect in their own organisation.

The construction of health care, laboratory and hotel facilities by the same design or contracting organisation provide good examples. They have features in terms of project content and client needs that are similar. Clients frequently emphasise the need for a good track-record with their type of company and building type. Putting financial packages together can also be part of the approach, for example in the large-scale steel bridge projects in the global market as contractor Trafalgar House, now part of the Norwegian group Kvaerner, undertook in the 1980s.

The third approach is an *innovative* one. A distinction is made between invention and innovation. Innovation requires bringing something new to the table in a strategic way. It is therefore more than problem solving. For example, as engineering consultants, Ove Arup are known around the world for their innovative approach and sometimes this leads to invention too. The cutting edge or 'guru' architects of the day are of this ilk. Interestingly, it is the design professionals that tend to fall in this category, few contractors offering *innovative* services in the marketplace. High profit margins are required because of the corresponding risks. The prestige, especially in building design, is seen by clients as a worthwhile price, particularly where corporate image or identity can be enhanced. Selling is carried out through non-company networks. Reputation is a key plank. The other is profile, prestigious journal articles and books being the media used by organisations. The client will always want to see the most senior person and therefore it is their responsibility to close the sale. This is not carried out through 'selling techniques' in an overt way, but is achieved through technical and professional means, convincing the client that they have the most appropriate solution.

Value orientation

The second variable is the *value orientation* of the company. Value orientation refers to beliefs and outlook. This is shown on the horizontal axis. There are two main options. These are shown below in Table 1.2.

The *paternal orientation* embodies a set of values of oversight and control, which provides definitive direction. At best this creates a 'family' atmosphere and security for employees. At worst it tends to be autocratic and sometimes overbearing. It largely depends upon whether management likes to see employees, and subcontractors to some extent, conform to their way or whether the oversight is there to provide a context in which people's strengths and potential can develop. The paternal orientation does not necessarily mean that a contractor is family controlled, although it may be. Conversely, a family-controlled business does not have to adopt a paternal approach.

The *practice orientation* is not about scale. Again, it is the attitude that is important. The value orientation is one of being in business to be an engineer, architect or surveyor. Being professionally successful yields a profit.

The *corporate orientation* embodies three prime criteria for existence – the bottom line, the bottom line and the bottom line! In order to be successful, the organisation has to perform its contracting or professional function efficiently and competently. Henry

Table 1.2 Value orientations of contractors and consultants

Value orientation	Contractors	Consultants
Option 1	Paternal orientation	Practice orientation
Option 2	Corporate orientation	Corporate orientation

Source: Adapted from Coxe *et al.* (1987); Smyth (1996).[16]

Ford stated that a business has to be profitable or it will die, yet in other words, 'when anyone tries to run a business solely for profit ... then also it will die for it no longer has a reason for existence'.[17] Under both options 1 and 2 in Table 1.2, financial success and operational success are requirements. The issue is one of emphasis; that is, what primarily drives the organisation.

Market position matrix

Combining the vertical axis and horizontal axis produces the market position matrix. Six positions are created (see Figure 1.4). Organisations tend to perform well when they are firmly based within one market position or when subsidiaries and divisions are set up to deal with services or businesses occupying different market positions.[18] There tend to be very few contractors occupying market positions *E* and *F*. In all cases, the tendency is for services and products to migrate to the top right-hand corner of the matrix, as yesterday's innovations are refined and applied elsewhere until such time as they become standard ways of doing business and delivering a service. Refreshing and developing the business through the injection of new ideas, techniques and management processes is necessary in order to maintain the current market position in the long term. This is one of the strategic marketing

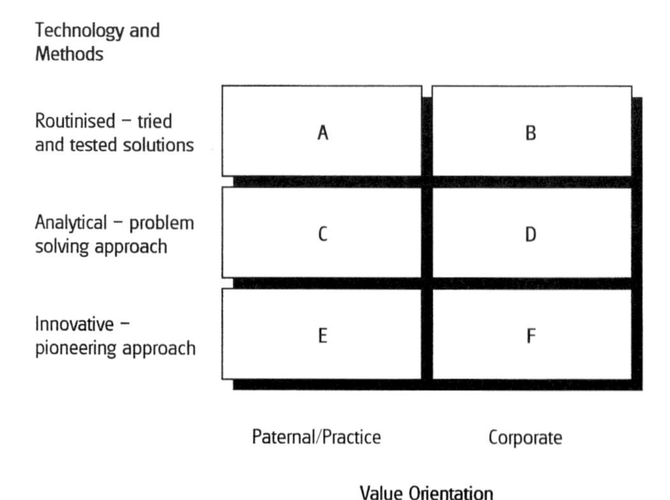

Figure 1.4 Marketing position matrix.

issues being developed in this chapter – the need for coherent, consistent and dynamic strategic marketing.

There are ways in which a more sophisticated approach can be adopted to analysing the market position. Project simplicity and complexity can be introduced to create sub-classifications of each market position. An architectural practice, specialising in community and arts projects and occupying an *E-type* market position, is increasing the size and complexity of projects, in order to cover the equipment costs and maintain profitability. Urban renewal schemes, involving liaison with local stakeholders and user involvement in design, have provided a bridgehead to being able to undertake larger projects. The reputation of the practice amongst peers and clients has created opportunities to increase project size. Smaller projects remain important as a way of training and developing staff. In this way, the practice is injecting new talent with appropriate experience into the organisation in order to maintain its market position in the longer term.

This process has the effect of imparting to staff its value orientation, as well as instilling the methodology and technological applications for the next generation.[19] A similar exercise can be carried out for geographical coverage, adjacent markets to the operating base or headquarters being treated differently to those at a distance,[20] perhaps outside the Far Eastern market for a Japanese contractor or outside Europe for a French contractor. However, this type of approach brings the analysis closer to the next stage, *market segmentation*, which will be looked at in the next chapter.

The typical key features for each market position are summarised in Figure 1.5. Assessing risk, particularly in the context of managing growth and indeed market decline, is very important. There is a relationship between risk levels and desired returns. Crudely, the lower the risk, the lower the profit margins. Consequently, the higher the pressure becomes to generate turnover. Low risk for the business can therefore lead to high-pressure sales. The reliance on existing clients and referrals is usually insufficient to maintain the status quo. New clients have to be found and the cost is high. This high cost should ensure that management keeps pressure on their sales staff and all those involved in selling. This is typical of the experience of *A* and *B* type companies, and those that have low levels of repeat business in *C* and *D* type organisations. Expansion of any business, when the rate of expansion is greater than overall expansion in the market will generate similar pressures.

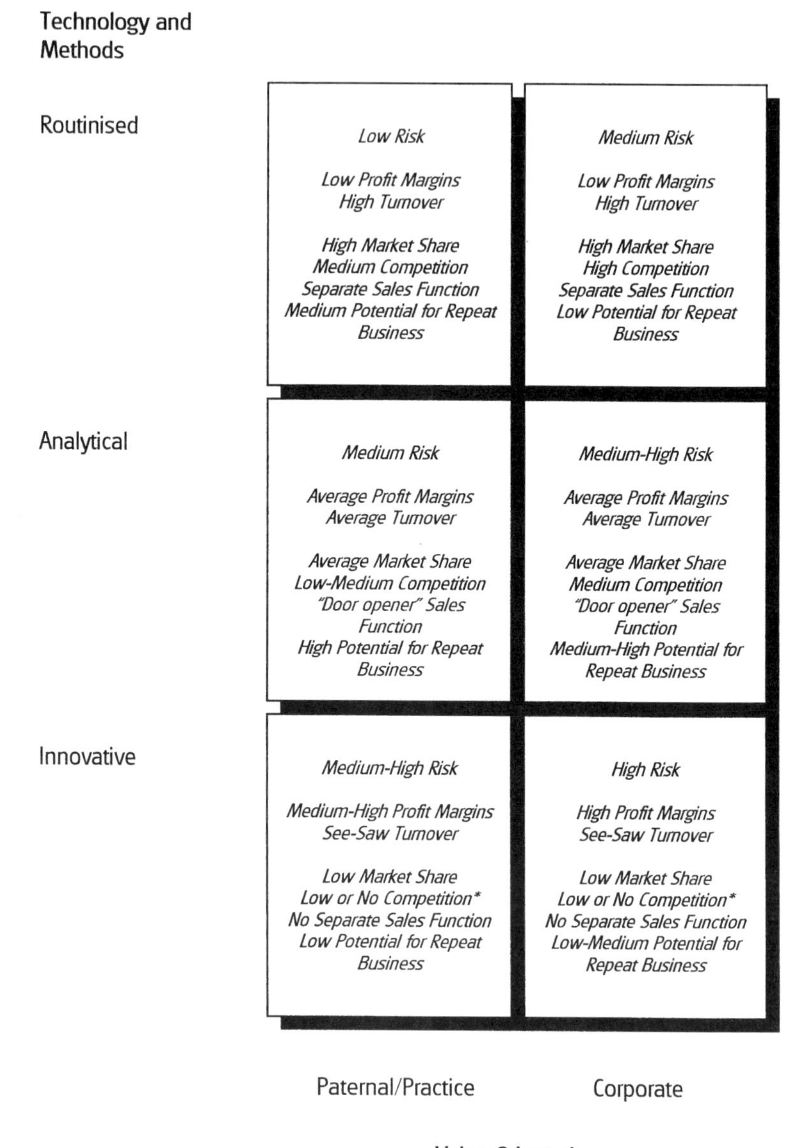

Figure 1.5 Typical key features of market positions.

There are a number of ways for beginning to establish the market position. A few options provide strategic alternatives:

- ❑ Ask key staff to locate the current position, giving reasons
- ❑ Ask existing clients, using it as part of the market research and perhaps include some potential clients too

- Ask consultants
- Bring in consultants.

The next task is to establish the preferred position, then, analyse how to get from where you are to where you want to be, by undertaking a detailed analysis of competencies, strengths and weaknesses.

Finally, it is important to determine how to reinforce your market position and protect it. This is a complex process and involves managing the marketing effort so as to ensure that differentiated services are created and stay ahead of the competition. This will ensure the potential for growing in the market.

Models for growth

There are basically three growth models:

- Expand in existing markets
- Expand in new markets
- Expand by take-over or merger.

The third option is favoured by *A* and *B* type companies, particularly the latter. This provides a route to achieve the first two. The *routinised* approach is imposed upon the acquired organisations and access to markets has essentially been bought.

The first two options are organic and tend to be pursued by organisations occupying other market positions. The risk levels are different and are shown in Figure 1.6.[21]

Because of the higher risk in diversification, contractors tend to offer a range of services in a *routinised* and standard way – more so than consultants. This leads to homogenisation of services and price competition and makes it hard work for the sales team and all those involved. Selling existing services in existing markets requires identifying targets of increasingly lower priority, where the strike rate will inevitably become less and less, as will the profit margin. Diversification is the antidote.

Diversification can be about offering new services in existing markets. New services can involve developing a brand new capability in the way many contractors have been doing with facilities management in recent years. It can involve transforming existing services in order to fit niche markets not currently being served.

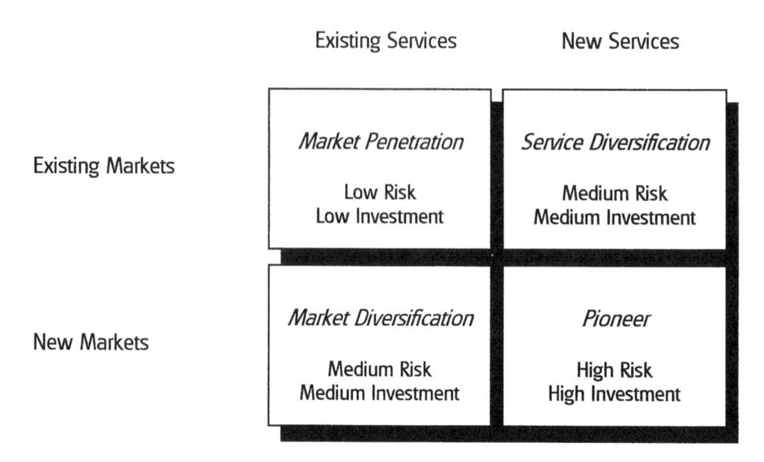

Figure 1.6 Risk analysis of organic growth.

Diversification can be about offering existing services in new markets. New markets can be defined in terms of expanded geographical coverage, where generic skills in *analytical* market positions can be applied to different building types or in new market segments. Organisations need to be certain that there is a demand for the products or services. US companies tried to sell 2 × 4 inch timber joists in the Japanese housing market. Yet this represented 7% of the market, Japanese housing construction tending to be post and beam construction, using 4 × 3 inch members. Canadian firms were more successful, as they paid attention to demand, backing this up with product quality and an after-sales service.[22]

The highest risk occurs where the new services are offered in new markets for the organisation. Pioneering is therefore occurring at two levels and the likelihood of failure is high. Careful and consistent client auditing can help monitor performance and this can be seized upon as an opportunity to get close to the client – a theme that will be discussed in subsequent chapters.

The inherent risk of any market position should be coupled with the risk of the growth models. Particular combinations therefore embody greater risk levels. These need to be managed and that is one function of the market planning – providing a framework for managing the chosen growth path (see the Paradox 5 in Appendix A). The risk is also a matter of timing. Doing the right thing at the wrong time, usually too late, can prove very costly (see ... *On the case* 5).

... On the case 5

A mid-sized international civil engineering and quarrying contractor decided to diversify its operations in the 1980s, selecting the leisure and housing sectors as areas of growth.

The contractor's analysis proved correct; however, competitors had identified the growth markets at an earlier stage and had built up both credibility and a track record, which acted as a barrier to subsequent entry.

The company had carried out little market research. The marketing director appointed to implement the diversification was therefore working against the odds from the outset. He wished to identify other markets, to get himself and the company out of its precarious position. He was essentially a sales person and not a marketer. He was also acting under increasing corporate pressure without the management support to undertake a plan review. He became an inevitable scapegoat.

The corporate outcome was a set of disappointing financial returns over the best part of a decade, despite a string of capable managing directors being appointed. An eventual withdrawal from building work and a later re-entry into speculative housing has been undertaken to complement the civil engineering core business.

This case illustrates the importance of timing of diversification.

There are factors driving organisations towards diversification. Internal factors include ambition and other career politics, overhead structures and financial accountability within the organisation. External factors include realising increased profits, market saturation, political and taxation issues. These issues are well documented in marketing texts and for construction.[23] Financing growth can come from external sources, such as loans and share issues. Finance can be internally generated from mature markets, the so-called *cash cows*. However, a cautious note has entered into diversification in recent years. The return to *core business* has been found to give managers better control and thus *diversification* should be seen as being desirable within bounds. It should clearly be distinguished from *differentiation* within the core business. The former focuses upon the organisation, whereas differentiation compares the organisation against what is on offer from the competition.

Company growth rates above those of the whole sector, or in particular segments, induce increased *market share*. This is a key indicator for organisations pursuing financial growth. Given the cost of pursuing new clients and thus of new markets too, there must be a balance struck between growth through diversification, growth through differentiation in existing markets, servicing existing clients and stagnation. Which growth model do most companies use? Table 1.3 provides an indication of competitive strategy across a number of sectors, setting this against price as a key variable.

Table 1.3 indicates that Japanese companies have pursued high-risk strategies, one that the Far Eastern stock market collapse of the late 1990s may have highlighted. Yet the British strategies at the other end of the spectrum contain high risks of losing markets to overseas competition, something which is underway at present in the British construction sector.[25]

Table 1.3 Strategic focus for growth

Option	US companies (%)	Japanese companies (%)	British companies (%)
New markets (entering emergent segments)	50	70	40
Existing markets (gaining market share)	73	83	53
Cost reduction & productivity gains	70	43	83

Source: Adapted from Doyle *et al.* (1992).[24]

Conclusion

This is an era in which certain nations will come to dominate the international construction market. Other national contractors will largely be confined to national or regional markets within their own countries. It is also an era in which the international players will seek to grow by undertaking the major contracts within a country. In other words, what has hitherto been the experience of developing nations will become the global norm. One way in which contractors and consultants can compete in this marketplace is to take marketing seriously in terms of plan formulation and implementation. This will form part of the process of shaking up the market.

Realism embodies what is possible in terms of aspirations, which in turn is dependent upon having both a good level of detailed knowledge of the market and planning to manage that market both effectively and efficiently. Organisations will do well to follow the 80:20 ratio between service and product impact and seek to *differentiate services* that match organisational strengths and client needs. *Differentiation* starts with the marketing strategy. For example, establish your market position or begin geographically with a global outlook or seek service niches. In this context, assess the market risk and choose a *growth model*:

- ❑ Market penetration
- ❑ Service differentiation
- ❑ Take-over.

It helps to focus on client benefits and to avoid price competition as the predominant issue unless a uniform and standard approach based on pure volume, high turnover and low margins is chosen.

Those that take marketing seriously will be better equipped to operate on the scale and territory they do and have aspirations to do. As the shake out is occurring from the global down to the local, contractors will need to be realistic in their aims. Managing those market aims and objectives is important. Determining the detailed management of the market is the subject of the next chapter.

Summary

1. The *purpose* of this chapter has been to:
   - ■ Recognise the need for marketing planning
   - ■ Formulate robust plans from the top of the organisation

- Implement plans consistently and coherently from the top to the bottom of the organisation.

2. The chapter has:
   - Established that there are different ways to approach the marketing process
   - Presented the market position matrix as one starting point
   - Acknowledged that there are risks associated with any one approach
   - Recognised organisational growth and profit growth as key objectives, which require increased market share and/or productivity as the means of achievement
   - Outlined that growth in market share is achieved through increased market penetration, diversification and/or takeovers, each of which has an attached risk profile.

References

1. Fisher, N. (1986) *Marketing for the Construction Industry: A Practical Handbook for Consultants and Other Professionals*, Longman, Harlow, p. 18.
2. Spencer Chapman, N.F. and Grandjean, C. (1991) *The Construction Industry and the European Community*, BSP Professional Books, Oxford.
3. Doyle, P., Saunders, J. and Wong, V. (1992) Competition in global markets: a case study of American and Japanese competition in the British market, *Journal of International Business Studies*, **23**, 419–442.
4. The sample is derived from questionnaire responses among delegates attending courses run by the Centre for Construction Marketing, Oxford Brookes University, Oxford, which runs short courses and training events, conferences and publishes research.
5. Leading Edge (1998) *On the Edge*, **9.1**, Leading Edge Management Consultancy, Hitchin.
6. MacDonald, M. (1992) Ten barriers to marketing planning, *Journal of Business & Industrial Marketing*, **7**, 5–18.
   MacDonald, M. (1995) *Marketing Plans*, Butterworth-Heinemann, Oxford.
7. Simkin, L. (1996) People and processes in marketing planning: the benefits of controlling implementation, *Journal of Marketing Management*, **12**, 375–390.
8. Dibb, S., Reiko, K. and Simkin, L. (1995) Marketing practice in management consultancies, *Journal of Management Consultancy*, **11**, Fall, 30–36.
9. Kotler, P. (1991) *Marketing Management*, Prentice Hall, New York.
10. Reynolds, A. (1986) Marketing strategy for wall partitioning in the UK, *European Journal of Marketing*, **20**, 68–79.

11. Kotler, P., Armstrong, G., Saunders, J. and Wong, V. (1996) *Principles of Marketing*, Prentice Hall, London, p. 78.
12. Valikangas, L. and Lehtinen, U. (1994) Strategic Types of Services and International Markets, *International Journal of Service Industry Management*, **5**, 72–84.
13. Smyth, H.J. and Stockerl, K. (1998) Strategic marketing planning by UK contractors, *Proceedings of the International Construction Marketing Conference*, 26–27 August, Leeds University, Leeds.
    Langford, D.A. and Rowland, V.R. (1995) *Managing Overseas Construction Contracting*, Thomas Telford, London.
14. Peters, T.J. and Waterman, R.H. Jnr (1982) *In Search of Excellence: Lessons from America's Best-Run Companies*, Harper & Row, New York.
15. Adapted from McDonald, M.H.B. (1992) *Marketing Plans: How to Prepare Them; How to Use Them*, Butterworth Heinemann, Oxford.
16. Coxe, W., Harting, N.F., Hochberg, H. *et al.* (1987) *Success Strategies for Design Professionals: Superpositioning for Architecture and Engineering Firms*, McGraw-Hill, New York.
    Smyth, H.J. (1996) Design and build marketing: issues and criteria for architecture selection, *Proceedings of the 1st National Construction Marketing Conference*, 4 July, The Centre for Construction Marketing in association with CIMCIG, Oxford Brookes University, Oxford.
17. Quoted in Reichheld, F.F. (1994) Loyalty and the renaissance of marketing, *Marketing Management*, **2**, 10–21.
18. Coxe, W., Harting, N.F., Hochberg, H. *et al.* (1987) *Success Strategies for Design Professionals: Superpositioning for Architecture and Engineering Firms*, McGraw-Hill, New York.
19. Smyth, H.J. (1996) Design and build marketing: issues and criteria for architecture selection, *Proceedings of the 1st National Construction Marketing Conference*, 4 July, The Centre for Construction Marketing in association with CIMCIG, Oxford Brookes University, Oxford.
20. Smyth, H.J. and Stockerl, K. (1998) Strategic marketing planning by UK contractors, *Proceedings of the International Construction Marketing Conference*, 26–27 August, Leeds University, Leeds.
21. Ansoff, H.I. (1957) *Strategies for diversification, Harvard Business Review*, **35(5)**, 113–24.
22. Czinkota, M.R. (1994) A national export assistance policy for new and growing businesses, *Journal of International Marketing*, **2**, 91–101.
23. See Hillebrandt, P.M. and Cannon, J. (1990) *The Modern Construction Firm*, Macmillan Press, Basingstoke.
24. Doyle, P., Saunders, J. and Wong, V. (1992) Competition in global markets: a case study of American and Japanese competition in the British market, *Journal of International Business Studies*, **23**, 419–442.
25. Stockerl, K.C. (1997) The importance of strategic marketing planning for the UK construction industry in a changing European business environment, *Proceedings of the 2nd National Construction Marketing Conference*, 3 July, Oxford Brookes University, Oxford.
    Smyth, H.J. (1998) The competitive stakes and mistakes: the position of British contractors in Europe, *Proceedings of the 3rd National Construction Marketing Conference*, 9 July, The Centre for Construction Marketing in association with CIMCIG, Oxford Brookes University, Oxford.

Smyth, H.J. and Stockerl, K. (1998) Strategic marketing planning by UK contractors in an international business environment, *Proceedings of the International Construction Marketing Conference*, 26–27 August, University of Leeds, Leeds.

Chapter 2

From Strategic Market Positioning to Selling

Themes

1. The *aims* of this chapter are to:
   - ❑ Strengthen marketing plans through development to a detailed level
   - ❑ Bridge the gap between strategy and practical implementation
   - ❑ Set the context for enhancing dynamic plan formulation and refinement.
2. The *objectives* of this chapter are:
   - ❑ To be equipped to identify segments and select niches in the market
   - ❑ To enhance the ability to target potential clients in each segment and niche
   - ❑ To explore sales as an enhanced and complex management function.
3. The *outcomes* of this chapter are to demonstrate that competitive advantage embraces:
   - ❑ Identifying client management as a key area of marketing and sales activity
   - ❑ Understanding the options for managing perceptions of marketing and sales inside and outside the organisation.

Keywords

Segmentation, Niche markets, Targeting, Sales process

Introduction

Children in front of the sweet counter take ages to select the sweets they wish to buy. Frequent changes of mind are evidence of the child learning to sift through the options and possible permutations. Parental patience may be running out, yet patience is essential. The exercise is as much educational as it is an indulgence of delight and expectation.

The construction culture is about *getting the job done*. Agonising over decisions and pawing over the options is not the typical style. There has been a certain amount of impatience with marketing; 'A basis for establishing a market position has been given, so it is done and dusted, and now we can get back to the real business' (see Paradox 6 in Appendix A). Yet, establishing a market position is only the first stage of selectivity (see Figure 2.1).

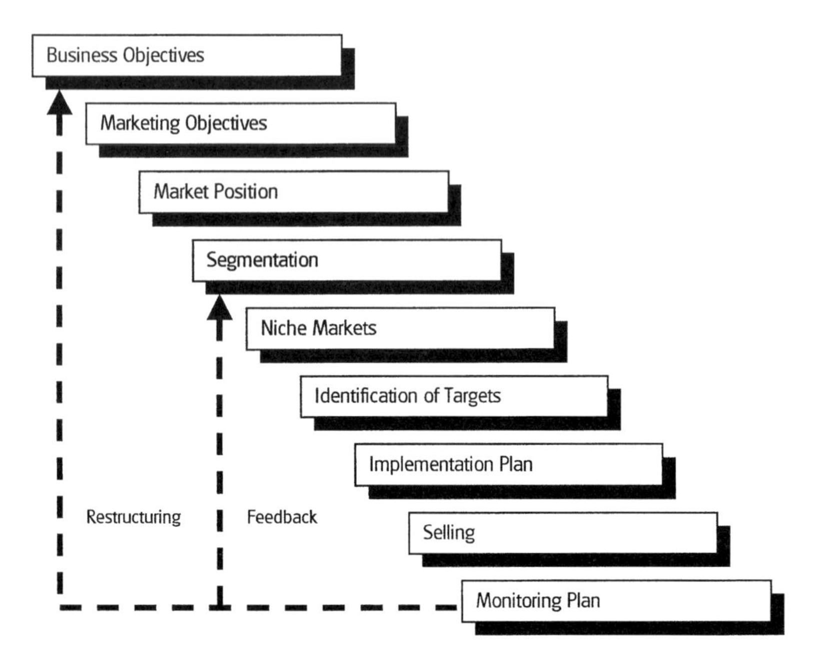

Figure 2.1 One alternative marketing planning process.

In marketing of consumer products, advertisers put together adverts that address very specific audiences. Over recent years, the people groups being singled out are smaller and smaller. The mass markets of the 1950s and 1960s have been transformed into highly targeted markets. Standard products are superseded by those designed or adapted for specific groups of people. This is the result of several forces at work:

- Producers penetrating new markets
- Competition in existing markets for standard products, as a result of an unhealthy emphasis on price competition or market saturation being the driving force
- Consumers forming high expectations and becoming more discerning and demanding.

Construction is no different. The dominance of public sector work that provided a backbone of standard projects has given way to greater reliance on a diversity of clients, whose building and infrastructure requirements have become more sophisticated and diverse. A standard approach is price dominated. Price has only dominated in construction for so long because of the highly fragmented nature of the supplier market and reluctance amongst contractors and consultants to differentiate their services from the competition. Service differentiation in existing markets and diversification into new markets is becoming the order of the day. This requires highly detailed analysis of markets and patience in exploring the potential of the organisation to serve particular market *segments* (see Figure 2.1).[1]

A sample of 133 responses spread across 30 construction and consultant organisations cited in Chapter 1 highlighted that 36% believed that they or their organisations needed to address the marketing processes of market position, segmentation, niche marketing and client targeting. This contrasted with only 3% believing that the financial issues of budgets, higher margins and pursuit of credit-worthy clients were the best way forward.[2] This data does not demonstrate that financial improvements are derived from addressing marketing, although this could be one reasonable conclusion to draw,[3] but it shows that improving marketing will benefit the position of the organisation as a whole. What are the stages in addressing the process of market analysis for segmentation purposes? These are set out in Figure 2.2.

Overview of market selection

As the process of market analysis is undertaken, choices and decisions are made. The market selection process takes into account three dimensions:

- ❑ Is there a market in that area?
- ❑ Do our strengths actually or potentially serve that market?
- ❑ How effectively can we manage that market and the clients within it?

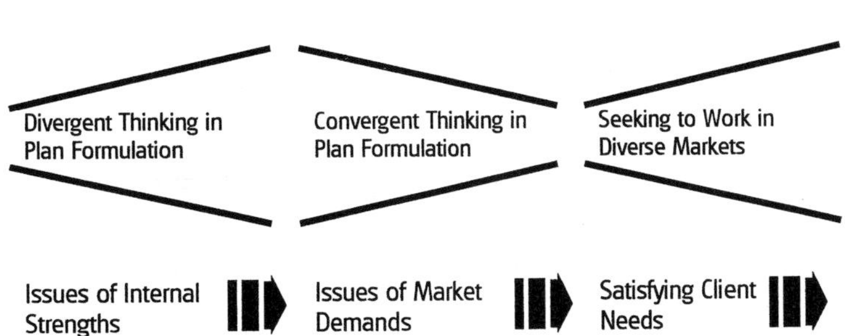

Figure 2.2 Process stages of market analysis for market management and service diversification.

Market area

Markets of growth in the sector are not the same as the best growth areas for the organisation. Several dimensions require scrutiny. These are:

- Existing markets – invest, harvest, divest
- Growth markets
- Competition
- Market entry – timing, risk, cost of entry, cost of subsequent withdrawal
- Internal capabilities – management strengths, skills, technologies, financial resources
- Reasons for entry – turnover, profit, commitment to client, type of work, short-term or long-term opportunity
- Financial returns.

All these issues are adequately dealt with in standard textbooks. The caveat to emphasise in a construction and consultant context is the time horizon. Many projects continue across the peaks and troughs of the economic cycle. Too little attention is given to how a project opportunity and the client will make out under dramatically changing economic conditions. Similarly, too few contractors and clients pay attention to key clients, who are regular building procurers, during recession or when there is a lull in their programme!

Management of market and client

The key to managing the market is to match organisational capabilities with market demand. There are a number of steps involved in this process:

1. Identifying strengths in the organisation for which there is a market demand
2. Managing the strengths to satisfy market needs
3. Managing the market, requiring the development of strengths to the changing pattern of market demand
4. Managing the market, which is an aggregate of individual client needs
5. Satisfying client needs, which is the management of the client
6. The effectiveness of client management, which is the extent to which the strengths meet client needs.

Thus, market management requires focusing upon parts of the market that are manageable. This therefore provides the platform for responding specifically to individual client needs – adding value and enhancing the service quality. The level of detail at which market analysis and management is conducted is getting smaller and smaller. The framework for focusing in this way is encapsulated in Figure 2.3.

Many organisations start with a *mission statement*, which in reality is a slogan, containing little of substance. Another approach is to position the organisation in relation to another familiar product (see ... *On the case* 6).

Establishing the market position helps to determine the segments open to each organisation. Each market position excludes other segments. Segments are simply categories of work that help the organisation to focus its efforts and thus begin to manage the market more effectively. How can organisations approach the segmentation process?

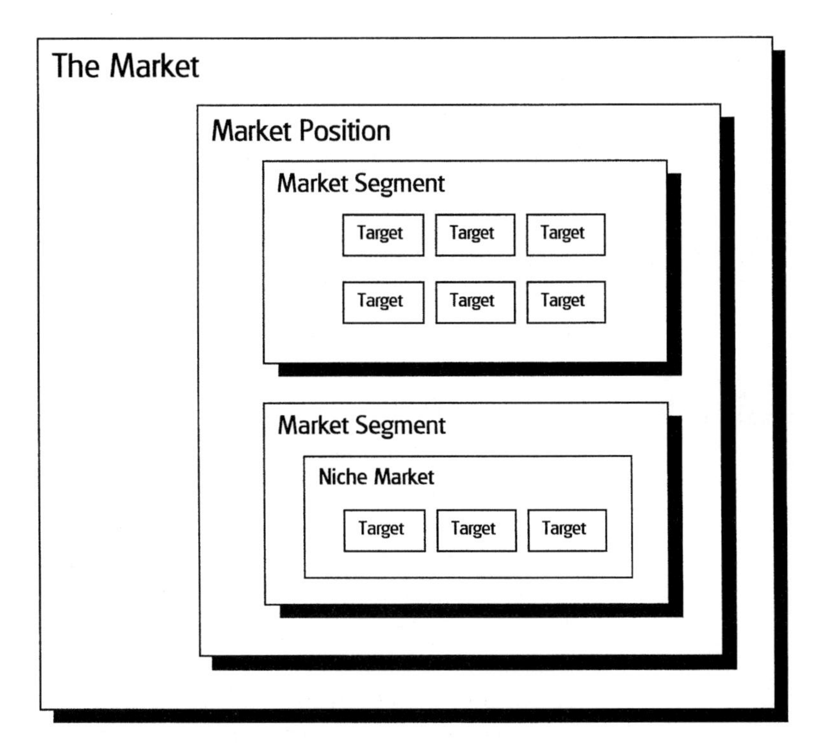

Figure 2.3 Market selectivity and management levels.

... On the case 6

A large architectural practice, employing around 100 people asked its clients to compare their service to brands of cars.

The result was that most clients rated the practice towards the upper end of the BMW range. This provided guidance on market position through the brand name of the manufacturer. It also began to provide some implicit data on market segments through the model selected.

This was subsequently followed up with a comparison of staff perceptions at director and associate levels, which showed a consistent picture of a *D-type analytical-corporate* market position. Interestingly, the managing director held a different perception, that of a much more innovative approach. However, this was where he wished to push the firm in the future. It helped to give direction for this approach; a recently formed R & D unit was aimed to refresh and sustain the *analytical-corporate* market position for the long term, rather than to compete against other *innovative* practices.

Segmentation process

Segmentation originates from micro-economic theory for price discrimination. It was a way of grouping things according to price sensitivity and was adopted into marketing theory and practice through the traditional *marketing mix* approach, creating four key variables, the so-called 4Ps:

1. Product or service
2. Place
3. Promotion
4. Price.

Price is just one variable: other variables come into play in terms of how to divide up the market. This is important to appreciate as organisation, endeavour to break the price domination for both their benefit and the benefit of clients in terms of quality of service and product.

The aim of segmentation is to divide up the market across two dimensions:

- Internal management of the marketplace from the viewpoint of most efficiently and effectively mobilising resources where they will meet the marketing objectives
- Identifying where the greatest market needs are for the product or service and grouping the targets into segments towards which the resources can be channelled.[1]

Most organisations focus upon the internal factors, rather than the market factors. The reason for this is that managers feel they have more control over the internal issues. However, such an emphasis may restrict the effectiveness of selling. It depends whether the organisation is primarily client driven or operationally driven. Both can be successful business objectives, providing the implications of this are realised and consistently and appropriately managed.

The traditional way of segmenting the construction market is set out in Figure 2.4.

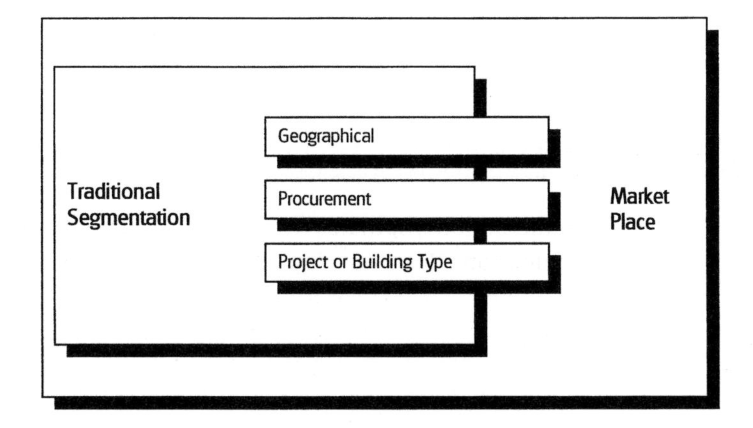

Source: Smyth, H.J. (1998) *Innovative Ways of Segmenting the Market: Practice Guide No. 1*, Centre for Construction Marketing, Oxford Brookes University, Oxford.

Figure 2.4. Traditional ways of segmenting the construction market.

Markets are organised around geographical coverage, the contract type or procurement route, such as design and build or partnering, and by the type of building experience with which the company has been involved.

Geographical coverage

Geographical coverage is a function of two key elements:

1. Size of organisation
2. Travel to site time and distance.

The size of the organisation is important. Large organisations increase geographical coverage by opening multiple regional offices. The limits of regional coverage tend to be determined by the second factor – travel to site time and distance. The ideal is often seen as a return visit within one working day. In Belgium and Germany, for example, regional issues give rise to the political necessity for having offices in each regional area.[4]

Geographical coverage is not generally seen as a market planning decision. It is perceived as an overall business decision and therefore *structural*, marketing tending to be organised mainly as a *process*. There is nothing wrong with this in itself, but it tends to inhibit thinking and thus the potential for differentiation of the service.

The international market tends to be treated differently. The North American and European contractors tend to treat parts of the world market as reasonably homogeneous. For example, the Far Eastern market for major projects has been dealt with in this way during the 1980s and 1990s. This is possible where the economies are developing. The more complex economies become, the more individual attention has to be paid to them.[5] Within the European Union, the market may become more harmonised over time; however, this will make marketing and hence segmentation more and not less difficult, due to the increased competition.[6]

Many of the world leaders in the construction industry have been making the step from being multinational or transnational to becoming global. In other words, instead of operating in a few countries and world regions, the company operates in any region or location. This requires a series of regional offices on a world scale and a single operational headquarters, with project work wherever the demand is located.[7] The demand tends to be in a major projects category or segment. Although this global approach does not theoretically fit with the structure of the construction sector,[8] a number of groups are competing against each other to become dominant in world terms. They tend to exhibit some of the following features:

- Diversification into conglomerate groups, frequently having strong characteristics of *vertical integration* with the opportunity to manage supply chains for individual projects
- Connections with financial and aid agencies so as to *leverage project finance* as a means to secure work
- Global *sourcing* of project inputs
- A desire to primarily pursue *market penetration* and build *market share*
- Approach to construction as a *mass market*
- A 'think global, act local' perspective.

The marketing aim is domination in the market for its own sake and, as a spin off, to be the most successful within the national location of the organisation (see ... *On the case* 7).[8] It is a geographical strategy (see Paradox 7 in Appendix A).

Another international player operating in similar markets uses the following criteria:

- Potential for negotiation rather than competitive bidding
- Availability of funding
- Cost of pre-qualification and cost of submitting
- Extent of capital expenditure required
- Availability of specialist plant
- Potential contractual requirements, payment terms and local laws
- Extent of competition.

The organisational strengths, measured against a potential major project include:

- Critical mass of work
- Scope for innovation
- Potential for using other vertically integrated businesses in the group, up- or downstream
- Technical edge
- 'Star' client status.

What steps can an organisation take to address the implications of geographical penetration and market competition? The global

... On the case 7

A large international conglomerate from Germany has a large multidisciplinary contracting organisation, which is going global.

The mission statement has been to move from being a 'master builder to system leader', providing multi-focused construction related services from design and finance through to maintenance and operation of constructed facilities.

The geographical strategy has been to enter world regions through take-overs and cross shareholdings, and to reinforce its presence in the home region through the same means. The company operates through associate companies in distant markets.

Strategic alliances and setting up new regional offices are supplementing the strategy.

Over 50% of turnover is generated overseas and over half its staff are engaged in international work. Of its international activities 86% are carried out through associated companies.

market is used below to show some of the issues. Consider these points:

- Evaluate the appropriate mix of businesses, *portfolio analysis*, in the context of the overall strategy
- Ensure that every element of the international strategy reinforces the home market position
- Integrate marketing into a business strategy for market penetration
- Build market share as the first objective, profit second, in times of expansion
- Trade-off profit against risk
- Be careful to take profits on the beginning of the upswing of the trade cycle and seek market share in remaining periods
- Control the supply chain through:
  - Networking in the financial and aid agency markets

  - Take-overs and alliances formation to achieve vertical integration
  - Having a local presence in key markets and regional centres
  - Developing strong head office control of markets in the home region and semi-autonomy for distant markets from the domestic base
- Build market share through take-overs and alliances within contracting or consultancy.

Procurement

Procurement is typically dealt with as a *structural* decision. This is more a matter for contractors, although some large consultants are aware of this issue in their organisation at a team level.

Whenever a new contract type, procurement route is identified, then the temptation has been to create a separate division or subsidiary. Contractors find it difficult to create a marketing culture in their organisations. Staff tend to feel comfortable with the 'way we've always done it'. The structural solution is a relatively easy and quick way to deal with procurement issues. It means that contractors can identify a set of key individuals who will deliver the new or particular procurement approach. The key individuals know they are accountable for this. They also become isolated and hence protected from others, who continue to do things the 'way we've always done it'. This can also inhibit cross-selling in the organisation, especially where the client is a naïve purchaser. It does inhibit different ways of looking at the market. The net result is that contractors all tend to adopt pretty much the same sorts of approach, albeit tailored to the procurement route, in much the same way. This tends towards a zero-sum game of intensive price competition.

Project or building type

The third traditional means of segmentation is project or building type. This puts the emphasis upon *track record*. It has become an arbitrary way for clients to include or exclude contractors at prequalification or tender stage. It is arbitrary because most contractors have a range of generic skills, which are transferable across projects regardless of building type. A prison shares common features with a hospital, which in turn has some commonality with a hotel. Not having built one building type does not mean an inability to build another type of project successfully. Having the generic skills and experience are the key competencies. This may be dependent upon market position to some extent.

Innovative segmentation

There are alternative ways of proceeding that have been introduced during the 1990s. The framework is set out in Figure 2.5.

Financial structure and corporate culture

Financial structure and corporate culture are located upstream in the business planning process. This recognises that segmentation could be heavily influenced by the business strategy of the contractor or consultant. It may therefore prove helpful to find like-minded clients. Matching the financial approach or corporate culture of contractor and client are therefore ways of segmenting the

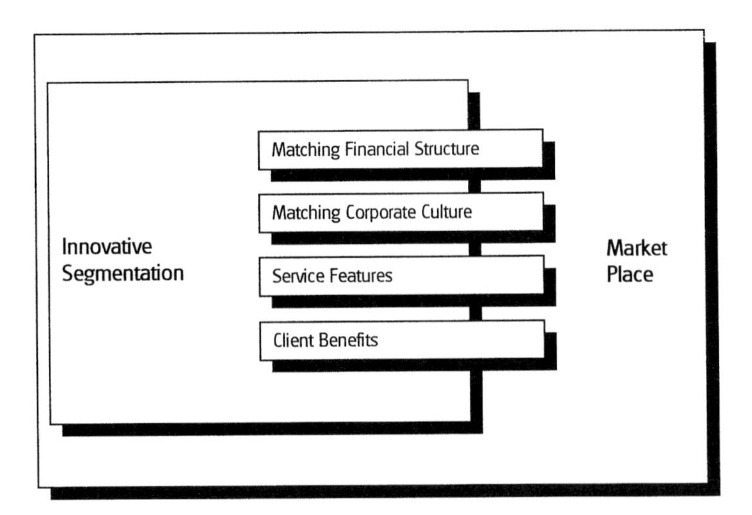

Source: Smyth, H.J. (1998) *Innovative Ways of Segmenting the Market: Practice Guide No. 1*, Centre for Construction Marketing, Oxford Brookes University, Oxford.

Figure 2.5. Innovative ways of segmenting the construction market.

market, by trying to identify those who have similar approaches to business. An affinity between the organisations may lead to being able to satisfy the client more easily and help facilitate long-term working relationships. Occupying a similar market position in terms of strategy may be another variation on the theme.

However this may not be as straightforward as it first looks. Like-minded organisations may not necessarily get on or work most effectively together. They may work in a complementary way at a very general level. At a level that is too specific, it does not necessarily induce an ability to work together in the short run, as the quote illustrates:

> *The evidence for this is based more on logic than practice. It is not certain that things do work quite this way. In a family context, to use this as an analogy, people often get on best when they are very different from each other. They share interests and an overall outlook, but the approach to specific issues is frequently contrasting. Indeed, it is said that the one curbs the worst excesses of the other and can plug the gaps their partner leaves. The outworking of this is frequently a source of conflict. Ideally it is a creative one and thus allows the partner to flourish in areas of strength. Perhaps another method can be to seek clients with which the cultures and structures dovetail rather than overlap.*[1]

The purpose of this approach is to largely ignore the types of development or project needs of clients, but to develop a strategy that looks at the overall corporate approach. Companies that match the criteria are targeted. The selling is usually geared to the higher levels in the client organisation. It is at this level that decisions can be influenced. It recognises that many clients have few people at senior levels with property or construction expertise and those that do are frequently afforded little status.

The approach requires the contractor or consultant to ignore the structural aspects of their business. People and resources are broadly clustered around particular client segments and resources mobilised towards targets where there are project opportunities.

Service features

Segmenting by service features is another method. It also looks upstream in the market planning process, particularly to the SWOT analysis stage. Looking closely at what the organisation delivers to the client, the aim is to sift the features into groups and match these features with the client types in the market.

This is usually comfortable for operationally driven organisations in overall terms. Yet it is also challenging in three positive ways. It prompts the questions:

- What are our real strengths?
- What do we actually do that is different from everyone else or most competitors?
- What features do clients really want?

Many organisations start the market planning with a SWOT analysis. One regional contractor of a national group started the planning process by meeting to determine the company strengths. They soon generated a list of 13 strengths. This is not uncommon, yet totally misses the point. Senior management tend to be unaware of their real strengths. Most organisations seldom have more than three genuine strengths. Longer lists are simply competencies, which they perform, probably as well as many others, as it was in the case of this regional contractor. If this is a characteristic of the planning process, then planning must be in the *weakness* box of the SWOT analysis. Strengths have to be seen as being well above the average. Strengths have to be offered in a market where few or no competitors can match the strength, as they are the basis for the strategic differentiation of services.

Once genuine strengths have been identified as existing or for development, the situation should be compared to competition in the market. A competitor analysis can be undertaken in parallel to ensure that the competition is manageable. Delivering service features that differ from the competition are clearly key in the last instant and this has to be balanced against the number of potential clients who desire those features, in order to create a market segment. A successful outcome is lower levels of competition and higher margins in the long run, providing the features are consistently delivered (see ... *On the case 8*).[9]

There are a variety of ways and means. A regional contractor in France segmented the market by *milieu*. A milieu has a geographical basis, but its heart is the economic–political network. The purpose was to use the *milieu* to gain market information on inward investment. The political network regionally and nationally is used to solicit referrals and to gain direct access to inward investors located overseas. This is the dominant segment for the contractor and therefore has become a means of organising its sales

... On the case 8

An international civil engineering contractor has operated in the marine market for many years. Even in that market it recognises the diversity of clients. It offers a range of services to differentiate itself and to enable the service to be tailored to client needs. Value engineering and specialist in-house design comprises a key value-added component. This is supplemented in the segmented approach with:

- ❑ Tracking clients over a long period, prior to known project programmes
- ❑ Introduction of a multi-skilled team at the problem definition stage
- ❑ Plant ownership
- ❑ Access to specialist labour.

A diverse contracting organisation brought in a management consultant to redefine its market segments. The overall approach became to focus upon the 80% of the service, which accounted for 20% of the costs, and to reduce emphasis on the core product, although accounting for 80% of the costs only constituted 20% of the benefits for the clients.

This was supplemented by a customer and potential customer audit.. This yielded the following segments that emphasised service features and benefits, described as:

1. Keep it simple
2. Efficiency seekers
3. Guardians
4. Worriers
5. Image seekers.

The first two are more conventional, price sensitive segments, the latter three looking for added value, the last one of which is potentially a high investment and high margin segment of a small elite group of clients.

effort. Differentiation can be reinforced through promotion and branding as Bovis have successfully achieved through the Bovis fee system in the UK. Even though the original concept came from a client in the 1930s, Bovis built on it. Organisations need to identify new service features or clusters of features for the evolving market. Of course, a feature may not necessarily be one that is particularly attractive to the client. Consultants and contractors frequently believe they know what clients want. Sometimes this is based upon some sophistication, sometimes upon arrogance concerning their expertise. A report commissioned by the CIOB a few years ago highlighted some areas of considerable mismatch between client and contractors in the UK.[10] Market research or audit data can be a useful source of confirmation and new information about client needs. It is an important and undeveloped area in contracting and among consultants.

Such approaches, combining innovative segmentation with market research, can be used to develop marketing commitments:

- At board level
- Investment in time
- Adequate development budgets
- Implementation: budgeting, training and re-organisation.

Client benefits

Every service feature has a corresponding benefit. For example, the efficiency of a car engine is a feature, the low fuel consumption the benefit. On a contract, being consistently within budget is a feature. The client benefit is the reduced borrowing and interest payments. The benefits of a feature may vary from client to client, as will their prioritisation. In many cases, it is the softer service elements which clients most benefit from, for example responding quickly to changes, helping the manager on the client side to secure promotion through successful completion or being able to trust the contractor.

This is innovative and challenging. It operates both upstream and downstream in the market planning process. It operates upstream as part of strategic market auditing and downstream as part of the monitoring plan for client satisfaction at the pre-qualification, tender and contract stages.

The aim is to identify the benefits that clients want. Market research and auditing are useful tools to supplement the day-to-day contact of the sales and contract staff. Benefits can then be clustered according to those that can be delivered by the contractor or consultant and those that occur commonly enough to make a segment. In practice, it may be that there are several clusters of benefits, each constituting a segment. However, these clusters are not likely to be as large a segment as classifying segments traditionally. A reasonable course of action would be to focus sales attention around the clusters – the defined segments – and act in a reactive or opportunistic way to the remaining market. The opportunism will need to be tempered with selectivity in terms of likelihood of success, in order to avoid diluting the targeted sales effort in the segments.

Taking the strategic steps towards addressing these issues may involve:

- Dividing up the market to most effectively penetrate the market, using the strengths of the organisation
- Allocating time, effort and costs providing segmentation is pursued with consistency and commitment, in order to secure an adequate return
- Organising segmentation carefully:
  - Structural organisation is easy to monitor, but is the least flexible
  - Innovative segmentation is more about *process*: creating new ways of tackling the marketplace can change the way of selling – either with greater input from senior management or through the sales and contract teams in the case of segmentation by features and benefits
- Creating time and financial budgets for policy formulation and implementation
- Developing genuine strengths that mark you out from the competition
- Creating feedback loops for refining the segmentation, using:
  - Market research and feedback from client audits
  - The sales and contract staff on a client-by-client and project-by-project basis for the service feature and client benefit approaches to segmentation
- Starting with *senior management*, the decisions being carried through to *sales* and *contract staff*.

Niche markets

Niche markets are a subcategory within a segment. For example, a consultant that specialises in healthcare work, may have developed particular expertise with clean room design. Healthcare and pharmaceutical clean rooms may constitute two niche markets. This is the traditional approach based around building type.

The identification of niches is easier using innovative approaches to segmentation. Taking client benefits as an example, the process of segmentation, through niche marketing, to the identification of specific targets can be seen as a series of screening exercises. For example, it may be found that many clients particularly want the following benefits:

1. High availability
2. Non-adversarial approach
3. Empathy with client culture and management approach
4. Good site appearance
5. Excellent safety record
6. Keeping projects within budget
7. Being flexible and co-operative in response to design changes
8. Rapid response to defects in the defects liability period
9. Maintaining the same contracts and site managers throughout
10. Advance warning to client and architect in a spirit of co-operation that production information is needed.

It may be the case that some of these commonly occurring features can be grouped into segments. Each segment may require a minimum of three of these desired features and benefits. Some combinations may occur that can be identified as niches. This process will be described with the aid of two illustrations (Figures 2.6 and 2.7). A screening process is used to identify where segments, niches and targets can be identified. A simplistic view of this is presented in Figure 2.6. Each client organisation is represented as a small box, the benefits they require being identified numerically from the list above.

The raw data from companies is presented at the top of the figure. Where several common benefits exist, these are screened into segments, as shown in the second layer of the figure. A further filter extracts niche markets. Some niches will occur within a segment and clearly will be given a high priority. Some niches will occur outside segments, yet there is sufficient commonality to select the niche as a secondary niche market. Other client returns will be discarded,

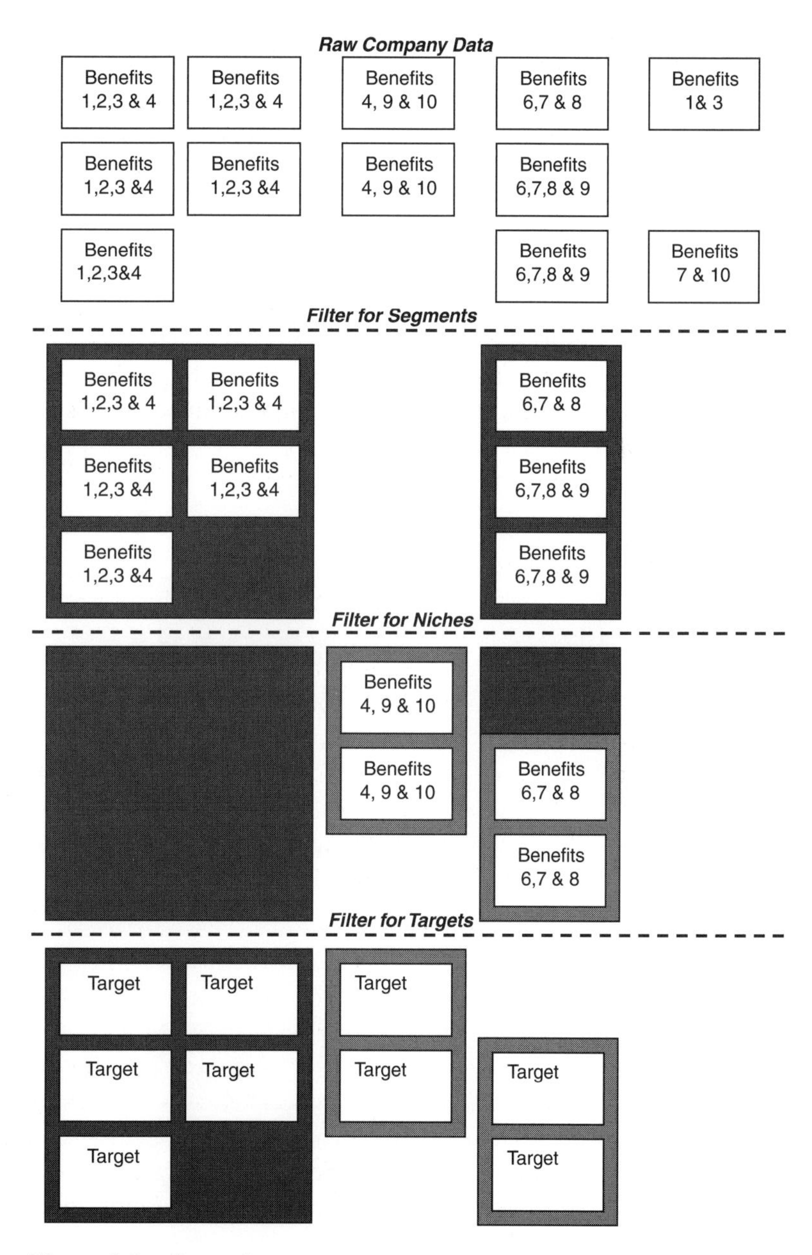

Figure 2.6. Screening process.

as shown on the right-hand side of the figure and therefore do not appear as sales targets shown at the base of the figure.

In practice, the process is more complex than shown. For convenience, the grouping of benefits makes segmentation easy in Figure 2.6. The reality is that more complex overlaps and layering occurs, as demonstrated in Figure 2.7. In this figure, the shadowed rectangles represent the benefits numbered above. The size of the rectangles represents the number of respondents.

The complexity arises in the following forms:

1. More than the minimum of three features and benefits for a segment
2. Niches occurring within segments
3. Three benefits occurring, yet insufficient numbers of clients to constitute a whole segment, yet enough for a stand-alone niche
4. Two benefits occurring for stand-alone niches.

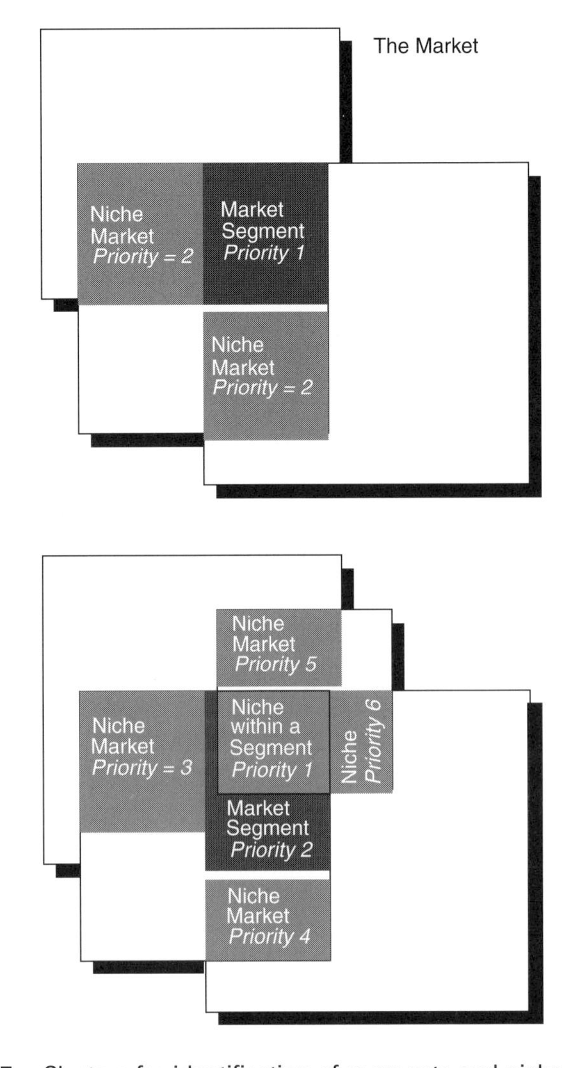

Figure 2.7. Clusters for identification of segments and niche markets.

While such overlaps may yield complexity, presented in this way it is easy to prioritise the data, hence develop layers of primary, secondary and tertiary markets.

While there are considerable advantages to be gained from this detailed level of planning, it should not be pursued for its own sake. Contractors and consultants have a long way to go before they come up against this, but as a pointer for the future niche marketing has its dangers. It has been claimed that North American and European organisations have frequently misunderstood client dynamics. US companies, such as Hewlett-Packard and Hyster, have withdrawn from some markets in favour of pursuing profitable niches. This has permitted Japanese companies to enter the volume markets with greater ease. With their focus upon building up long-term market share, re-entry by US firms has been well nigh impossible.[11] The aim is not to 'throw the baby out with the bath water'. A dual approach of niche marketing, combined with some broad and perhaps traditional segments can prove very effective.

Targets

The data from market research, audits sales, team and contract staff can be combined in order to provide a screening system. In practice, the combination can be highly complex. This leads to identification of targets, again in priority. This is illustrated in Table 2.1, the more benefits required that can be delivered, the higher the priority.

One note of clarification may be useful at this point. The prioritisation stated here is from the point of view of internal market organisation. It is in order to prioritise and monitor the sales effort. It is not necessarily a reflection of how clients might prioritise each of the listed benefits. Contracts directors and site managers should not take their cue from this information. The data can be used as a basis for then identifying the most sensitive issues in client management, but that is a separate issue, as it comes later in the marketing process – an issue of client management, which particularly pertains to relationship marketing.

Clients may be the ultimate targets; however, they are not the only sales target. There is a considerable reliance among contractors on advocacy, such as the design team and recent clients. In fact, there are several other avenues (see Table 2.2). Each avenue

Table 2.1 Targeting through screening

Primary market			
Priority 1: Niche market and segment			
Client No. 1	Client No. 2		
High availability	High availability		
Non-adversarial	Non-adversarial		
Empathy	Empathy		
Within budget	Within budget		
Same managers	Same managers		
Design changes	Design changes		
Priority 2: Segment			
Client No. 3	Client No. 4	Client No. 5	
Non-adversarial	High availability	Non-adversarial	
Empathy	Empathy	Empathy	
Within budget	Within budget	Within budget	
Defects liability period	Defects liability period	Defects liability period	
Secondary market			
Priority 3: Niche market			
Client No. 6	Client No. 7	Client No. 8	Client No. 9
High availability	High availability	High availability	High availability
Same managers	Same managers	Same managers	Defects liability period
Tertiary market			
Priority 4: Niche market			
Client No. 10	Client No. 11	Client No. 12	
Empathy	Empathy	Empathy	
Advance warning	Advance warning	Advance warning	

should be pursued with reference to targeted clients, although of course, opportunities may also arise through third party advocacy and referrals.

Table 2.2 Targeting using the six markets framework

Markets	Parties
Clients	Current and recent clients
Potential clients	Current direct targets
Referral markets	Management contractors, project managers, repeat business clients and past clients
Influencers	Design team organisations
Supplier markets	Subcontractors, material and component suppliers, labour subcontractors and agencies
Internal markets	Application of marketing internally and using employees to promote and secure work

Sources: adapted from Payne (1991) and Thompson (1996).[12]

Implementation

There is frequently an *implementation gap* between marketing and sales. Part of the reason may be the inadequate formulation of a marketing strategy and plan. The consequence is insufficient information on which to act or know how to take forward, especially if those responsible for implementation in the field are different personnel. Part of the reason is that successful marketing plan formulation addresses *what* is to be done, but rarely addresses *how* to do it. The advantage of following through the innovative ways of segmentation is that it already begins to point the way towards implementation. Those responsible for selling are already equipped with important data as to what matters to the client. They can then concentrate on why is it important to them and hence how to address the issues in the sales process and once a contract is won. This is a facilitative role. It is seeking further depth of knowledge of the client and is as much about briefing other members of the contracting or design team as it is about moving towards closing a sale.

A facilitative approach appears to be open ended, therefore how do those involved in sales and those managing or overseeing the sales process know whether it is going well? There has already been some prioritisation through discriminating between primary, secondary and tertiary markets. Further prioritisation is required at the level of individual targets. This goes a considerable way in bridging the implementation approach. Indeed, more monitoring can paradoxically create more room for manoeuvre (see Paradox 8 in Appendix A). How can this be done?

The aim is to set this up in a way that can lead market managers to develop a *sales time budget* for each prospect. A quantitative output is therefore useful. How should time be allocated overall? The first step is to prioritise clients. This may include some *sales experimentation*. This is a key part of market research. It aids the ability to see where the market is unfolding, provides information from the bottom and feeds this into top-down strategic planning. In other words, this refreshes planning with raw market data from the sales and contract staff view of the market (see Figure 1.3 on the Four Market Views). It is recommended that sales staff should be highly targeted in their activity, yet 10% of their time should be *experimenting* on new market contacts. These are selected by looking at niches and segments where the organisation is not currently active, selecting those organisations that exhibit change:

- ❑ Management change
- ❑ Take-overs and mergers
- ❑ Growth
- ❑ Restructuring.

The type of indicators can be gleaned from the financial and management press for commercial organisations, from newspapers and other audio-visual media for government and public bodies. *Sales experimentation* is treated as a *segment* in its own right and thus in the same way as any other for the targeting exercise.

To make this exercise relatively simple to explain, it is assumed that the same criteria required by clients are those that the organisation can offer. While many organisations, especially contractors, have been of the 'we can do that' school of marketing, in reality differentiation is beneficial and increasingly a requirement. It shall be assumed that the organisation has strengths in certain areas. To recap, the list of client requirements is repeated below, with the contractor and consultant organisational strengths highlighted:

1. High availability
2. ***Non-adversarial approach***
3. ***Empathy with client culture and management approach***
4. Good site appearance
5. Excellent safety record
6. Keeping projects within budget
7. ***Being flexible and co-operative in response to design changes***
8. ***Rapid response to defects in the defects liability period***
9. Maintaining the same contracts and site managers throughout
10. Advance warning to client and architect in a spirit of co-operation that production information is needed.

Four items are highlighted, reflecting what would be a radical set of deliverables for most contractors! Therefore, these are the criteria with which the organisation would select clients. For the purposes of illustration, an arbitrary weighting on a 1–5 scale can be given to each, which in practice would be a reflection of actual strengths of the contractor or consultant, perhaps gleaned from a client audit.

If the first five clients from Table 2.1 are used, then an overall weighting can be calculated. This is shown in Table 2.4. Then each potential client is given a rating as to how well the client profile matches the needs of the organisation. The rating is derived,

Table 2.3 Hypothetical weighting of internal strengths

Weighting	
Non-adversarial approach	2
Empathy with client culture and management approach	3
Being flexible and co-operative in response to design changes	3
Rapid response to defects in the defects liability period	1

using a Likert scale, from a low score of –2 up to a high score of +2.

The criteria used could include a number of different matters. The rating is given as to how well each fits the marketing objectives. For illustration purposes, let us say that the following issues are important:

- Financially stable
- Size of development programme
- Potential for long-term repeat business
- Potential for facilities management work
- Prompt payers.

Table 2.5 is set out as an illustrative assessment of how the potential clients may rate against each of these five issues from the viewpoint of the contractor or consultant.

A score for each target can be calculated for each client and potential client. The score is calculated using the following formula:

$$\text{Score} = \text{Weighting} \times \text{Rating}$$

From this, a ranking is then derived in order to prioritise the sales effort. It should be recognised that these figures are fairly subjective, based upon people's evaluation. However, having figures is better than having nothing, as they are much more than a 'gut feel'.

Table 2.4 Hypothetical weighting for each client

		Client weightings				
	Weightings	No. 1	No. 2	No. 3	No. 4	No. 5
Non-adversarial approach	2	2	2	2	–	2
Empathy	3	3	3	3	3	3
Design changes	3	3	3	–	–	–
Overall weighting		2.66	2.66	1.66	1.0	1.66

Table 2.5 Rating for potential and existing clients

Degree of fit	Rating
High	+2
Good	+1
Tolerable	0
Poor	−1
Intolerable	−2

The figures need updating regularly. This is part of making the market strategy a dynamic and working document. If they are updated regularly then a pattern tends to emerge and the market shifts can be monitored, which lends further confidence to the data and helps improve the evaluation on which the figures are based. It also helps to argue against individuals and senior management demanding fickle change. It aids a consistent and coherent approach.

More than this, it is a vital component in the efforts of management to move away from the personality culture towards systems and processes that are owned by the company. The 'little black book syndrome', whereby sales staff are hired and fired upon the basis of their contacts, starts to be overcome. Although contractors and consultants have databases, they are underused and frequently the key information remains the property of the individuals rather than the company – *the little black book syndrome*! Addressing this in the way suggested above has the benefit of addressing a number of marketing paradoxes (see Paradoxes 3, 8 and 9). A *learning organisation* is in creation for marketing and sales.

Table 2.6 Hypothetical score for potential and existing clients

		Client									
		No. 1		No. 2		No. 3		No. 4		No. 5	
Weighting		Rating	Score	Rating	Score	Rating	Score	Rating	Score	Rating	Score
A. 2.66	×	+1	+2.66	+2	+5.32	0	0	−2	−5.32	+1	+2.66
B. 2.66	×	+2	+5.32	+2	+5.32	+2	+5.32	−1	−2.66	−1	−2.66
C. 1.66	×	+1	+1.66	+1	+1.66	+1	+1.66	0	0	0	0
D. 1.0	×	0	+1	+1	0	−1	−1	+1	+1	0	0
E. 1.66	×	+1	+1.66	−1	−1.66	−1	−1.66	+2	−3.32	−1	−1.66
Totals			12.3		10.64		4.32		−9.98		−1.66
Ranking			1		2		3		5		4

Having established a score, that score and ranking can be used to allocate time against particular clients over a period. A year may be a reasonable period, with quarterly reviews of sales progress and the scores, hence time budgets. The experimental segment identified earlier will have a constant time budget of 10% of the sales time and effort.

This exercise helps to bridge the gap between the marketing strategy and sales in the planning process. It also has the powerful potential to move selling away from the *little black book syndrome* toward the more facilitative approach. How? In order to make the scoring exercise valuable, then the subjective evaluation that gives rise to the weighting and rating figures needs to be based upon sound information. That information can only be obtained if those engaged in selling:

- ❑ Draw in the client, rather than push the service
- ❑ Make the intangible needs of the client more explicit and hence more tangible to you as well as them, which is particularly important for the infrequent purchaser.

In this way, you get to know their needs. If this information is shared within the contracting or consultant organisation, hence becoming owned by the company, then the particular combination of your service features can be mobilised for their benefit – your *unique selling proposition* or *USP* for the client. The facilitative sales role is therefore optimised when the needs at each stage of the sales process are identified and acted upon. Selling is therefore a highly skilled process, when conducted properly and in an appropriate way for the organisation. It is also a continuous one.

Some organisations, such as many of the large contractors, have established procedures along these lines, especially those operating at an international scale (see ... *On the case 9*).[13]

Sales as a continuous process

If selling is a process, how can it be practically linked back to marketing? Selling is itself part of the competitive advantage a contractor or consultant has in the market. It is part of the overall experience that the client receives. It is important that it is both effective and sensitive to the client. This is conceptualised in Figure 2.8. Selling is presented as a continuum. How far those involved with selling proceed along that continuum depends

... On the case 9

A major international contractor, which is part of a larger conglomerate, undertakes a ranking and weighting process for countries in which it will operate in the international market. This is carried out according to their regional competitiveness and the attractiveness of each country. The contractor admits that while subjective in approach, this gives a fairly clear visual representation of country or project attractiveness, and therefore is utilised in comparative analysis. The visual aspect of its analysis is to plot the results onto a matrix:

Country Attractiveness	Low	*Organisation's Competitive Strength in Country*	High
High	Market Entry		Joint Venture
		Project By Project	
Low	Project By Project		No Interest

This contractor operates on a global scale, its market being major projects by international standards. The criteria used for assessing the attractiveness of a country will vary for different contractors.

upon the type of service on offer. Crudely, a great deal depends upon the market position of the organisation. It also depends upon whether the organisation is driven by clients or by operational criteria; in other words, 'are you in business to build, to be an engineer or to serve?'. The continuum should not be interpreted in absolute terms. There will be exceptions. It may be a useful guide – to try to locate your organisation on the spectrum and where to place it for the future. Clearly, the more scattered your

Figure 2.8. Concepts of the sales process.

approach across the continuum, the more closely the sales effort needs to be evaluated. In tactical terms, those involved in selling can use the continuum to inform their actions. The further to the right-hand side of the continuum, the more a team and cross-functional approach is needed. Those in dedicated sales positions become door openers and sales co-ordinators or managers, rather than closers of the sale. They may become *account managers* too, an important and neglected issue in contracting, which will be discussed at a later stage.

This begins to raise the stages in selling, which is the second issue. The key stages in selling can be summarised as:

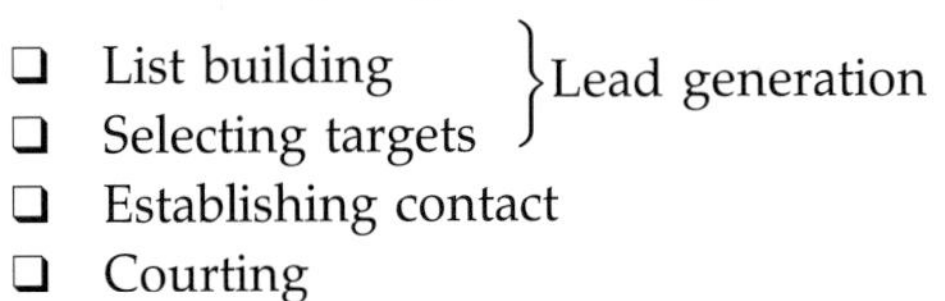

- List building } Lead generation
- Selecting targets }
- Establishing contact
- Courting

- ❑ Door opening for project opportunities
- ❑ Pre-qualification
- ❑ Shortlisting
- ❑ Tendering or bidding for consultancy work } Closing sale
- ❑ Post-tender negotiations
- ❑ Project
- ❑ Post-completion
- ❑ Repeat business.

List building and selecting targets

List building is the identification of potential clients from other published sources. The generation of such lists and selecting them is part of the targeting process for each segment. This may or may not be carried out by those selling. It may be carried out by those in marketing; however, involvement of both perspectives is important.

This raises a further point. The marketing and sales roles have been presented as distinctive. Indeed they are, yet the roles are often carried out by the same person. As this chapter links marketing through to selling, it is appropriate to look at this in more detail.

Marketing has been seen primarily as a cerebral function, requiring analytical competencies, whereas selling is perceived as a 'seat of the pants' activity. In other words, selling is responsive rather than proactive.

The sales skills are an ability to respond:

- ❑ *Intuitively*
- ❑ *By thinking on one's feet.*

Having the same person responsible for co-ordinating and spearheading both activities can lead to a kind of schizophrenia (see Paradox 3 in Appendix A). Alternatively, it can lead to a neglect of one or the other, usually the marketing function. In contracting, this dilemma frequently culminates in the worst combination:

- ❑ Operational bias – the sales effort being driven according to the needs of the organisation, rather than the state of and profile of the market, which puts unrealistic demands on those selling
- ❑ Sales bias – those selling often find themselves working in a vacuum with no back up from senior management and also an absence of support from other functions

- Rigid approach – either selling being pragmatically based around notions such as, 'we got by last year on it', or a marketing approach that stifles the development of conceptual competencies among those responsible for sales: marketing being seen as 'king' and selling as 'servant', rather than the two acting in harness to create a dynamic marketing and sales operation
- Lack of strategy – a reactive approach where the sales effort is diluted and disparate, usually lacking any real targeting and frequently being under-funded.

However, there are changes afoot in other sectors. These are two-pronged:

- A challenge to integrate marketing and sales in order to serve the customer more attentively
- A challenge as to whether a separate marketing and sales function is ideal, in order to serve the customer directly and consistently throughout their life as a customer.

These have yet to be substantially posed within construction and consulting organisations. The restructuring during the early 1990s was not grasped as a way of reconstituting marketing and selling in a coherent way. It was an opportunity to 'convert' many organisations to be client orientated. Nonetheless, marketing and selling activities are likely to merge for several reasons:

- Flatter management structures, which encourages both the reduction of the number of departments and encourages cross-functional working
- Faster changing markets
- Internationalisation of markets or working over larger geographical markets in ratio to the management and sales resources
- 'Top-down' marketing can be complemented by a 'bottom-up' sales response in order to refine the strategy
- Developing a marketing culture so that everyone is involved with marketing and selling.

All of these aspects help to plug the implementation gap between marketing and selling.

The other process involved in this stage of the selling process is lead generation. Identifying leads to pursue is important. In the way that segmentation and targeting have been portrayed in this chapter, lead generation is a matter of seeing which targeted clients are 'hotting up', in the sense that a project is on the horizon or imminent. This would lead to a re-evaluation of their rating and hence score. Other organisations that lack a strategy may see sales as being primarily driven by lead generation. There may not be much selectivity or it may be undertaken in a more pragmatic way. There are dangers in this. Sales people are hired and fired by their ability to generate turnover. There are two sources of sales from the viewpoint of the individual:

- Their own contacts – *little black book syndrome*
- Following leads – opportunism.

The former has been raised (see Paradox 9 in Appendix A), the latter can mean that the sales person pursues leads that generate turnover, rather than those that are in the best interests of the organisation. Short-term turnover is necessary, but there can be trade-offs:

- More profitable projects may be better than sheer turnover
- Low turnover projects can open the door to longer-term repeat business.

Most sales staff merely carry out the job as perceived by them. Few will take initiatives beyond that. The paradox at work here is that career interests and politics dominate over the accountability as an employee (see Paradox 10 in Appendix A). In order to test the approach of the organisation in selling, use the following criteria:

- Is the sales effort looking inward? If so, the concerns are:
  - Performance – pay levels and incentive schemes
  - Politics – taking the praise, yet passing the buck
  - Position – job retention/promotion.
- Is the sales effort looking outward? If so, the concerns are:
  - Performance – quality of product and/or service
  - Promises – delivering the goods!
  - Pleasure – job satisfaction derived from client satisfaction.

Establishing contact and courting

The strategy will inform the timing of this. If selling is being conducted with a production or operational bias (see the left-hand end of the continuum in Figure 2.8), or if the organisation is sales-led, then contact is likely to be initiated as a project becomes public knowledge. A repeat business client will be the subject of regular contact, even between projects, because a client orientation has been adopted. A new client may be contacted well ahead of a project, being singled out as a valuable client for the organisation through the segmentation and targeting processes.

What approaches can be adopted to selling? There are basically two:

- ❑ Hard sell
- ❑ Soft sell.

The *hard sell* is often perceived as:

- ❑ Hard work, more effective
- ❑ Slick, with more hype and less accuracy
- ❑ Unpleasant being on the receiving end!

The *soft sell* is often perceived as:

- ❑ Being beguiled into buying something you do not really want or need
- ❑ A form of manipulation.

In reality, the *hard sell* is pressurised and the later tends to be more subtle. The hard sell tends to be used more where there are standard products and services on offer. The features are vigorously sold to the client. The softer sell may involve a process of helping to define the product or service. Where there are problems to be overcome, briefs tend to be developed – a more facilitative approach. Both can be conducted professionally. Both approaches can be conducted with factual accuracy and with integrity.

How can a choice be made between the hard and soft sell approaches? There are several dimensions:

1. *Market position*: A sales force offering a standardised routinised approach would be more appropriate for the hard sell, whereas the analytical problem solving approach would more appropriately use a soft sell approach.
2. *Presentation skills*: The hard sell depends upon talking about and through the offer – selling the features, identifying the benefits and overcoming objections. The soft sell depends more on listening and other counselling skills before pushing home the offer. The hard sell uses demonstration skills, such as showing how a similar project was designed and constructed, whereas problem-solving skills are more pre-eminent in the soft sell approach.
3. *Inputs and outputs*: The hard sell is about taking, usually the briefing and/or tender documents, whereas the soft sell contributes ideas and tries to influence the course of action not just the final decision about the supplier of the services.

The innovative-pioneering approach mostly uses referral markets and press promotion. The analytical problem-solving approach is the most complex, especially in the soft sell, although this is where the opportunities for competitive advantage are greatest. Listening and counselling skills are used in the soft sell approach. This is so that you and the client can tease out and develop the client's requirements. It is also valuable market research, which has two main functions:

1. Solicit information to tailor your offer to match the needs of the client
2. Solicit information that is of generic application across several clients, perhaps a distinctive segment or niche.

Courting involves getting close to the organisation at two levels:

- ❑ Decision making unit (DMU)
- ❑ Corporate organisation.

The first level concerns personalities as much as information. The second level concerns the policies and actions that shape the organisation and hence the client development and construction requirements. The organisation needs to be mapped in order to understand the corporate and DMU. After all, we have to ask, 'Who is the client?' There are many interest groups and values within any organisation and these are going to be expressed through the DMU. The DMU will endeavour to represent the cross-section of views, as well as those from their own professional and managerial viewpoint. Many contractors and most consultants fail to identify who the key decision-makers are and their criteria for operation. The source of key decisions is usually easier and many of these are announced publicly in the relevant media.

It is important not to be fickle about courting. If the romantic analogy is pursued, someone is not going to expect the relationship to develop unless the contact is maintained, 'for better and worse' according to marriage vows. Why then should, say, a developer client believe that a contractor is interested in a long-term business relationship when the contractor goes completely cold during the winds of property recession. Similarly, in romance the seeds of suspicion can be sown if someone comes on too strong too early. Their motivations may be questioned. A contractor or consultant therefore has to court the client with the consistency and sensitivity demanded by the stage reached in the sales process. This is part of the facilitative skill of selling, especially in the soft-sell approach.

Door opening for project opportunities and pre-qualification

When the project opportunity arises, the client needs to be in a state of receptivity to your organisation. Whether taking a short-term or a purely sales-led approach, or whether adopting the longer-term courtship, the person in the sales role needs to be able to mobilise the resources and personnel necessary to get to the next stage: pre-qualification.

Pre-qualification may combine soft and hard sell, depending upon the way in which clients conduct this stage. Some go straight to shortlisting, while others have elaborate procedures and presentations to complete. The sales person needs to be proactive at this stage in order to offer more than competitors, even if the client decides not to respond to all of them, such as visiting past projects. There are a number of initiatives that can be taken; however, the more preparation that has been conducted in the courting stage, the greater potential for influencing the criteria by which the contractors or consultants pre-qualify. The more the client is understood, the easier it will be to decide what the options are to be proactive at this stage. There will be more brought out about understanding the client in other parts of the book.

Shortlisting and tendering

Many of the points raised above apply equally to these stages too; however, these are overtly hard sell stages in the process. Yet, there is a need to be very cautious. The danger is to steamroller the adopted line. The client can change direction at this stage for several reasons:

- ❑ Power in the DMU can shift, hence changing the criteria for selection
- ❑ Being close to a major budgetary decision can mean changes in emphasis
- ❑ Factors important to the organisation have not been made known to the DMU
- ❑ Price becomes more dominant and budgets can be re-appraised, sometimes drastically.

Many changes are unpredictable or are hard to gauge, yet insufficient knowledge as to what makes all members of the DMU 'tick' is bound to hamper ability to see that there has been a shift at all, let alone to respond to it. This is a neglected area of selling and usually requires a highly co-ordinated effort down the hierarchy of the marketing and sales organisation, as well as across functional departments.

Another neglected area is the decision on the final price, especially the profit margin – if there is one! Those involved in selling are frequently excluded from this process or meeting. This is short-sighted. Leaving the sales people out encourages contractors to become too price dominated in their decisions, because a partial picture is used for evaluating the tender. The sales people can contribute vital information, as well as helping to advise on what the client is expecting and will stand. The sales people can also gain insights from others, especially estimating the competition. The contribution the sales people can make depends upon how knowledgeable they are about the client, which leads onto the next point.

The other and most frequently neglected area is to ask the client what their criteria for selection is. This is vital. Many clients do not have a list written down. Is this not the least you should expect? Otherwise, and it all too frequently happens, the criteria are set verbally during the client meeting of selection. This throws it wide open to internal power play in the client organisation. This sort of power politics or personal prejudices can completely scupper the hard work put into the bidding process, which in a few cases can mount to expenditure into six figures (in US dollars or sterling).

If there is no set of written criteria, offer to help the client work out a set as part of the service. Be careful not to try to skew the criteria in your favour. Simply helping can give competitive advantage. You also have a tremendous opportunity to learn more about and what is behind the project. Some valuable insights could be gained.

Post-tender negotiations

Closing a sale or 'clinching the deal' can be a straightforward matter for many products and services. In construction it is a long and drawn out process, culminating in the negotiated contract or the post-tender negotiations. The sales people still need to be heavily involved at this stage. Changes and amendments can be made to secure the contract. This can be in the light of the price given, or new information can emerge at this stage. What is important here is that the promises made at the courting, pre-qualification and tender stages are carried through with consistency.

How many times does a sales person promise that the contractor works harmoniously with clients, or the design and build projects are constructed to a high quality specification? Yet when it comes to the tender and post-tender stages, the legal department or consultants want the contract clauses to be tightened up, with a consequence of upping the stakes of adversity, the estimating department paring down the specification, and so on. Of course, if the contractor cannot nor has any intentions to deliver on these matters, they should not be promised in the first place. The sales person can use this stage to integrate their approach to selling and carrying out projects more closely at this stage.

Project

In most contracting organisations, the marketing and sales people move on to the next opportunity at this stage, if they had not already left the scene long since! Partnering deals and framework agreements have gone some way to improve matters on this front, yet there is still a lot to do to improve continuity of selling across the board.

Monitoring what was promised and assessing the performance against that is a marketing matter. One way of looking at this is that:

- *Contract terms* specify the minimum levels of performance
- *Marketing* sets out the target for what can be achieved to satisfy the client.

There could be an internal system of auditing performance for this.[14] It will be seen that some contractors are developing such systems in Chapter 3. Those involved in selling have the other main task of monitoring this aspect. However, this too is seldom carried out and fed back into the market planning process. The book will look at how this may be achieved from scratch or in a more systematic way towards the end in Section III.

The simple point that needs to be made here is that a satisfied client can be the direct or indirect source of new work. Concerning a client directly providing repeat business, it is estimated for all industries that it is five times more expensive to find a new client than it is to keep an existing one – a lesson still to be taken on board by contractors.

Post-completion and repeat business

Consultants tend to be much better at keeping close to the client than contractors during and after a project. The design team usually has a lead consultant as the client representative with direct contact. The contractor frequently has the design team or project manager as a buffer between them and the client. And yet, when client expectations are at their highest towards the end of the project, that is precisely the time contractors tend or appear to take least interest in client projects. The 'A' team of contract staff has long since gone to win the next job and the 'B' team is now removed, replaced with a new site manager to finish things off. That new team, and the 'B' team for that matter, does not have a knowledge of the project history to resolve issues, does not know what the client expectations are, and in many cases has little in terms of systems and corporate culture to inherit. There is plenty of paper work and procedures, but project after project is run with the personality culture of the *street-fighting man* (see Paradox 1 in Appendix A). By the time handover comes, the client is frequently frustrated with the project.

Consultants do not always perform particularly well against this measure. Once the project is complete, they move onto the next. Keeping in touch with the client is seen as less of a priority than launching the next job or relieving the pressure on a stretched design team.

Handover, defects liability, final account should all be stages of opportunity to seek the next job. Even if those handling often thorny issues are not the key decision-makers for the next job, keep the decision-makers informed of progress. Even if it is not that positive at times, better the decision-maker hears it from you than from colleagues. Provide it as part of the after-sales service and try to ensure that others do their best to address the problems, such as sorting out a defect quickly and rapidly. Quite small things that go wrong, and are noticed by key people in the client organisation or among building occupants, may spoil the chances for repeat business in the future. Independent commentators summarised this:[15]

> *A commitment by the contractor to the client must be developed. It is noted that performance of a project after completion may ultimately be recognised as more important than possible overrun on time or cost (p. 77).*

Addressing some of these issues will improve the service level. It is an indictment of the industry to be able to say that success in these areas can yield considerable competitive advantage. It is hoped that this will not be the case within a decade. For, although taking measures in this direction requires time and effort and hence has a cost implication, the industry cannot afford to neglect them any longer. The investment is small compared to the additional cost of attracting new clients.

It is fully appreciated that not all clients have repeat business opportunities. However, the market works at several levels (see Table 2.2). Clients, even infrequent ones, can act as *influencers* and along with the design team, project managers and others, give third party recommendations in *referral markets*. Suppliers can also act as promoters of the organisation.

Client and project management

Selling as a continuous process has been shown to encroach upon contract management. Similarly, those involved with delivering the services can be valuably employed in the pre-contract stages. Project management therefore involves client management, and good selling is about client management from 'cradle to grave'.

It is this type of overlap that has contributed to the debate as to whether there should be a separate marketing and sales function. Should marketing and selling be endemic to the culture with everybody playing their part within traditional functional roles? In international markets there tends to be more integration because of the resource issues. The relationship between project and marketing has been characterised as an interface within a holistic and seamless picture (see Figure 2.9).[15]

From an industry sample,[16] 68% stated that there were areas of sales from courting through to repeat business where considerable improvements are to be made. The percentage is from a total of 107 responses. Nearly 24% identified courting specifically as an area for improvement and 8% saw the need to improve marketing and sales during the contract period. The courting and contract stages were the two greatest areas of concern amongst the industry sample. Clients appear to have similar concerns. One major property developer, running their own project management arm, said that his organisation expected of contractors:

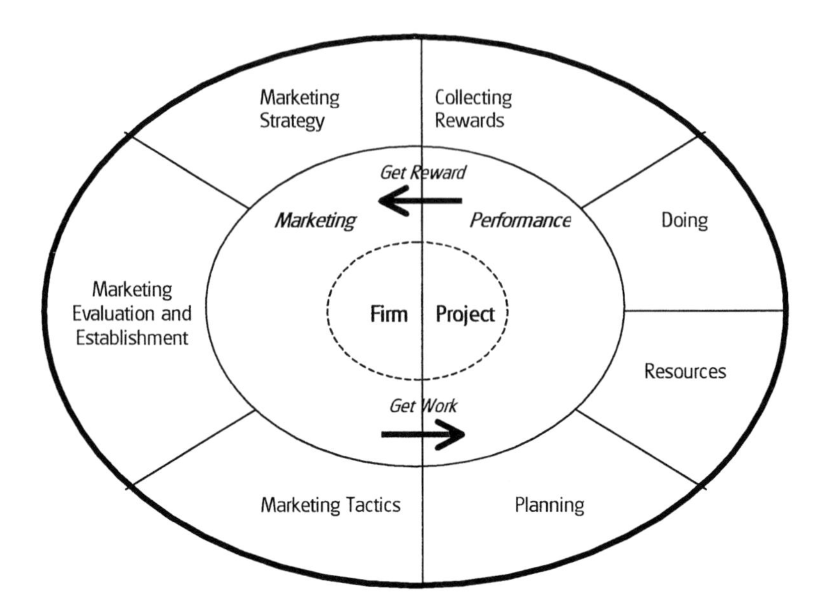

Source: adapted from Langford and Rowland, 1995

Figure 2.9. Relationship between the marketing and sales process and project management.

- ❑ Professionalism
- ❑ Tremendous emphasis on selling building skills
- ❑ Someone who has taken time to understand the project.

The client wants to be sure that the contractor can 'read their real needs'. This starts at the sales stage, specifically courting.

Conclusion

I used the analogy of forming and maintaining relationships: if you love someone you want to see them and tell them all that is happening. Indeed, the ultimate goal is to share together what is going on through marriage. This chapter has been about choosing partners – not blindly as a dating agency would do, but through careful selection. Clients take time to select consultants and contractors that suit them. That does not mean that the match is equal. Should not contractors and consultants take an equal amount of time selecting their clients?

The analysis has gone on to look at some of the key aspects of the sales process. Gaps in the client–contractor relationship have been identified. It is said of marriage breakdown that it starts with a lack of *communication*; the problems then follow – usually erosion of trust and lack of commitment. The same is true of business relationships. Poor communication exists at several levels between:

- ❑ Marketing and sales – the implementation gap
- ❑ Marketing and project management.

Hence, there is bound to be poor communication with the client because the parties act in a partial way, for they start off with a handicap. Contract and site managers have arranged marriages with their clients, which is not to say it cannot work, it just takes a higher level of commitment and trust than is currently exhibited in construction. The marketing and sales personnel are highly promiscuous and cannot be trusted to stay loyal. They are on to the next romantic engagement before you know it. There is no stability in this. It is costly for everyone in the long run. Careful selection of clients is the key.

To be able to consistently grapple with selection of and working with clients, some sales systems are needed. The subject of sales systems is the focus of the next chapter.

Summary

1. The *purpose* of this chapter has been to:
   - Identify the management implications of adopted market positions
   - Relate market positions to the way in which sales offers differ to clients
   - Provide the options to decide on the method of market segmentation
   - Identify specific client targets
   - Introduce selling as a continuous process from courting to securing repeat business.
2. The chapter has:
   - Identified the levels for differentiating services
   - Introduced some of the important areas to focus upon for improved sales and performance.

References and notes

1. Smyth, H.J. (1998) *Innovative Ways of Segmenting the Market: Practice Guide No. 1*, Centre for Construction Marketing, Oxford Brookes University, Oxford.
2. The sample is derived from questionnaire responses among delegates attending courses run by the Centre for Construction Marketing, Oxford Brookes University, Oxford, which runs short courses and training events, conferences and publishes research.
3. However, there is a professional bias in the sample towards marketing.
4. Scoubeau, C. (1997) The behaviour of some important Belgium construction firms, *Proceedings of the 2nd National Construction Marketing Conference*, 3 July, The Centre for Construction Marketing in association with CIMCIG, Oxford Brookes University, Oxford.
5. See for example Daniels, J.D. (1987) Bridging national and global marketing strategies through regional operations, *International Marketing Review*, **4**, 29–44.
6. Guido, G. (1991) Implementing a pan-European marketing strategy, *Long Range Planning*, **24**(5), 23–33.
   Smyth, H.J. (1998) The competitive stakes and mistakes: the position of British contractors in Europe, *Proceedings of the 3rd National Construction Marketing Conference*, 9 July, The Centre for Construction Marketing in association with CIMCIG, Oxford Brookes University, Oxford.
   Smyth, H.J. and Stockerl, K. (1998) Strategic marketing planning by UK contractors in an international business environment, *Proceedings of the International Construction Marketing Conference*, 26–27 August, University of Leeds, Leeds.
7. Siehler, B. (1998) Different approaches of European contractors to be a global player, *Proceedings of the 3rd National Construction Marketing Conference*, 9 July, The Centre for Construction Marketing in association with CIMCIG, Oxford Brookes University, Oxford.
8. Porter, M. (1986) *Competition in Global Industries*, Harvard Business School.
9. Vincent, S. (1998) Selling in specialist markets, *Proceedings of the 3rd National Construction Marketing Conference*, 9 July, The Centre for Construction Marketing in association with CIMCIG, Oxford Brookes University, Oxford.
   Pratt, J. (1998) Re-segmentation – a new route to better margins, *Proceedings of the 3rd National Construction Marketing Conference*, 9 July, The Centre for Construction Marketing in association with CIMCIG, Oxford Brookes University, Oxford.
   Cova, B. (1996) Construction marketing in France: from reaction to anticipation, *Proceedings of the 1st National Construction Marketing Conference*, 4 July, The Centre for Construction Marketing in association with CIMCIG, Oxford Brookes University, Oxford.
10. Fellows, R. and Langford, D. (1993) *Marketing and the Construction Client*, CIOB.

11. Doyle, P., Saunders, J. and Wong, V. (1992) Competition in global markets: a case study of American and Japanese competition in the British market, *Journal of International Business Studies*, **23**, 419–442.
12. Payne, A. (1991) Relationship marketing: the six markets framework, Working paper, Cranfield School of Management, Cranfield.
    Thompson, N. (1996) Relationship marketing and advocacy, *Proceedings of the 1st National Construction Marketing Conference*, 4 July, The Centre for Construction Marketing in association with CIMCIG, Oxford Brookes University, Oxford.
13. Ostler, C.H. (1998) Country analysis: its role in the international construction industry's strategic planning procedure, *Opportunities and Strategies in the Global Marketplace – Proceedings of the 1st International Construction Marketing Conference*, 27–28 August, University of Leeds, Leeds.
    Hand, P.W. (1998) Cast the net wide – but use a wide mesh, *Opportunities and Strategies in the Global Marketplace – Proceedings of the 1st International Construction Marketing Conference*, 27–28 August, University of Leeds, Leeds.
14. Bean, M. (1997) Developing and supporting a trial performance measurement system, *Proceedings of the 2nd National Construction Marketing Conference*, 3 July, Centre for Construction Marketing in association with CIMCIG, Oxford Brookes University, Oxford.
    Smyth, H.J. (1999) Performance audits and client satisfaction, *Proceedings of the CIB Symposium on Customer Satisfaction*, September, Cape Town.
15. Langford, D.A. and Rowland, V.R (1995) *Managing Overseas Construction Contracting*, Thomas Telford, London.
16. The sample is derived from questionnaire responses among delegates attending courses run by the Centre for Construction Marketing, Oxford Brookes University, Oxford, which runs short courses and training events, conferences and publishes research.

Chapter 3 Sales Systems

Themes

1. The *aims* of this chapter are to:
   - ❑ Translate the implications of the sales process outlined in the previous chapter into *sales systems*
   - ❑ Address the attitudes of all staff to selling
   - ❑ Set the context for enhancing dynamic plan formulation and refinement.
2. The *objectives* of this chapter are:
   - ❑ To create a sales framework from the *top down*, which encourages *bottom-up* feedback
   - ❑ Learn from past selling experiences for future opportunities in the context of the *learning organisation*.
3. The *outcomes* of this chapter are to develop a corporate approach to all sales activity:
   - ❑ Developing a marketing and sales culture throughout the organisation
   - ❑ Moving from an individually 'owned' sales approach to a corporately owned system.

Keywords

Corporate ownership, Sales system, Learning organisation

Introduction

What we do not want to do is to invent or develop systems for their own sake. Just because everyone else has them does not mean that they are right. Building something formal can be time consuming and cumbersome to operate. Is a systematic approach to selling better than the 'taken for granted' or common sense approach that has worked for so many years? Is there a danger of destroying perfectly good informal systems? These are important questions.

If the answer to these questions is to stick to what most organisations have, then we will produce a sales process that is a 'chip off the old block'; that is, we will do what we have always done with periodic and incremental changes. If the answer to the questions is that something more comprehensive is needed, then there are some important issues to explore.

In many ways, corporate culture has become more individualistic in recent years. Industry generally has contracted out activities that are not seen as core and staffing levels have been reduced. The bringing in of management consultants places reliance on an external opinion – an informed one, yet one that comes from a small number or one person. In many environments and departments individual performance monitoring, frequently related to the pay structure, has become the norm. This latter point is commonplace in sales environments. At one extreme, a reliance on teamwork means that those that lack motivation hide behind the results of those that do not. The net result is under-performance. On the other hand, performance and incentive-based systems tend to erode teamwork, as they set one against another. They also set short-term goals ahead of long-term ones. Expediency results in short-term actions at the expense of long-term goals. The outcome is that competitiveness comes to mean posturing over very few differences within any market position or segment.

More fundamental advantages are initiated upstream. In the search for competitive advantage, and perhaps the development of differentiated services, regular reviews of current activities are needed. The starting point for this chapter is to keep an open mind. This chapter can act as just such a review. Marketing is not only a means to create competitive advantage, but can itself be one source of competitive advantage, as the sales approach can give a competitive edge.

For the next generation of staff in construction a developed sales system will be taken for granted. They will use and interrogate the sales system to serve them and the organisation without challenging the need and the way in which it works. The issue, of course, is this: the next generation can only do that if the system is introduced in the first place. Here lies the problem. Many of us do not actually want to use a more systematic approach. We are quite happy with the way things are. How reasonable is this – logical reasoning or discomfort, fear, laziness, even rebelliousness? We muster all the best and reasonable excuses of work pressures. The real problem is the mentality of the *street-fighting man* (see Paradox 1 in Appendix A). While our motivations and hence our resistance to change may differ, the basic mentality is the same. It is this mentality that leads to the *personality culture* where the organisation does not *own* the way in which business is conducted in any coherent and consistent way. It has dominated construction.

In selling, the personality culture of the *street-fighting man* manifests itself in a number of ways. A good example, already mentioned, is the *little black book syndrome*. This is where sales staff are hired and fired on the basis of the contacts they have. The assumption is that these yield work for the new employer. The irony is that this recognises the absence of systems in construction generally. The assumption only works if the clients in the *little black book* are loyal to the sales person and not the organisation for which they worked – personal, rather than corporate ownership of contacts.

In many other sectors, customers tend to focus on the organisation and not the sales person. Sales people tend to stay around 5.5 years, whereas average customer–supplier relationships are 11 years in duration.[1] Yet the construction sales person and others often seem to keep as much information secret as possible. Many contacts may not even find their way onto the computer database and certainly with as little other information as possible (see Paradox 9 in Appendix A). The worst case of this individualistic approach to selling is that sales people stop being proactive, becoming entirely reactive, as Maister has found with many consultants:[2]

> *... at many firms, marketing is reactive, i.e. responds to an external impetus such as an RFP. Accordingly, available marketing time is quickly filled with activities initiated by these external stimuli.*

Secrecy and reactivity falls short of objectives to maximise sales potential in the best market segments. Employers have taken on the sales people in order to do the best for their organisation. The very least the employer owes themselves is that all those involved in selling feed information into a system that is 'owned' by the contractor or consultant.

How can this tendency be reversed?

- ❑ Shift the role of sales people away from secrecy and information ownership to a more dynamic one of facilitating and managing selling across an organisation
- ❑ Create career opportunities for sales people to progress into marketing and other areas of management.

This can be achieved by setting up an efficient sales system in order to:

- ❑ Improve sales effectiveness with the current staff profile
- ❑ Continue to maintain existing and established contacts, regardless of the coming and goings of personnel.

As a spin off, it will also help to encourage a more systematic approach to contract and site management. This is largely absent at present, as ... *On the case 10* demonstrates.

... On the case 10

A medium-sized developer let a design and build contract for a phase of a retail park. The architect and contractor worked well together. The client was pleased and decided to give repeat business to the contractor–architect team. The architect maintained the same design team and key project personnel on the next phase – director, associate and project architect. The contractor selected an entirely different contract and site team. There were several consequences to this decision:

- ❑ A new learning curve was needed between the architect and new contractor team, and the potential efficiencies were lost
- ❑ The same understanding and spirit of co-operation did not emerge between the new set-up
- ❑ The project was not completed in the successful fashion of the first phase and the client was not pleased.

This case demonstrates that many clients are in effect hiring contract teams, rather than contractors. The project management process is dictated largely by personalities and team style, rather than by the employer. What management function do contractors therefore offer over a consultant project manager, especially where there is little direct labour employed or strong subcontract and supplier alliances?

Sales objectives

The overall need for long-term objectives in marketing was set out in the previous chapter. In practical sales terms, short-term objectives lead to sales that may not be in the best interests of the organisation. How does this manifest itself? If a sales person is required to achieve certain sales levels or targets in order to secure a living wage, or if their job security is dependent upon that alone, then the pursuit of immediate sales will overrule longer-term sales. The essential groundwork – for example, the work that will secure pre-qualification or secure the information from which competitive advantage can be created – for a £10m or $15m contract that will not come to fruition for 9 months – is neglected. Preference is given to a £2m or $3m contract, which will fall into the next reporting period. The agenda of the sales person is dominating in this example. The fault lies with the contractor or consultant, not the individual, because the performance system embedded in the organisation has invited this response. There are many examples, falling into the following categories:

1. Short-term, low-turnover work pursued at the expense of long-term, *high-turnover* work (see hypothetical illustration above), with potential consequences for market share in the segment or overall marketplace
2. Short-term, low-margin work pursued at the expense of long-term, *high-margin* work
3. One-off projects pursued at the expense of clients with long-term *repeat business.*

Of course, a balance between performance-based criteria and teamwork is needed. Chapter 13 will look in more depth at how such a balance can be created. At this stage, it is important to gauge and establish the necessary sales systems. A balance is more likely to be created if the following guidelines are taken into consideration:

- The market and the marketing plan, which encapsulates the ability and approach to market management, drives the sales effort (rather than the share and power structures of the organisation)
- The sales process is *top down* and *bottom up*, so that a dynamic is created between all management and sales levels
- Signs, pointers and mistakes are learnt from in a way that such lessons are 'owned' by the organisation and not just the individual – the *learning organisation.*

Using a balance of performance- and teamwork-based guidelines, a list of sales objectives can be produced, all of which are designed to improve the level and quality of sales:

1. Respond quickly to opportunities, based upon good market information
2. Target potential clients, based upon segment and client information
3. Obtain access to new clients through knowing their needs
4. Obtain further client information to enhance the sales response
5. Overcome objections
6. Mobilise a unique selling proposition or your competitive pitch
7. Get price and profit margin at a mutually beneficial level
8. Close the sale.

All sales objectives depend upon communication of information and building of relationships as a means for successfully delivering a service. Section II will consider in greater depth how these objectives are put into action during the selling process. Relationship building is dealt with in particular in Chapter 7. At this stage, it is important to understand how these objectives fit into the totality of the sales process.

Grease lightening

The song in the musical *Grease* talks about turning an old crock into a racing car. In the same way, if something can be created out of nothing or informal systems can be converted into ones of substance and importance, then knowing what a system is can be helpful. *System* is one of those words where we know what we mean, but which is often hard to pin down in practice. What is not meant is a series of proforma and procedures. Contractors and the design team have more than their fair share of these. Nor is something totally abstract required. It is something in between.

The *Concise Oxford Dictionary* puts it this way:

- *System*
  - Complex whole, set of connected things or parts, organized body of material or immaterial things

  - Method, organization, considered principles of procedure, (principle of) classification
- *Systematic*
  - Methodological, according to plan, not casual or sporadic or unintentional, classificatory.[3]

Sales system

A system is proposed, one with a light touch – one that is dynamic, which oils the wheels rather than rusts the sales process. What are the building blocks of such a system? These are shown in Figure 3.1. Some of the elements have already been identified, yet notice that this figure, like many in this book, is not presented as a flow-line diagram. It is presented in layers, one fitting into a larger whole. This is intentional, in order to move away from a linear conception, emphasising the integrated and holistic nature of many processes in practice. The holism is about establishing *systematic staff relationships*. That is to say, requirements are specified to staff as to when and how they relate. Where holism is lacking, the design of systems can build in integration.

There are new elements to add to those introduced in Figure 3.1, summarised below:

- Selling as a market research exercise feeding back into the marketing plan
  - *Discovering needs* of targeted (potential) clients
  - *Experimentation* on a selected number of clients undergoing sector or corporate change
- Corporate ownership of data for effective selling
  - *Database*
  - *Reporting* and sales feedback
- Management for effective client satisfaction and repeat business
  - *Client management*
  - *Project management* and *service delivery*.

Figure 3.2 links these systematically together. The arrows show the movement of information: *information exchange in order to inform action*. The vertical arrows pointing down are a *top-down* management approach, which comprises three management processes:

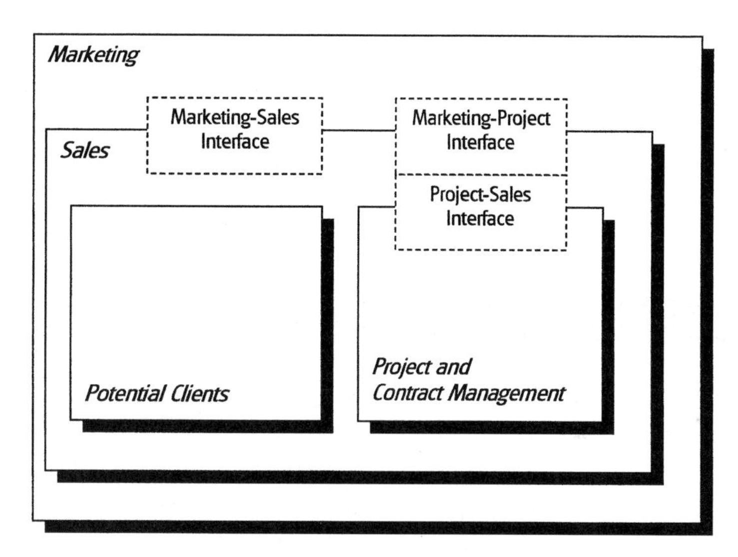

Figure 3.1. Building blocks of the sales system.

1. Communication of strategic planning aims and objectives – to be embodied into the sales plan for implementation of strategy
2. Communication of market data
3. Monitoring cross-functional co-ordination and action concerning sales progress.

The vertical arrows pointing upwards are part of the *bottom-up* approach and also comprise three main processes:

1. Feedback about specific clients for response on existing contracts from the perspective of satisfying the client for referral and repeat business opportunities
2. Generic feedback that is of relevance across a number of clients, which will contribute towards refining the marketing and sales strategy
3. Feedback from experimentation in potential segments for strategic plan development.

This is necessary to create dynamism in the marketing and sales strategy. These management processes are captured in terms of the movement of top-down and bottom-up management in Figure 3.3. This dynamism must be made to work from the top and from the bottom; thus, all senior management should be familiar with the marketing plan to a high level of detailed recall. Those responsible for marketing should have the plan on their desks the whole time in order to be guided by it and to be refining it. As was stated in Chapter 2, there should be long-to-short-term goals. The long-term strategic objectives provide an immutable framework. Therefore the majority of the refinement comes from the marketing–sales interface:

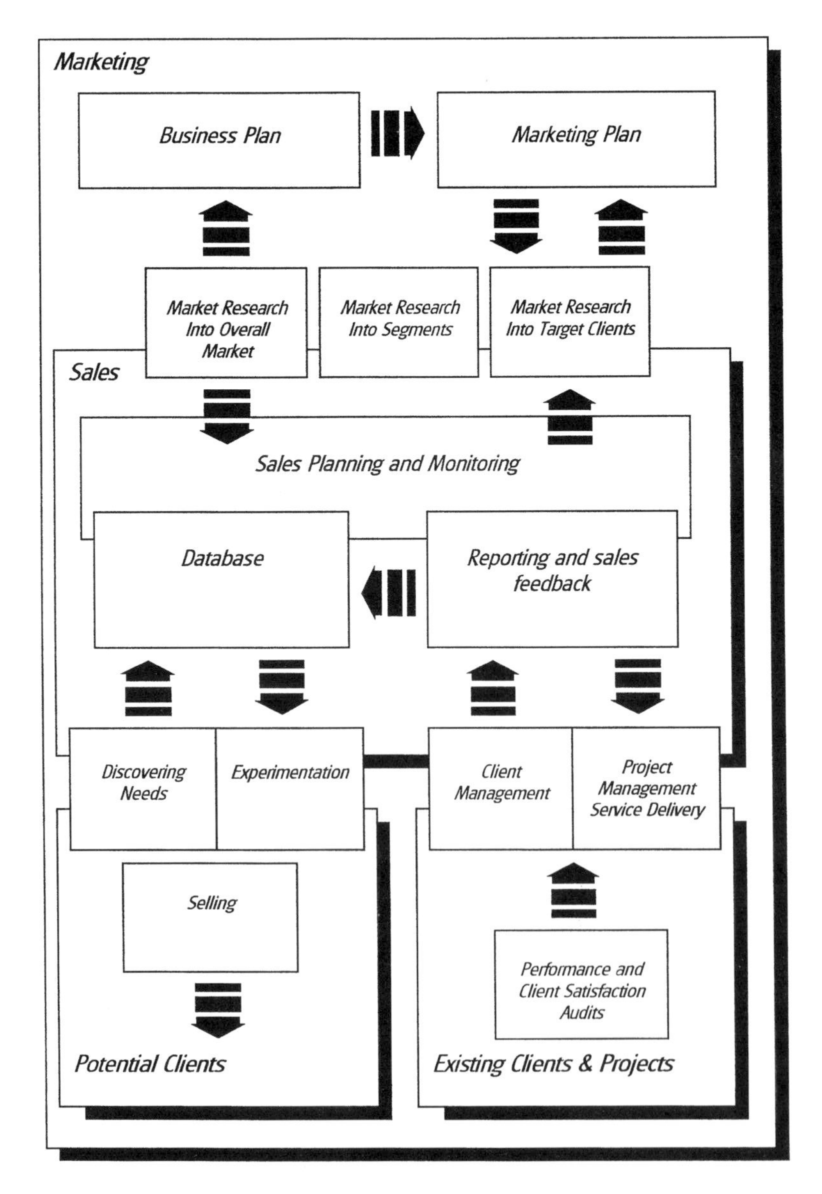

Figure 3.2. Elements of the sales system.

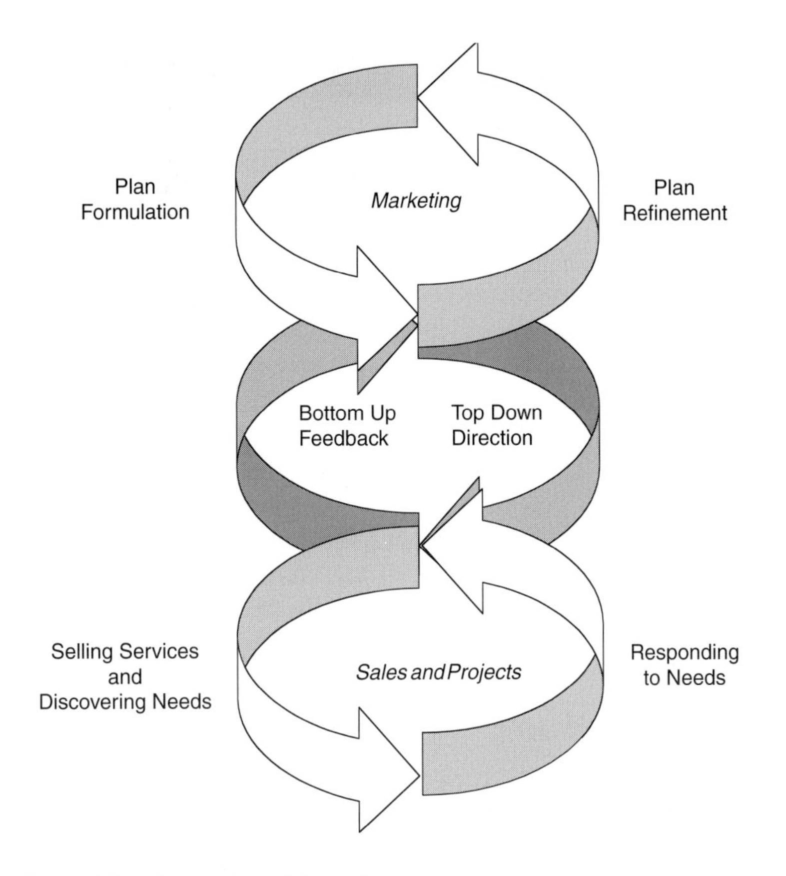

Figure 3.3. Dynamics of the sales system.

- ❑ Refining detailed objectives
- ❑ Subtle changes in segmentation
- ❑ Changes in targets and areas of experimentation
- ❑ Shifting of prioritisation of existing targets
- ❑ New information on targets.

There should in parallel be a system for the collection of more substantial data that goes beyond that needed for day-to-day refinement, yet is valuable for an annual review and for the periodic review of the overall long-term strategy.

Sales procedure

The key information sources have been identified (see Figure 3.2 and Table 3.1). Other sources can be added. Some sources are internal to the organisation and some are external.

Most organisations only use a few of these means. The informal nature of most 'systems' leads to partial use or an *ad hoc* use of sources. Not every organisation should set its sights on substantially tackling every one of these. The first task is to improve what already exists and then to add to it. However, all the elements need to be put into place, so that the selectivity can be in line with other marketing and sales policies within the organisation. It is this aspect that makes the *system* and the proof of it will be whether people use it in order to make it effective. Given the mentality of the *street-fighting man,* organisations need to be realistic as to what can be achieved at any one time. The long view is therefore needed to improve and develop the system,

Table 3.1 Key sources of information in selling

Source	Origin	Level	Purpose
Published material: journals, newspapers	External	Marketing	Strategic segmental information
Internet and other media	Sales		Tactical client information
Market research into overall market	External	Marketing	Strategic sales information
Market research into segments	External	Marketing	Strategic sales information
Market research into target clients	External	Marketing	Strategic sales information
		Sales	Tactical client information
Business and peer network	External	Sales	Tactical client information
Database	Internal	Sales	Strategic and tactical sales information
Reporting and sales feedback	Internal	Sales	Strategic and tactical sales information
Client and project management	Internal	Sales	Tactical sales information
Performance and client satisfaction audits	External	Project Sales	Repeat business and service improvement

which is a test as to the commitment that management have of the long-term nature of marketing in contracting.

The elements need to work together. For example, a database should be kept up-to-date with a broad range of data about the client, so that it is both current and has depth. More extensive information may be carried in memos, minutes and reports from the reporting and feedback mechanisms. It is this requirement that ensures information is owned by the organisation, rather than the individual. Yet there needs to be a verbal review and exchange process carried out in parallel in order to maintain the dynamic. Otherwise, if key sales staff leave, too much of the information is cold to the remaining staff, whether the lead or client is 'hot' or not!

The real danger, as opposed to fear factors, is that such systems become cumbersome – a bureaucracy where sales people spend their lives desk bound, form filling, rather than getting out there and selling the services. This is indeed a danger. Two things need to be said about this.

First, an appropriate targeting of clients and potential clients, preferably with time budgets as discussed in Chapter 2, means that the sales effort is concentrated on fewer and more productive targets. Capturing adequate information about clients is part of making the sales effort more effective and hence ensuring that a greater number of the targets are indeed productive for the organisation. In addition, no other main function such as accounting or personnel could function properly on the level of information that marketing and sales currently operate upon. In part, accountancy and human resource legislation impose requirements in most countries. This is simply to ensure that all organisations meet minimum standards of practice. In this case, we are looking at practices that will help to install a competitive edge for organisations.

The second point is that a balance needs to be established. Enough needs to be kept to meet the marketing plan objectives. Clearly, the demands will increase over the years as competitors play 'catch up'. It will not be necessary to systematically record information from every source listed above at the outset; however, it will be necessary to widen the scope over the years in response to the competition. For larger organisations, such intensification comes sooner rather than later. The upshot is that information gathering requires careful management. It requires an overview of how this is evolving over the long term. It also needs day-to-day oversight to ensure that the 'tail is not wagging the dog'. A systematic approach cannot be confined to proforma and databases, but is something with which everyone concerned is engaged.

What sort of information can be recorded? Each source is briefly reviewed below to provide some indication. There is no claim of being comprehensive here. This is impossible given diversity of market positions and modes of differentiation covered in this book. In any case, part of the competitive edge is finding aspects and types of information that will prove useful to the organisation for effective or optimum response, given the type of services being offered.

Taking *published material* first, the purpose of collecting information is two-fold. Information about sectors or segments is important. This includes segments or clients that are part of the sales experimentation. The other aspect is specific information on clients. What is being looked for? Some information is general information on the size and stability of the potential client. Some clues as to their market position is useful. There may be a fair amount of information about their development programmes and approach to procurement, especially for clients with substantial repeat business. The information may come from:

- ❑ Annual reports and accounts
- ❑ Promotional material produced by the client
- ❑ Stockbroker and other independent financial reports
- ❑ Specialist journals, such as those concerned with management or the sector in which they operate
- ❑ Newspapers
- ❑ Construction trade press.

All this information is good background reading, so that the initial approach to the company is an informed one. Many contractors and consultants fail to get under the skin of their target organisations. This information helps to build up a good initial picture. The purpose is not to be able to go into the target organisation with 'all guns blazing' as to how much you know about them, but to be able to ask the pertinent questions so that you can to get to know even more. Clients will always be impressed if it can be demonstrated that you have taken an interest in them. Just knowing the right questions to ask can be sufficient to do this. If you are asked a question, which allows you to demonstrate some of your knowledge, then you will really be in a strong position. The strong position is not to exhibit your sales prowess, nor to dominate them, but to create a secure footing for a stable and trusting relationship. This is always important and is the overriding factor if a relationship marketing strategy is being pursued.

One other kind of information is important to record for organisations from this source. Clients undergoing structural change can often have changing building and estate requirements. The situation may be one of growth; it may be restructuring in the face of a shrinking market or adverse conditions. There may simply be key personnel changes, where a new management wishes to undertake substantial corporate change, which again can lead to changing building requirements. These situations do produce a state of flux in clients. Their requirements may be changing, but they may also be changing their criteria for selection or indeed the type of contractor or consultant, which they will employ. Changes at the personnel level of estates and building management may also yield changes as the 'new broom' wishes to assert authority. All these are opportunities for a contractor and consultant to get a foot in a door, which may have hitherto been closed.

In some cases, the next project may be sewn up, but the long-term construction requirements should be considered.

Verbal reports and file notes and reports are the best way to capture the unfolding picture for each target. The higher the priority of the target, the more information is needed. Those in primary markets for the organisation will receive more attention than those in secondary or tertiary markets (see Table 2.2) and those to whom the organisation is responding opportunistically; then a bare minimum is required, usually on a 'need-to-know' basis. A file is therefore needed for each client. Files can be linked back to database fields.

A database record is needed for each client too. Some of the essential information would be recorded here, enough for anyone to pick up the important factors and know whether they need to refer to the file notes or not. A stand-alone sales database is used by most companies, although there is a great deal of scope for linking this with an accounts system, especially where there is emphasis placed upon repeat business.

Market research into the overall market is the next source. The main purpose of soliciting this information is to inform marketing issues. However, there may bc specific information relating to clients or types of client that is worth putting onto the database. Market research into target clients will certainly yield specific information. It is likely that some will be detailed and therefore a file note is required.

The business and peer network is an important source of information. It is a source that is more difficult to manage. Those invited to corporate events and client lunches can frequently learn information that is not of their immediate concern, yet is of immediate concern to others in the organisation. This information is frequently lost. Recording notes in file or database format is valuable. Those, who obtained the information, should record it. Even vigilant people often pass on a quick verbal report to those concerned and the nuances can be lost.

The two main internal sources are the database and the written reporting and sales feedback. These should be comprehensive. Databases are frequently out of date and are used partially by staff. The fields of information are frequently ill-considered in relation to the market information needed. Most branded databases are more suited to other types of sales environment than those in construction. There will be further analysis of databases in a subsequent chapter.

The information needed must help to build up a profile of the client. It can be divided into a number of categories and headings:

- Client information
  - Leads
  - Development programme
  - Size and type of organisation
  - Key financial data
  - Key senior personnel data
  - Types of products, services or activities and their approach to managing their operations, especially those that may impinge upon construction and contract requirements
  - Profile of the decision making unit (DMU)
- Sales information
  - Initial contact details
  - Progress and stage of courting reached
  - Previous projects and their history in sales and construction terms
  - Organisation contact details: sales and contract personnel.

Client and project management information implies a link between sales and contracts. This is a neglected area of selling by senior management in construction. It implies that a sales database is not confined to dedicated sales personnel or simply located in that department. The intranet is an important means of creating common access to data in large companies, as well as controlling data inputs. All clients need to be managed and therefore an organisation must identify someone in overall charge of the client at any one time. Each individual contact in a client organisation must also be 'owned' by someone in the contractor or consultant organisation. Different contacts may be owned and managed by different people, but the lines of responsibility are important in the management of the system and hence the management of the sales process. It is a marketing function to set this up.

Performance and client satisfaction audits concern repeat business clients. Capturing performance data for sales purposes is clearly an important part of ensuring that the client is being managed and that past performances are improved upon for the next opportunity.

Summarising, put in place a sales system that is owned by the organisation and not by the individual. Design the system so that it relates back to the marketing plan. Design it so that it can be easily monitored and that it is effective, yet not too bureaucratic for all those selling, the test being, 'Will I as a senior manager learn to use it?' Do not see selling as separate from the project. Selling is very much part of it; it is simply the front end. At an aggregate level, projects, and hence the selling, must be driven by the marketing plan.

System output

Having this information means that it can be reviewed and used. Review focuses upon the use of information as it is pushed up the organisational structure to inform amendments to and refine the marketing plan. Using information for sales purposes involves the sales person taking it and applying it to the next sale.

Captured information can be reviewed regularly to refine the marketing plan or to see what sort of mistakes occur. There may be a pattern to the type of clients being won or lost that is more precise than currently defined in the plan. At this stage, marketing is shifting from information capture towards using that information to make changes and refine decisions. This is the essence of the *learning organisation*. Experience is no longer encapsulated in the minds of individuals, but is shared in and across the organisation in a relevant and appropriate way. Hence, a corporate experience is built up.

Attitudes of sales staff are important. They will need to be inducted into having procedures and thus feed information into the system. They need to trust their management in this facilitative situation. Their management therefore must reflect these requirements in employment packages and monitoring procedures. Achieving this will mean that those involved in selling, whether they are dedicated to this function or not, can benefit more from the experience of others. This has a positive effect upon the organisation. It is essential as services become differentiated and as the change in the marketplace continues to speed up. Change is the norm and staff competencies are no longer to do with having the data – *the little black book syndrome* – but knowing what to do with it in both general and specific terms. The construction market in this sense will increasingly demand the facilitative competencies among their staff with high capabilities to adapt.

Therefore, greater integration of marketing and sales is likely to take place in this changing market environment. The requirement is to be able to take on both the cerebral function of analytical marketing competencies and the more intuitive 'seat of the pants' approach of sales, discussed in the previous chapter (cf. Paradox 3 in Appendix A).

In the sales process, the information is useful to existing personnel. This is two-fold. First, sales staff, who have tended to be 'seat of the pants' in style infrequently review what they are doing to a sufficient level. The mere act of having to record information and data does have the effect that it will be mulled over. If marketing and sales staff regularly review the information, then further reflection takes place, which will aid the next sales moves.

The second aspect is that the information is no longer owned by the individual, but by the organisation. If the sales person leaves the organisation, someone else can use the information. Even if a sales person remains with the organisation, others may need access to it and indeed input further information. Most contractors do not operate as advertising agencies do, where the account handler is the common way of managing a client. Instead, responsibility for the client tends to pass from one person to another. Sales, estimating, legal, contract personnel and management all have access to sales information. This is important in delivering to the client what was promised, that is to meet their expectations and is important if there is potential for repeat business.

The output has so far been described in terms of individual clients and project opportunities. This focus may appear to be at odds with the emphasis placed upon the big picture. Yet, it is at the detailed level that important data is gathered to assess the overall situation. Of the four market views (see Figure 1.3), the focus is on the processes that are internal to the organisation and involve the sales and contract staff at the bottom of the management structure. The overall picture is generally qualitative. The system is needed to solicit that information and to make sense of it.

The information can also be quantified and that needs to be carried out too. Indeed most contractors and consultants tend to start with this and work down to the project level – very much the *top-down* approach to management. Contractors are much better at the collection and use of such data. It can be aggregated at various levels – total turnover, individual segments, niches and core or key clients. New and repeat business distinctions are also useful and could be used more frequently. A balance has to be struck between having a sea of information and being comprehensive. Periodic reviews break down information according to market position and the direction in which the organisation is developing the market and other dimensions, in order to shed new light and give a different angle on the business.

Below are provided some of the key data requirements. Again, the emphasis will change from organisation to organisation, depending upon the business and marketing plans. They have been grouped into categories:

- ❑ Sales performance measures
  - ■ Clients actively courted to *pre-qualifying stage*:

$$\frac{\text{Number of pre-qualifications of courted clients}}{\text{Number of courted clients}} \times 100\%$$

    This is an important measure of the overall performance of the sales effort. Monitoring this is necessary to gauge how well sales staff are serving the organisation. Two things must be borne in mind. Selling is a team effort, cutting across several functions. Rarely is a sales person fully responsible for bringing in a job, but this measure is more effective than turnover secured, which is frequently used. The second factor is that pre-qualifications are also dependent on the health of the marketplace, so a sales effort can only be as good as the context that effective demand permits.

- *Conversion rate* to tendering stage:

$$\frac{\text{Number of tender opportunities}}{\text{Number of pre-qualifications of courted clients}} \times 100\%$$

  This is an effective measure of the teamwork in contracting, bearing in mind the caveats above.
- *Strike rate*:

$$\frac{\text{Number of contracts won}}{\text{Number of tender opportunities}}$$

- *Value performance*:

$$\frac{\text{Value of contracts won}}{\text{Value of tender opportunities (estimated or known winning bid)}}$$

- *Market share*: This is an important figure to be able to gauge, yet is the least easy to calculate. The data for individual segments and for the total market is difficult to monitor and 'best guesses' may be the only realistic measure of the total value of work at the time. Even historical statistics are not collected in a very usable form. However, these limitations are not an excuse for not doing anything. Monitoring the press, use of the 'grapevine' in addition to data on tender bidding can provide a basis to come up with reasonably accurate 'best guesses' over time. Historical analysis should be undertaken and the results compared. This is more relevant to contractors than the even more highly fragmented consultant markets; however, certain segments, such as the major international projects market, could be assessed in this way.

$$\frac{\text{Value of contracts won}}{\text{Total value of contracts}}$$

- Competitor performance
  - *Intensity of competition*:

$$\frac{\text{Total number on tender lists}}{\text{Total number of tenders}}$$

    This figure is useful to compare and contrast against the strike rate.

- *Competitor success rate*: Taking each competitor that wins a contract instead of your organisation:

$$\frac{\text{Number of contracts won by named competitor}}{\text{Total number of tenders bid by your organisation}}$$

- *Named competitor's market advantage*:

$$\frac{\text{Number of contracts won by named competitor}}{\text{Sum of rank order on tender list for your organisation across total number of tenders bid against the named contractor}}$$

❑ Financial performance measures

- *Average tender profit margin*:

$$\frac{\text{Tender profit margin}}{\text{Number of contracts}}$$

  Historically, margins have been low in construction. It is important to monitor these and couple them with some oversight of the range of margins between projects. This may help to set the margin for a tender, especially if there is a pattern to high and low margin work across segments.

- Demands on *working capital*, to take account of cost of specialist equipment, components or contract terms and contract risk:

$$\frac{\text{Maximum working capital requirements}}{\text{Contract value}} \times 100\%$$

  If the project is a large one in proportion to overall turnover, then it may also be necessary to calculate the percentage of working capital required for that project out of the total working capital employed. There are two issues to be considered. The first is that tying up a considerable amount of capital in one project may close down opportunities to work for other clients. In this case, the priority rating given to this client or project within the marketing strategy will be important. The second is that contracting and consulting are both cash flow-based businesses. Indeed, survival during recession can depend upon keeping overheads at a low level.[4] Minimising the investment in new equipment and so on can be part of carrying a minimal asset base and keeping the organisation as liquid as possible.

- *Average contract profit margin* (at practical completion and final account):

$$\frac{\text{Realised profit}}{\text{Project cost}} \times 100\%$$

- *Turnover secured*:

$$\frac{\text{Turnover secured}}{\text{Target turnover}} \times 100\%$$

  This measure needs to be gauged against the period of measurement. The portion of time to secure the remaining work will indicate the extent to which the organisation is on target. A shortfall or being ahead of expectations may be a sign of sales performance or the setting of unrealistic targets in the current market. Assessing the likelihood of meeting the target depends upon prospects. If there is four times the turnover required in the pipeline for the period and the strike rate is 1:4, then there can be a high degree of confidence among management. However, some work may come 'out of the blue'; in other words, the pipeline of prospects is an inadequate indicator. This is the case in some markets and at certain times. It may therefore be helpful to keep records of how much of the work historically secured has come 'out of the blue' and then this can be factored into the pipeline assessment. This is particularly relevant for consultants. The pipeline can be divided up as follows:
  - Pre-qualified work, speculative work often being involved especially for architects
  - Work being tendered
  - Post-tender negotiation stage
  - Estimation of projects coming 'out of the blue'.
- *Profit secured*:

$$\frac{\text{Profit secured}}{\text{Target profit}} \times 100\%$$

This type of information is elementary, yet many consultants and contractors do not have it to hand. The very same organisations would be offended if it was suggested that they did not take great interest in their business. Most organisations can tell you in detail the outputs in terms of products or service, particularly in accountancy terms. Chefs will say that good cooking depends on putting in good ingredients or as the computing adage goes, 'rubbish in, rubbish out'! The marketing and sales ingredients are a key step to competitive advantage (see ... *On the case 11*).

A multidisciplinary design consultant went into receivership, one reason being that they did not take an interest in the inputs, although they had very detailed analyses of accountancy outputs. A database of clients and contacts did not exist. They had relied on referrals during boom years. When it came to more stringent times, there was a lack of understanding as to why clients came back and where the organisation made their profit. They had charged on an hourly rate. Once the organisation appreciated

... On the case 11

A large firm of international engineering consultants that is financially driven uses a strict regime of financial control. At group level, the data focuses upon operating performance, secured work for the next 3 months, orders in the pipeline and the cash position. As 80% of the work comes from repeat business – a very high proportion – and most of the remaining work is through referrals, then this traditional focus is reasonable.

The chief executive believes *cash management* is vital because the individual business units can then direct attention to the client. One business unit head identifies staff costs as the key to the success of unit operations, it being 60% of total costs. Provided that threshold is not exceeded, then the factors that are perceived to deliver high levels of repeat business can be developed. These factors are:

- Developing the unit's 'knowledge economy'
- Staff 'ownership' of knowledge and clients
- Developing a skills database
- Tackling problems and issues through task forces
- Getting close to clients.

that clients had become more price sensitive, the organisation had no idea how long or how much it cost to undertake a design project. The organisation had accountant systems for shareholders and legal purposes, yet were unable to support sales requirements in a fast-changing market.

A balanced system

In the presentation of the system, it has been argued that the flow of information in the organisation is two-way, *top down* and *bottom up*. It is acting upon this information that leads to a systematic use of the information. Actions involve taking initiatives, making decisions and then implementing these. These actions are from the top and the bottom of organisations, people at both ends having the responsibility to act. This is 'empowering' and is thus quite different from delegation, or worse still 'passing the buck'. In this sense, everyone has some management responsibilities. The system is there to provide a framework or context in which people can operate. As a framework, it provides the parameters in much the same way parents that set down boundaries for children as to what they may and may not do. The procedures within also act as disciplines. It is important not to make a sales system a procedural one alone, or sales staff will vote with their feet. So, in designing a system, what is the balance of responsibilities? Figure 3.4 tries to provide some visual guidance.[5] It can be read in relation to responsibilities and scope for action. It can therefore also be used in setting time budgets for staff. It is presented in terms of market position and management structures vary *across* different market positions. Clearly, the flatter the management structure *within* a market position, the greater the scope for empowering staff and encouraging a *bottom-up* sales approach.

Alternative and complimentary approaches

Other approaches to setting up a sales system include the use of:

- Matrix management and mentoring
- Diagnostic review groups or workshops

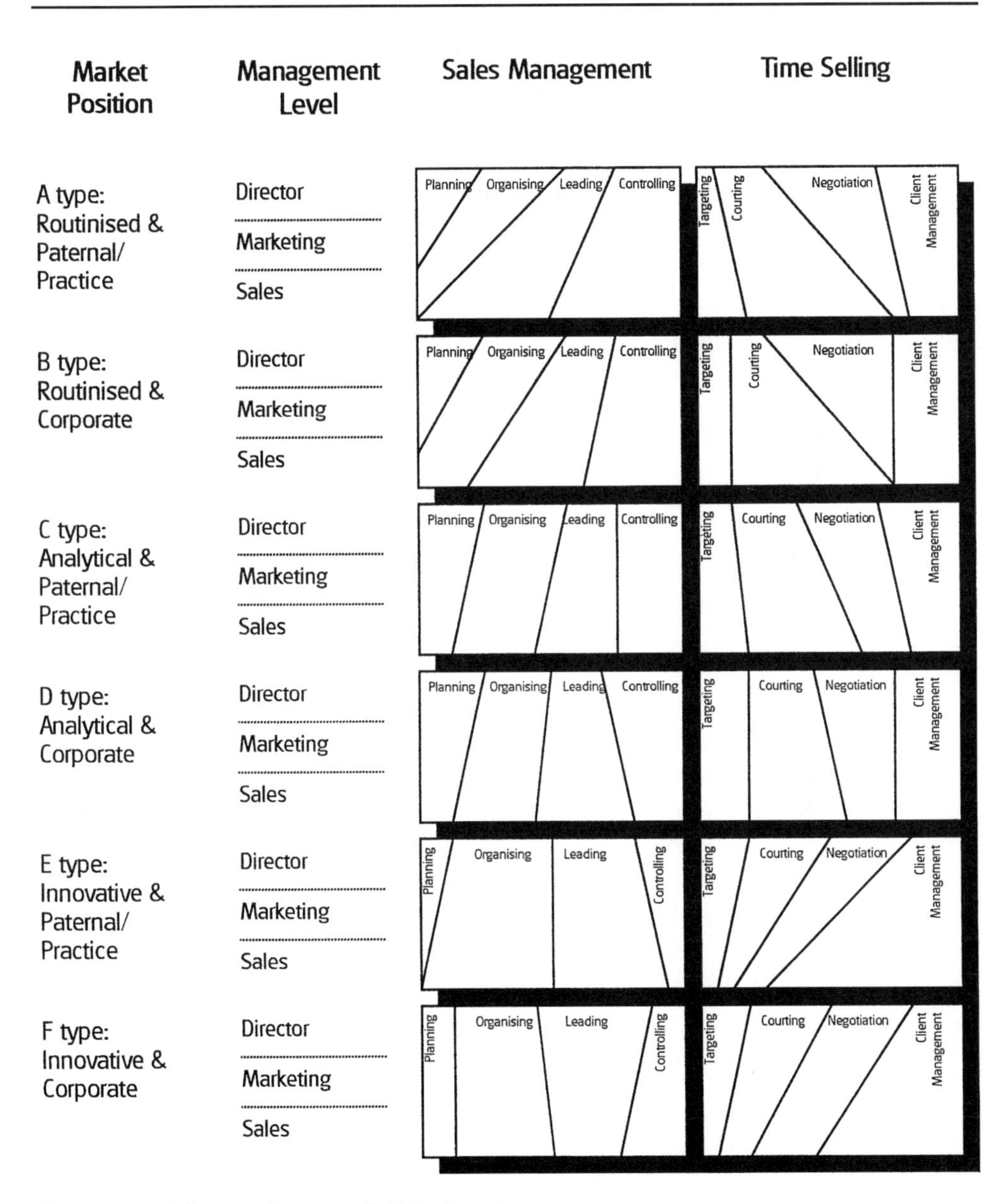

Figure 3.4. Time and responsibilities in sales.

Matrix management and mentoring

This approach has the advantage of being continuous, yet the disadvantage of being potentially time consuming, especially in contracting where people may be located on site, as well as in the office. The main benefit of this approach is that people report in two directions. Accountability to a functional or departmental line manager will remain. Another manager is chosen. One manager must directly or indirectly have marketing and sales as part of their brief. It is ideal as a process, therefore, for those in estimating, legal departments, contract and site managers, for understanding the complexity of their role in the sales process.

Diagnostic review groups or workshops

This process may be regular yet not continuous, capturing snapshots of the sales position. However, it has two distinct advantages. First, senior management can more easily monitor how the process is going and what impact it is having upon the culture. Consequentially, it also helps to demonstrate management commitment to the sales system and to encourage a positive response from all staff. Second, the focus is more task orientated, which fits with construction more neatly, yet many actions and tasks can only be decided by people sharing information and interjecting with their side of the story, so issues have to be reconciled.

What is being explored through both these alternatives is in essence *internal marketing*. A system in itself is without meaning. It is people who give it meaning and the structure of the system helps to evolve the meaning. These alternatives both insist that personnel are brought together to develop their understanding. It has been expressed this way:[6]

> *Internal marketing is a relationship development process in which staff autonomy and know-how combine to create and circulate new organisational knowledge that will challenge internal activities which need to be changed to enhance quality in marketplace relationships (p. 354).*

Marketing and sales effectiveness can be improved through restructuring. The development of processes as complementary or alternative systems can be a major move towards a more client-orientated approach.

Conclusion

In the last chapter, courtship and marriage was used as an analogy for the client–contractor or consultant relationship. It was said that communication was an important ingredient in cementing and maintaining a stable relationship. It was also said that sales personnel have tended to be quite 'promiscuous' and therefore the relationship has to be between the organisations, rather than solely between individuals. A sales system is a means to achieve this and this chapter has tried to sketch out some of the possibilities and to explore one or two avenues in greater depth. The aim has not been to provide a fully comprehensive account, but to stimulate improvement in this hitherto neglected area. After all, every stable family is different.

To pursue the family analogy, marriages are more likely to be successful if the family from which each partner comes has itself been stable, providing love, security in an environment of trust and good communication. The family systems for this include a variety of things, such as talking at mealtime, playing games together, going for a walk, having a family meeting at the weekend, and so on. In a corporate environment, the organisation needs to provide that same secure base in an environment of trust, in order for individuals to begin to cement lasting sales and contract relationships with clients. The sales system is simply a way of ensuring that this happens, complimenting all the informal systems that may be robust in the short term but in themselves do not deliver enduring relationships on a corporate level.

Another way of looking at systems, as we have briefly explored, is that they are the oil of *internal marketing* within the organisation. There must be effective support systems.[6] These must be *top down* in order to guide the marketing and sales effort, with senior management lending and hence demonstrating their commitment to the process and the underlying system. It must be *bottom up* in order that senior management are in touch with what is happening on the ground, so that they can initiate responses to client need and hence improve sales and contract effectiveness in meeting client need.[7]

This chapter may have been quite demanding for the contractor or consultant that is running just to keep still with the existing workload. The idea of adopting new systems may be an anathema or a shift too traumatic in both adopting the change and maintaining the new systems. Organisations occupying routinised market positions will find adoption easier than others (see Figure 1.5), yet need it least, for the risks are lowest in their market position. It is those in remaining positions that need this type of analysis most. A balance needs to be struck between the organisational flexibility for any market position and efficiency of operation (see Figure 3.5). How can it be approached in a realistic way? Undertaking project change, in the broadest definition of projects is the easiest level at which to achieve change. The more complex the change and adaptation, the more a system is needed in order to facilitate structural changes and evolution of the organisational culture.

The systems can be introduced incrementally as long as the whole picture and the integration of the components is constantly borne in mind and communicated as a reminder at each stage of introduction and implementation. Monitoring progress is important:[8]

> *Organisation and people often behave like a large block of rubber in respect of change; they change under pressure, but when the pressure disappears, they revert back to their old habits.*

In order to ditch old habits, the encouragement of a *learning organisation* approach is helpful (Table 3.2).

Achieving change requires managing expectations. This means overcoming resistance and cynicism through building up the expectations people have for introducing the necessary systems:

$$\text{Motivation} = \text{desire} + \text{expectation}$$

Therefore, the first step is to have a desire for change amongst the

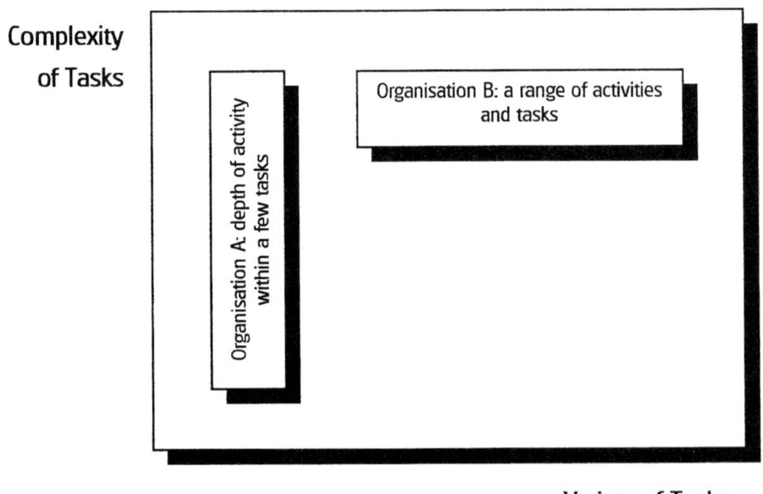

Source: adapted from Markham, 1987.

Figure 3.5. Analytical tools.

Table 3.2 Learning organisation characteristics

Focus	Traditional organisation	Learning organisation
Structure	Hierarchy	*'Bottom-up'* management and networking
People	Managers who control	Managers who facilitate
Staff who know	Staff who learn	
Goal orientated teams	Process orientated teams	
Skills	Adaptive	Creative
Performance measurement	Finance dominated	Multi-dimensional and balanced
Strategy	Rigid	Evolutionary

senior management. High motivation breeds high levels of commitment:

$$\text{Commitment} = \text{implementation} + \text{results}$$

The methods for achieving such results are based upon four possible lines of approach:

1. Internal PR
2. Training
3. Demonstration through senior management coming alongside key staff and helping them in a 'hands-on' way
4. Command.

That is the chronological ideal. Moving too quickly to the fourth way will prove unproductive. Taking such steps helps to make the process continuous. This permits systems to evolve and be dynamic in the setting of a learning organisation. In other words, there comes a shift in attitudes that fundamentally shift, the organisation, rather than a set of procedures and proforma that stultify initiative and choke productive activity. Systems are effective when they in turn make other processes effective.

Summary

1. The purpose of this chapter has been to demonstrate the importance of creating robust sales systems appropriate for the market in which the organisation operates. This incorporates:

- Structures and processes that are designed for a particular market position and segments
- Systems that are set in place from the top down
- Systems that are refined from the bottom in the context of a learning organisation
- Processes that encourage the feedback of market information to ensure that the marketing plan and sales practice is dynamic and evolves in relation to changes in the market and organisation.

2. The chapter has:
   - Set the scene for more effective selling through the corporate ownership of the sales process
   - Created an understanding for the need for cross-functional links.

References and notes

1. Donay, P.M. and Cannon, J.P. (1997) An examination of the nature of trust in buyer–seller relationships, *Journal of Marketing*, **61**, 35–51.
2. Maister, D. (1989) Marketing to existing clients, *Journal of Management Consultancy*, **5**, 25–32.
3. Fowler, H.W. and Fowler, F.G. (eds) (1964) *The Concise Oxford Dictionary*, Oxford University Press, Oxford.
4. Smyth, H.J. (1985) *Property Companies and the Construction Industry in Britain*, Cambridge University Press, Cambridge.
5. This has been adapted and further developed from the work of others, namely:
   Coxe, W., Harting, N.F., Hochberg, H. *et al.* (1987) *Success Strategies for Design Professionals: Superpositioning for Architecture and Engineering Firms*, McGraw-Hill, New York.
   Neale, R. (1995) *Managing International Construction Projects: An Overview*, International Construction Management Series no. 7, International Labour Office, Geneva.
6. Ballantyne, D. (1997) Internal networks for internal marketing, *Journal of Marketing Management*, **13**, pp. 343–366.
7. Ballantyne, D., Christopher, M. and Payne, A. (1995) Improving the quality of services marketing: service (re)design is the critical link, *Journal of Marketing Management*, **2**, 25–28.
8. Markham, C. (1987) *Practical Consulting*, Institute of Chartered Accountants in England and Wales, London, p. 79.

Chapter 4 Market Vehicles

Themes

1. The *aim* of this chapter is to develop the ways and means of working in market segments:
   - ❑ Refining the selection of targets
   - ❑ Operating in the international market
   - ❑ Marketing management from head office in relation to regional offices and operational divisions.
2. The *objectives* of this chapter are:
   - ❑ To select the methods for market penetration, especially in distant locations
   - ❑ To identify the appropriate marketing management approach for serving markets.
3. The primary *outcome* of this chapter is having the ability to choose the right strategy to penetrate markets.

Keywords

Targeting, International markets, Operational divisions, Market penetration and market management

Introduction

The old joke about two sales people from two separate shoe manufacturers is that their managing directors simultaneously sent them out to two distant countries to assess the market for shoes. They both thoroughly investigated the market. The first returned a bit downtrodden and said, 'It's terrible, nobody wears shoes'. The second reported back, 'It's great, *nobody* wears shoes!' The point is that both had the same experience, both had the same information, but they came away with different perceptions. While there may be many reasons why individuals have different perceptions of the same situation, the crux of the matter is that their managing directors need to clearly brief their sales people as to what they are seeking in a given market.

Both perceptions could be equally correct. The first company may wish to generate a good level of exports with little up-front investment and commitment, in which case the response would be entirely appropriate. The second company would need to be in for the long haul, extolling the benefits of wearing shoes and hence creating a demand. This may yield high levels of sales, but only in the long term.

Conventional wisdom holds that the sales person must get out onto the streets and wear out their shoe leather! That is the way to be an effective sales person. If you cover the ground, then trial and error will determine where the market is located. There is nothing wrong with this and it can be effective. However, for complex services there needs to be more interaction at a number of levels and from different angles. That may not happen in distant markets, because the sales effort is located remotely from support at head office. While communications are a vital link and can provide an important source of inspiration and guidance, geographically distant markets can pose particular problems for cultural, political and economic reasons. There can also be 'distance' between operational divisions of an organisation. Selecting the best vehicle for selling in different geographical locations is important.

International markets are an obvious case when looking at geographical distance, although this is not the only geographical application. Within one region or country or even between locations within one country there can be strong differences. Understanding the culture is vital. Appreciating the subtleties of the operating environment sometimes can only be grasped by being there. Management decisions about targeting new and existing clients are therefore affected by this (see ... *On the case* 12).

This chapter therefore considers the way in which geographical and structural concerns affect the *vehicle* used in reaching the market. This is about internal issues that affect the delivery channel for appropriately selling services. It relates to *place* under the 4Ps of the marketing mix.

Selecting the vehicle

Contractors and consultants can set in place strategies and systems to manage the market in their local or domestic markets. Managing the marketing strategy and selecting the appropriate market targets in distant or remote locations can be more difficult. Travel to site and back can be a determinant of market segmenta-

... On the case 12

In Belgium, it is necessary for national contractors to have two headquarters, one in the north and the other in the southern part of the country, in order to have national coverage. Each of the regions requires a physical presence of the contractor within that region as a condition for prequalification.

Local policies can exert a powerful influence on repeat business opportunities. One large contractor working for a local city authority in Britain was complying with local and ethnic labour content policies on their contract. Further staff were required during the contract, so an advert was posted on a site office window facing the street. A staff member of one of the subcontractors wrote beside the poster in the dust on the window, 'wogs need not apply'. That contractor consequently was not permitted to undertake any further work for the authority and did not do so over the next 10 years, which included a boom period when a great deal of public sector work was commissioned.

In both cases, management is needed, whether it is for strategic and structural decisions or for tactical monitoring of daily activities. Getting it wrong has severe consequences. In misreading the culture there may be no feedback, whereas in the latter case the feedback could not have been more dramatic.

tion along geographical lines. Expansion into new geographical markets with the same services and in the same market segments has been addressed by setting up regional offices. This is a common strategy with which all are familiar and carries a medium risk (see Figure 1.7). Selecting an adjacent geographical area is the usual process. Market diversification by geographical area does not pose many issues under such circumstances. The same market strategy and segmentation processes can be used. The local sales effort can then, using the guiding principles set down centrally, identify their client targets in that market area. This can be illustrated very clearly through a case study where the process became built into government policy in Britain during World War II (see ... *On the case 13*).[1]

... On the case 13

In Britain, government policy favoured contractors that could operate nationally during World War II. Regionally based contractors were unable to compete effectively. National contractors undertook large contracts for government. At any one time they were permitted to undertake up to 60% of their annual turnover. If this was exceeded, they were not permitted to undertake additional work until the current workload dropped to 40% of their turnover.

National contractors were permitted to set up regional offices and for those operations to be discounted from the national totals, providing the contracts were under £25 000. This permitted national contractors to compete on an equal footing with regional contractors and hence to increase their market penetration.

Furthermore, when the national workload of contractors exceeded the 60% limit, that office could be designated a regional office. One British contractor had a Newmarket office with national contracts of £270 000 in March 1943. The turnover of the regional office was £71 000. The contractor asked for the Newmarket office to be discounted from national operations and that office was therefore suspended from taking on more work. However, it freed up the contractor to increase the national workload once again. In effect, the contractor increased its total turnover by £200 000 without affecting its national operations.

Therefore in adjacent markets, the procedures and processes are quite straightforward, whether market penetration is institutionalised, as in a war economy, or led by demand and strategic management.

In distant markets, the head office does not have the same level of direct contact. This has implications for market penetration and the selection of market targets according to the marketing strategy. Entering a new geographical market can take one of several routes:

- Follow the client
- Franchise
- Local agent
- Alliance with local partner
- Joint venture with local partner
- Acquisition
- Traditional route – set up a regional or local office.

Follow the client

This is the option of least commitment. There is no allegiance to any one geographical market, the organisation following the client wherever they go. This can work at any scale, but has taken on particular significance with many large organisations operating on a global scale. One international engineering consultant undertakes contracts for Coke on a global scale, their operations following the client wherever they have construction requirements. The requirements are frequently in high-risk locations from a contracting and consultant point of view, although not from the viewpoint of processing fast-moving consumer goods. In this instance, the risk is borne by the client and the contractor or consultant has chosen to work for that client because they have a 'blue chip' status. It may even be the (temporary) office of the client, which the contracting and design team use as their administrative base.

Sometimes the client can be internal to the organisation. The advertising group, WPP, conducts cross-referrals meetings among the diverse set of companies in the promotion and advertising industry. This can result in obtaining business in another location, working out of the sister company office for meetings, and so on.

Franchise

The franchise is the next option of least commitment. It is seldom used in construction for several reasons. The first is that barriers to market entry have been low historically, although one of the purposes of marketing is to increase the entry costs. The second reason is this: the capital asset base is low in construction and for consultants. The third reason applies more to contracting than consultants, the investment in training being low, too.

The ideal type of operation for franchising is a routinised service where the initial outlay and training is high (see Figure 1.5). Once this is in place, the corporate culture is in place and the marketing and sales process will follow. Identifying appropriate market targets will follow with relative ease and without the need for a very 'hands-on' approach.

Local agent

Selecting a local agent can have considerable advantages. It offers the benefits of someone or an organisation that understands the local context in all its ramifications. There is still a cultural issue, for that does not mean that they understand your organisation and its objectives, especially if they are agents for a number of different organisations, for they will have developed their own style and approach.

Matching the corporate culture as much as possible, for example, by using the market position matrix, is important (see Figure 1.5). There are still requirements for the contractor or consultant to understand the local market:[2]

> *Most of those offering their services will present themselves as able to provide influence as well as contacts.*

Such claims need to be substantiated. A variant of the agent is the 'sponsor', who is a prominent figure. In some markets, such as in parts of the Middle East, a contractor or consultant has to have a registered local sponsor in order to undertake any work.

The local agent will need induction. Both sides should be prepared to invest in the relationship. The further removed the service is from the routinised approach, the greater the induction requirements. On the other hand, innovative approaches to business are less suited to the use of an agent.

Alliance with local partner

An alliance will be made where a local organisation agrees to bid for projects outside its capability involving the international contractor in some aspect of the project, so that local capability is enhanced. This type of arrangement can be undertaken by signing a memorandum of understanding. When a project is secured, then a contract will be signed. One organisation may take the commanding position, the remaining one acting as a subcontractor or subsidiary partner in essence. A more equal relationship can be established through a joint venture agreement for the project.

This type of organisation will seek projects that suit them. Therefore, their market position and approach to segmentation needs to dovetail with your organisation.

Joint venture with local partner

A joint venture may be project-based or it may be a long-standing agreement. There needs to be mutually exclusive benefit for both parties. The usual benefits are that the local partner provides understanding of and access to the market, whereas the incoming organisation provides resources and expertise, including the sourcing of components and labour on the international market, as well as project know-how.

A joint venture needs to have a similar market position; however, the segments in which they operate need not be the same, especially if the joint venture is on-going and not project-based. As the venture is designed around mutually exclusive benefits, then pursuing different segments is possible. This model is suitable for working in any adjacent geographical market. They need not be distant.

Acquisition

This is the quickest way to establish a local base. Through increasing resources, expansion can be rapid. However, the corporate culture in another organisation can be very different in any context, but especially in an international setting. Acquisition in distant markets may be easier to integrate more rapidly than in the home market because of the advantages yielded from the relative autonomy given to local management.[3] The downside of acquisition is that in both contracting and consulting the primary resource is the people. They have the contacts, relationships and expertise. Take-overs and mergers can create insecurities and lead to people re-evaluating their careers even before the new management does. If staff vote with their feet, then making the expansion effective can be seriously hampered.

Traditional regional or local office

Setting up a local presence is far from straightforward. Some of the previous options may have been used as stepping stones. However, when a local or regional office is set up, it is vital that an appropriate marketing approach is adopted. Assuming all the hard investigative work on statistics and thorough networking has been carried out in order to establish the desirability of the market, the strategy has to be fulfilled and the right targets selected.

The key issue concerns how much central control is exerted from the head office. It has been found in other sectors that the level of headquarters control affects the success of operations; however, the requirements for control change according to how close or distant the market is located.[3] This is illustrated in Figure 4.1.

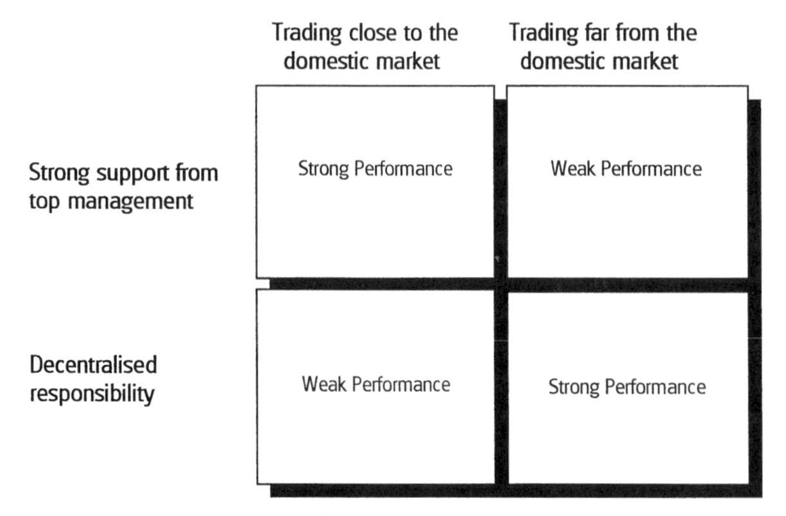

Figure 4.1 Implications of control from headquarters on regional performance.

It has been generally found that organisations that exercise close control of new offices and operations located adjacent to the home market perform well, as do those offices located in distant markets, where a high degree of autonomy is permitted. There is every reason to suppose that construction conforms to this general pattern. The most successful global contractors implement this dual strategy (see ... *On the case 14*, cf. ... *On the case 7*).

Different nations have different cultures and this type of approach is more suited to some national cultures than to others (see Table 1.1). Organisations will perform better in distant markets that typically show a preference for:

- Entering emerging market segments
- Winning market share
- Quality and reliability
- *Top-down* and *bottom-up* communications.

These characteristics are exhibited by many Japanese organisations,[4] contractors having been reluctant to withdraw from markets, even during recessive times and having invested for the long term. Much of the gained experience has proved useful and is imported into the domestic market to enhance existing construction practices.

Organisations will perform poorly in adjacent markets that typically show a preference for:

- Short-term profitability
- Cost reductions, rather than improved productivity
- Low levels of service and support
- Low levels of teamwork.

For example, this has been shown to be the case generally in British industry[4] and shown in British contracting in terms of the low levels of strategic planning.[5] There has been an absence of commitment to the European market, resulting in a loss of market share of over 18% in a 3-year period among the industry leaders.[6]

The conclusion is that the strategic set up affects the ability of those responsible 'on the ground' to select the best clients and sell the services of the contractor or consultant. A dual management approach is needed if geographical expansion is seen as necessary or desirable – strong support in markets adjacent to the home base and decentralised responsibility in distant markets. The management style that comes most naturally to contractors is

... On the case 14

A dominant and diverse international contractor from France has locations in over 80 countries and a workforce of nearly 100 000 and has ambitions to be a global player across its business interests. It operates at three levels of control:

- In the domestic market cross-shareholdings offer structural protection
- In the European home market outright ownership and strong direction from the head office is provided
- In distant markets, such as the Far East, expansion has been undertaken, through acquisition of majority shareholdings. The original intention had been to co-ordinate local and central activities more closely than has been the case in practice. For example, originally the plan had been to change the name of local companies to conform to a single global image. The local names were retained, however, due to their strong reputation with clients in their established local markets. Greater autonomy was found to be necessary and appropriate.

the decentralised one. It is proving insufficient in order to operate internationally and compete on a global scale. This could also be replicated on a smaller scale for national contractors within their own country.

Natural divisions

The emphasis has been on the international or geographically distant markets. There can be a great deal of distance between different divisions or companies within a contracting organisation. This is particularly the case in construction. It was shown in relation to segmenting the market that many 'marketing' decisions have been historically addressed as issues of structural organisation. The setting up of a separate design and build or management contracting organisation are good examples of this. Not only can this act as a barrier to being able to effectively advise a potential client and to serve a repeat business client with diverse needs, it can also create distance across the group.

The solution for effective diversification is not always a structural one. The convenience from a management point of view is that the central or headquarters management function can be at 'arm's length' from the division. The same dynamics and issues that have been raised across geographical distance also apply across the structural divisions designed around different procurement routes.

In the international case, there is a strong imperative to adopt a dual management approach. The same imperative is not repeated across different services. However, managers need to be aware of the implications for certain sorts of organisational structures and to decide whether this provides the best fit in reaching the markets being sought by the organisation. Using structural solutions for market management is perfectly justifiable when carefully considered in relation to client service and market targeting. It cannot be justified, as is all too frequently the case, when it is selected for the convenience of management. The management steps to a considered approach are to hold board meetings to review organisational structures, namely:

1. The relationship of regional offices to the headquarters
2. The relationship between operational divisions and market management.

Critically evaluate the degree of support given – should you be offering strategic guidance and a hands-off approach or strategic management? Decide what changes need to be made and how to introduce these. One way of evaluating the structure is to use an action matrix. A hypothetical review is shown in Table 4.1.

In Table 4.1, the organisation has clearly recognised a lack of strategic direction and a need to more highly value those responsible for selling. In general, there must be accountability. People need to be put in charge of various aspects of business development. There will be office directors and divisional directors and those in charge of individual segments. They are required to be experts in their segment. They may come from a marketing background or be in marketing; however, their brief is to understand market segmentation, the selection of potential and existing clients

Table 4.1 Hypothetical action grid for market management

	Actions				
	Main board	Divisional heads	Head of marketing	Marketing personnel	Sales personnel
Strategic evaluation					
Guidance	Review	–	–	–	–
Control	–	Restructure	Oversee	Train	Monitor targets
Strategic actions					
Induction programmes	–	Marketing	–	–	–
Training	–	Marketing	–	Develop strategy	–
Communication	Improve	–	–	–	Feedback
Tactical processes					
Monitor local Market targeting	–	–	Co-ordinate local & HQ	–	–
Monitor divisional Market targeting	–	–	Merge marketing across group	–	Empower sales people

to be targeted. Their personnel are required to furnish them with necessary information, to supplement their overview of the market.

Conclusion

Market targeting has involved undertaking the segmentation of the market and going through the interim steps until sales targets have been effectively identified. The precise purpose of this chapter has been to examine this in relation to business structures that can act as impediments to successful segmentation and targeting. This is true in any business environment, but is especially pertinent in construction, due to the decentralised nature of production and the fragmented nature of the marketplace, where client courtship can be pursued over a long period of time. It may therefore take quite some time before misjudgements about sales targets filters through, and even then the nature of the problem may not be clear. The steps outlined in this chapter can help to organise and structure the sales effort. Each organisation must have the right vehicle for selling.

Summary

1. The purpose of this chapter has been to demonstrate:
   - Market penetration where structural barriers exist
   - Appropriate strategies for market management
   - Ways to commence reviewing and managing complex issues of geographical and operational distance.

References

1. Smyth, H.J. (1985) *Property Companies and the Construction Industry in Britain*, Cambridge University Press, Cambridge.
2. Langford, D.A. and Rowland, V.R. (1995) *Managing Overseas Construction Contracting*, Thomas Telford, London, p. 84.
3. Madsen, T.K. (1989) Successful export marketing management: some empirical evidence, *International Marketing Review*, **6**, 41–57.
4. Doyle, P., Saunders, J. and Wong, V. (1992) Competition in global markets: A case study of American and Japanese competition in the British market, *Journal of International Business Studies*, **23**, 419–442.
5. Stockerl, K.C. (1997) The importance of strategic marketing planning for the UK construction industry in a changing European business environment, *Proceedings of the 2nd National Construction Marketing Conference*, 3 July, Oxford Brookes University, Oxford.
6. Smyth, H.J. (1998) The competitive stakes and mistakes: the position of British contractors in Europe, *Proceedings of the 3rd National Construction Marketing Conference*, 9 July, The Centre for Construction Marketing in association with CIMCIG, Oxford Brookes University, Oxford.

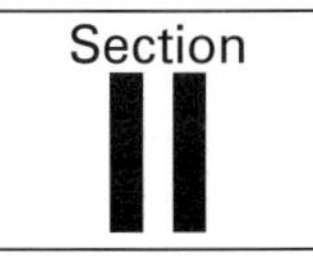

What is Selling?

This section contains eight chapters. Its emphasis is on selling and therefore it focuses upon the implementation of the marketing plan. The section covers the traditional *marketing mix* approach and the approach of *relationship marketing* to selling. It also looks at some of the alternatives available within the traditions, especially within relationship marketing. *Managing the sales process* is therefore a primary theme.

Hype or professionalism?

What is *sales*? In economics, it is an exchange process between two parties. In legal terms there is a consideration, which could be money or goods. Demand for products and services may be high in terms of what people wish to have, but it is far more constrained in terms of what can be afforded. On the supply side there is pressure to realise a healthy profit. Therefore, the customer wants to make sure they get good value in the exchange and the supplier does not want to erode overall profitability. Some tension is therefore built into the sales process from the start, which can lead to some distrust.

Selling is about oiling the wheels of the exchange process. Neoclassical economics assumes that the customer has perfect knowledge. In reality, that is seldom the case. Part of the sales process is to make sure the potential client knows as much about your service and product as possible, at least on the positive side. That is simply improving the level of knowledge in the marketplace. The onus is on the customer to check out whether the supplier is providing something that meets their requirements.

It is perhaps at this point that the idea that oiling the wheels of selling is just a slick operation, run by 'slimy people who would sell their own mother'. As in any professional activity, there are those who do not conform to the professional ethics either in law or a spirit of integrity. However, most do. It is natural to stress the positive. They are borderline areas. What about running down the competition to make your own products or services look better? Many people do that. If you wish to put your house on the market for example, it does not take long to find an agent who is prepared to knock the opposition. That does seem to work for some. It feeds the minds of those who have already partly made their mind up and consequently hear or believe what they want to hear. It can also work on those who find the process a hassle or stressful and are looking for someone else to help make their minds up for them. It also works better in markets where the service or product is very standard, the suppliers being driven towards it because the level of competition is high. While this may work in the short term, it does not work in the longer run, where organisations are reliant on reputation and referrals. Contractors have fallen into this trap at times and it merely adds one more layer to the equation that contractors are not to be trusted. It helps give the sector a bad name. From the viewpoint of the discerning customer, running down the opposition shows a lack of respect for others. Will the client be seen as on the opposite side of the table once a contract is signed? The same lack of respect or contempt for others may result in a very bad experience for the client or customer. Running down the opposition is an immediate warning sign: 'Stop! Do not do pass this point!' In any case, a good service provider should have some positive things to promote.

Therefore, selling should be a professional activity, as professional as anything else. Those in sales are simply trying to persuade the customer of the value of their product or service. Certainly there can be times when promotion borders on manipulation. Association is frequently used in this context, where sexual or lifestyle imagery is used in advertising to create a positive feeling or desire for the product in the mind of the customer. However, a great deal of research has been undertaken about the impact of this type of advertising. The results have shown a variety of things. The level of recall is not that high. People often remember the imagery or story line, but fail to recall the product. Many who recall the product line stick with what they have always bought, even though they found the advert entertaining. Some adverts become a talking point, but it does not necessarily mean that everyone is manipulated into a purchase. On the other hand, advertising works, otherwise companies would not use it. It comes back to the adage; '50% of advertising works, but we don't know which 50%.' What we can say is that the absence of advertising means that a brand or product can quickly drop from view in the mind of the customer. It would appear that a great deal of advertising and promotion, indeed all selling, works at the level of *being there* or perhaps more accurately *being seen to be there*. A good campaign may then act as a persuasion, providing someone wants to purchase something.

There are essentially two levels at which this works:

- ❑ Effective demand
- ❑ Latent demand.

Effective demand is where a client has secured the finance and wants a building. It is then just a matter of choosing the means and supplier. The sales effort concerns securing the design or construction contract. *Latent demand* is where the desire for something is pent up, but has yet to become a market reality or where the client does not even know they need a building. In the first case, the sales effort is about getting into position for when the demand becomes effective, so that your organisation is leading the pack, or better still doing something, such as arranging the finance, so that the demand does become effective. In the second case, the issue is *opportunity cost*. For example, a manufacturer may believe that the best way to increase efficiency is to provide a new piece of very expensive equipment, although it could be shown that a slightly reduced expenditure on the building to make operations more efficient could have as good or better results. The demand therefore switches from expenditure on plant to a building contract.

For example, an international cosmetics plant had a manufacturing and packaging operation on the same site as its warehouse facility. The high bay warehouse had fallen into partial disuse due to just-in-time production and delivery adoption. One choice considered was to improve production conditions. This option was pursued, locating one of the production lines in one of the high bays of the warehouse.

It was designed to give a feeling of a sports and leisure centre, with some of the key advertising imagery used around the bay. Instead of going down a dingy corridor to a rather squalid room during breaks, an elevated platform was constructed in the bay with coffee- and tea-making facilities and comfortable easy chairs. Staff felt valued and productivity apparently went up 25% initially. It fell back a bit later, once the novelty wore off, but a real improvement was secured over the long term. This example demonstrates the opportunity cost of investing in buildings.

Natural gifting?

Some people are naturally gifted at selling. Those using hype may seem that way, but they are not always amongst the most gifted. Many rattle on and do not listen. The best sales people are usually good listeners; in fact, they are often listeners first and foremost. This ability, coupled with counselling competencies, is very important. Communication may be important, yet technical expertise can be vital too. Responding to the details of client needs in terms of technical and service requirements is essential in the business-to-business environment of contracting and consulting. This is an issue of *hard sell* versus *soft sell*.

There are a number of issues that contrast the two approaches, *hard* and *soft* sell. While they are essentially about techniques, there are a number of misconceptions leading people to convince themselves that selling is for experts! The misconceptions depend upon perception:

1. Selling is only for the naturally gifted – while it is true that some are more gifted than others, many of the techniques in all forms of selling can be learnt and the gifting gives an important edge on top!
2. Hard sell is better than soft sell – either hard sell involves harder work and is more effective or it is slicker, with more hype, even though it may involve less accuracy and it can be unpleasant to be on the receiving end!

Bear in mind that either approach can be carried out professionally, with factual accuracy and with sincerity. There are important considerations in making the choices between the approaches:

- *The market position*: A sales force for offering a standardised routinised approach would be more appropriate for the hard sell, whereas the problem-solving practice-based architect would more appropriately use a soft-sell approach through one of the partners.
- *Listening or presentation skills*: soft and hard sell emphasised, respectively
- *Demonstration or problem-solving skills*: hard and soft sell emphasised, respectively
- *Briefing or education skills*: hard and soft sell emphasised, respectively.

The house-building industry has a more standardised approach compared to business-to-business parts of the construction industry,

... On the case 15

In the UK, volume house-building industry selling new homes could be characterised as a hard sell operation by trained sales negotiators through show home and estate agent 'retail' outlets.

Sales staff are not experts on construction, content or performance of the product. They are experienced at closing sales and negotiating around price, extras such as carpets and white goods, plus financial incentive packages, for example cash back, part exchange and discount schemes.

It has been found that prospective purchasers were interested in performance criteria if they were 'educated' about these. Focusing on energy features research found that purchasers would pay more, including a profit premium, if sales staff informed them about the benefits and any likely protection of the added value in the resale price. It showed how a standardised/routinised operation could benefit from a soft-sell approach, if training and support could be given to staff.

It has been further found that there is a relatively untapped market segment for the second-time homebuyer in the UK, among those in their late 20s and early 30s.

but even here there are always lessons to be learnt from different approaches and drawing technical expertise (see ... *On the case 15*).[1]

Many naturally gifted sales people have a very narrow view of selling. They merely carry out the job as perceived by them. Few will take initiatives beyond that. In other words, they tend to look outwards and get their rewards of recognition from clients wanting what they are offering. This involves a focus upon:

- *Performance* – quality of product and service
- *Promises* – delivering the goods
- *Pleasure* – job satisfaction derived from client satisfaction.

When they look inwards they tend to 'look after number one' and to 'mind their backs' with their employers. This covers concerns about:

- *Performance* – securing pay levels and incentive schemes
- *Politics* – accountability and passing the buck
- *Position* – job retention or promotion.

A more serving disposition may be unfashionable, but it works.[2] A serving disposition is about drawing people in, rather than pushing information out as an effective way to sell. Serving is also conceptually linked in with the service management approach of relationship marketing. There is close correspondence between this service orientation and employee morale plus job satisfaction. The theoretical implications for practice are:

- All staff need to have or develop part-time marketing skills and capacity in order to enhance the ability of the whole organisation to serve
- Full-time sales staff need to be selected according to the market position
- Full-time sales staff need to be selected according to the sales approach.

Relationship marketing is therefore also a management strategy.

Internal issues

Communication is central to selling. If clients and client representatives do not hear what the contractor has to offer, then selling cannot commence in earnest. Communicating the offer to the client is one thing. It is underpinned by the effectiveness of communication within the contracting or consultant organisation. Effective communication is hard work. Its importance is to create an internal environment in which selling can flourish. It is precisely this aspect that is neglected. Therefore, this section is not so much to do with *how* to sell. The emphasis is upon the sales environment within the organisation.

Sales communication within the organisation and the management action that follows is a neglected area in the construction industry. It is a neglected area because selling has low status. Selling generally has been perceived as an individualistic activity outside the mainstream of operations at best and at worst as some sort of activity that lacks integrity and is therefore unprofessional, selling being something of a career cul-de-sac. This is paradoxical because selling is the lifeblood of a commercial organisation (see Paradox 11 in Appendix A).

Although some aspects of sales techniques will be looked at in the next section, effective sales management ensures a competitive edge. The consequence of failing to address and to manage the sales environment is that this contributes to the trend of a contractor selling the 'same service' as the next contractor. Thus, it is all about price from the start, rather than at the finishing line. This is because those involved in selling have insufficient back-up. Inadequate back-up comes from their own staff, not only at top management level, but also horizontally across the various areas of expertise that can be called upon to add value and differentiate the services in sales communication. So, sales communication becomes limited. It is carried out on a 'need to know basis'. The perception of what people need to know is very narrow because of the low levels of status and support it attracts.

In this context, *selling* is defined as those who are dedicated to selling and they need the support of others, especially in market positions 'C' to 'F' (see Figure 1.5). These staff can feel isolated and operate more or less autonomously from the organisation, sometimes feeling more affinity to their client organisations than their own. If the sales people spend too much time at their desks, if it is not because of bureaucracy; it may well be that they feel isolated within their own company and want to be there in order to sense support and belonging.

Selling also embraces those who sell as part of their other functions, the so-called part-time marketer[3] or part-time sales person. For these staff, selling is often perceived as some extra activity that has to be squeezed in, frequently out of normal hours, a message implicitly conveyed through the attitudes of senior management. Indeed, many of those in management do not understand the role of sales and how it relates back to the marketing function.

The answer is to give managers the opportunity to develop and understand the marketing and sales functions. The answer is certainly not the commonplace one of giving the function a false status by calling sales people 'marketing managers', or as one major British developer[4] put it:

> *Scratch a business development manager and you'll find a sales manager underneath.*

... On the case 16

A highly successful multi-disciplinary surveying practice (employing around 150 staff across four regional offices) which sees marketing as fundamental to their business, faces a problem concerning sales. As the marketing manager puts it:

Selling is what separates partners from the rest of the staff.

Yet there is little preparation among prospective partners for their promotion in understanding the nature of marketing and sales:

Selling comes first and as they develop their role they realise they can't just sell and marketing provides direction.

So by trial and error they begin to appreciate the need for selling and how this relates to strategic business decisions. The marketing manager is helping the partnership to address the issue, yet it is recognised that this is a long-term objective.

Management tend to compensate for a lack of management by adopting self-policing or motivational approaches, such as incentive pay schemes. Yet employees rank pay as the third or fourth most important issue in employment exit surveys.[5] Most sales people want to feel wanted. Yet the relative isolation of sales staff has a side effect of creating tensions and sometimes conflicts with other operational staff. This in turn undermines support across the organisation and frequently from top management too, who more readily identify with the operational side of the business.

When dealing with communication on a 'need to know basis', the lack of understanding about the necessity for managing the sales process is that management do not know what communication and supportive actions are actually needed. This section addresses this management gap.

Communication can itself be viewed in several ways. It has been sketched out in this book in terms of attitudes, leading to certain behaviours. Behaviours, which are more than individual ones, become processes around which organisational structures emerge. What we say and how it is communicated is therefore very important. It is more than the verbal content. Some say that we listen to only 10% of what someone says, the rest is communicated by the setting, the means, tone of voice and body language.

External issues

The consequence of poor communication is that sales people, working in relative isolation, also fail to communicate effectively. Communication is reduced to the lowest common denominator. Hence, the sales process becomes culturally and structurally inbuilt as selling to the lowest common denominator – *price*. What is being suggested is that the failure to manage selling has reinforced the marketplace forces, such as fragmentation and a lack of service differentiation, and contributes to the many reasons why price has historically dominated selling. This in turn reinforces the lack of differentiation between competitors. Sales is the point of client contact and so if price dominates in the sales process by default, the client becomes educated by the contractor to connive with this. In other words, while there are good market reasons for price being a strong factor, the contractor and more recently the consultants, have been encouraging their clients to adopt such procurement approaches. The way out is differentiation. It has been expressed in the following way in terms of what it means to some contractors:[5]

> *Contractors have traditionally taken the narrow view of business development as the art of anticipating tender lists and getting their name on them. Nowadays, with the trend towards partnering and negotiated work, most medium-sized and larger contractors recognise the need for a more sophisticated approach to identify promising market sectors and ensure they are positioned to exploit them.*

What this seems to suggest is a move away from price domination, which is part of the 4Ps of the marketing mix, and entails moving towards the relationship marketing approach. There is choice as to how far the consultant or contractor moves. Differentiation can be achieved through emphasising other factors in the 4Ps, such as place, product or promotion, or for that matter any of the many 'Ps' that others have added to extended lists. While it has been possible to survive with a poor marketing and sales interface, it has failed to produce a seamless strategy. With such a hole, plus the additional hole – the sales management issue discussed here – it may not be possible to sustain a position in the marketplace in the emerging environment. And it is definitely not possible to sustain a long-term and effective sales effort in relationship marketing without managing the sales process, which includes relationship marketing tools, such as partnering.

How are these issues going to be addressed in specific terms in this section? Chapter 5 looks at the nature of sales and selling in more detail, particularly the marketing mix approach. Chapter 6 also looks at the nature of sales and selling from the relationship marketing perspective. In the two chapters, attention is paid to the models that can be used for managing clients. In the previous section, an emphasis was put on the strategic development of the business and marketing plan and how that helped to shape sales. In this section, the opposite end of the spectrum is considered, for the way in which clients are managed also affects the way in which selling is undertake and by whom. Of course, the two ends of the spectrum meet each other in the corporate management of an organisation and thus in the way the sales process is set up.

Chapter 7 looks in some depth at the nature of the product and service being sold in construction and by consultants. The steps and stages of the sales process are unpacked in terms of the more detailed objectives, mechanisms and outcomes being sought. Different procurement routes will be added into the equation towards the end of the chapter, spelling out the implications for selling.

Chapter 8 moves to a more distinctive emphasis upon relationship marketing, looking at how organisations can effectively sell their services through business-to-business relationships. It provides a base line for Chapters 9 and 10, the focus being the role of adding value to the product and service respectively in the creation and maintenance of competitive advantage. The move is towards differentiation in these chapters, where organisations are refusing to compete on the same ground as everyone else.

The overall outcome is to practically explore the management of differentiation in the sales process.

References and notes

1. Dibb, S. and Wensley, R. (1987) Energy efficient house design: the analysis of customer choice, *Reviewing Effective Research and Good Practice in Marketing* (ed. R. Wensley), Marketing Education Group, Warwick University.
   Smyth, H.J., Stallwood, P. and Thompson, N. (1997) *The Market for Energy Efficient Homes: Research Report No. 1*, Centre for Construction Marketing, Oxford Brookes University, Oxford.
2. Schneider, B. (1994) HRM – a service perspective: towards a customer-focused HRM, *International Journal of Service Industry Management*, **5**(1), 64–76.
3. Gummesson, E. (1991) Marketing-orientation revisited: the crucial role of the part-time marketer, *European Journal of Marketing*, **25**(2), 60–75.
   Gummesson, E. (1990) Making relationship marketing operational, *International Journal of Service Industry Management*, **5**(5), 5–20.
4. Knutt, E. (1997) The Scouts, *Building*, 21 February, 43.
5. Murdoch, A. (1998) Keep your sales team sweet, *Management Today*, January.

Chapter 5 Marketing Mix and Sales Promotion

Themes

1. The *aim* of this chapter is to investigate the approach to promoting the services of contractors and consultants using the traditional marketing mix.
2. The *objectives* of this chapter are:
   - ❑ To identify the implications in an economically fragmented market
   - ❑ To investigate the implications for geographically diverse markets
   - ❑ To explore whether company size has implications for selling.
3. The primary *outcome* of this chapter is to ensure that appropriate selling approaches are adopted.

Keywords

Marketing mix, Product, place, promotion and price

Introduction

Many people in the industry treat the client as remote in their attitudes. Many things that happen would not happen and many things would not be conducted in the same way if the client was right there watching and listening. It is a good *acid test* to ask, 'Would we do it this way if the client was here?' If we would not, then we are disregarding our clients and being driven by our own attitudes. This will not bring back the clients next year. The overall attitude and enough of what goes on is picked up by and indirectly fed back to the client for them to know that the poor image of the industry is fully justified and that the contractor is not to be trusted to take their interests to heart.

This makes the efforts of the sales people an uphill task. Their hands are being tied behind their backs before they even start. Is it not an indictment and a profound insult that clients are telling contractors how to run their affairs?[1] More is needed, but what? Contractors and consultants frequently swallow the myth that marketing *is* selling, and so the sales effort is further handicapped through lack of specific marketing and general management support. The false management solution to this problem is usually:

- ❑ More of the same – sell harder
- ❑ Replace the sales person – they are no good
- ❑ Get rid of the sales people – they are just a costly overhead and the work will come in anyway.

Several independent and industry studies have confirmed this as the dominant attitude.[2] An old sales adage is apposite here:

We sell what we produce, not produce what will sell.

This is an operationally driven approach of 'take it or leave it'. It assumes that the client wants what you are selling, regardless. It is only attainable when operating in the following environments:

- ❑ Monopolistic market position
- ❑ Cartel or Mafia-type market economy
- ❑ Homogenous market, where supply is heavily constrained
- ❑ What is supplied is a perfect match with demand in an unchanging economic climate.

The construction market is the direct opposite to this. There are numerous operators for whom the barriers to market entry are low, competition is very intense and the market is far from homogenous in the types of buildings, types of services and geographical coverage. Contractors have historically paid little attention to what it is that the client wants. Many consultants have been guilty of this to some extent too. Therefore, there can be a focus on the operations or a focus on the client. This sums up the two main approaches to selling:

- ❑ 4Ps of the transactional approach embodied in the traditional *marketing mix*
- ❑ Business-to-business approach of *relationship marketing*, based upon service management tl ieory.

The marketing mix advocates would say that the approach permits a great deal of customerisation. This is true; however, the emphasis is about getting the balance of the ingredients right in the recipe. The overall choice of recipe is not questioned at all. In one sense, what is being emphasised here is the need for a good marketing strategy. It is about getting sales onto the right footing. This is important, for selling has its own problems. Figure 0.3 sets out some of the key conceptual distinctions between the marketing mix and relationship marketing. In this chapter, the focus is upon the sales implications of the marketing concepts. How does that translate into the sales process?

Marketing mix

It is generally said that 75% of all business is lost in the very first contact.[3] Therefore, how the sales process is structured and carried out is crucial to the *sales performance*. The level at which the previous section has addressed the setting up has been through the business plan and strategic marketing. Now the job is to structure the sales process properly. It begins with the first point of contact. That may be the receptionist, it may be at a lunch set up by a consultant. It may be through a *cold call*. Whatever the contact, organisations do not perform very well. This is usually because those involved are too quick off the mark. The receptionist or telephonist wishes to hand the prospect on as quickly as possible. The technical person answers a different question to the one they were asked. The sales person is wanting to get in there quickly and blows it in the process. There is a real need to listen and make sure that the client knows they have been heard, understood and that an unhurried and considered response is going to be given, which reflects the question first asked. Many potential clients do not actually pose a direct question. They are only a prospect, because there are some issues to be explored. Listening and reading between the lines and gently using a series of open questions can help to uncover the issues. The sales process begins here and the steps are outlined in Figure 5.1.

Clients do not always know what it is they want. Even if they do, it may well be at a very general level. This is particularly true in the case of consultant clients, as selling occurs from higher

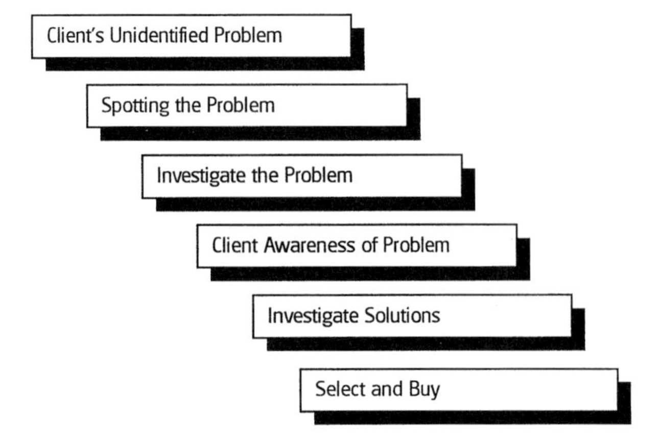

Figure 5.1 Sales stages from the client perspective.

up the supply chain. Of course, it is their expertise that helps to pin down the requirements in the briefing and concept stage, however, more time spent listening and teasing out the issues at the early stages can reduce uncertainty and reduce the number and extent of changes downstream.

Some clients, especially those who are experienced procurers are clearer about their requirements, but should always be seeking to learn about subtle changes in the external environment.[4]

The corresponding view to Figure 5.1 from inside the organisation is set out in Figure 5.2. Here the sales person is charged with the initiative of identifying generic and specific client needs and taking the initiative to respond to them. This is not merely for the individual. They are not able to cover every problem in terms of identification, analysis and solution. Hunting in pairs is often a good technique to supplement the efforts of the individual,

... On the case 17

The development director of an international property developer states:

> *If someone comes to me to market their services, I want to be able to have a useful session gathering market intelligence from them. It has to be somebody who takes responsibility for the company business – it's a bad idea to send a marketing director.*

In this case, the director wants someone who sees the business in the round and is not just after the next sale. The cryptic comment about 'marketing' directors is evidence of the real function being selling and that they are unsupported in the role.

This is substantiated by one of the largest and most stable developers in Britain:

> *Clients hear the same old thing all the time. They should be offered something different. But whoever is representing the company needs to have sufficient standing to go back and say, 'this is the action we need to take'.*

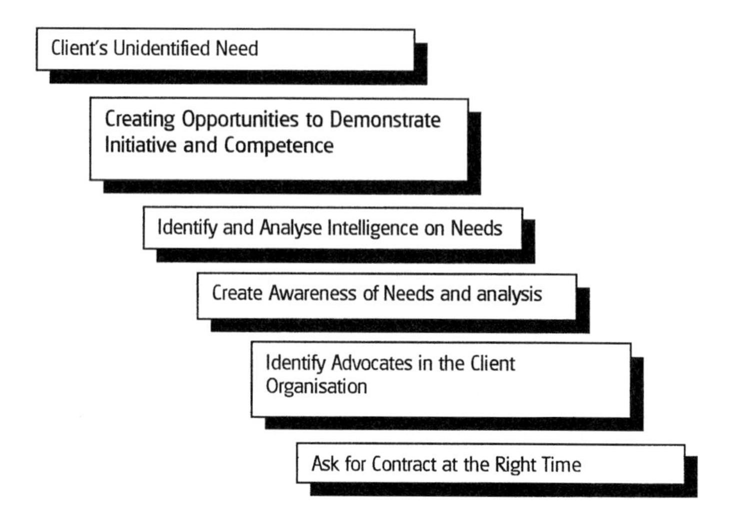

Figure 5.2 Sales stages from the supplier perspective.

especially for core or key clients. Complementary skills and contrasting styles can pay dividends. However, a wider constituency frequently needs to be drawn into the sales process. The part-time seller or marketer is relevant here.[5] Some companies have gone a step further to instigate loose sales teams or networks. This is ideally suited to the project environment of contracting and consulting. In one sense, it merely brings the project start date forward to embrace the commencement of selling, once the sales person has acted as a door opener and solicited some initial interest. It is also very suited to situations where technological issues are to the fore. Commentators have stated:[6]

> *This opportunity occurs in an environment where buyer's expectations of sellers are escalating, and marketers are faced with the challenge of differentiating and extending the utility of their products. Further, the needs of buyers are increasingly becoming complex, requiring buyers and sellers to rapidly exchange information.*

This is precisely the situation found in construction and consulting and team selling can yield a competitive and profit advantage. Like project teams, the sales team should be shaped around the needs of the customer and not around the operational drivers of the supplier organisation.

The idea may be a good one and it may help to launch a contractor or consultant into a client orientation or towards creating a marketing culture that penetrates the organisation until it becomes a matter of common sense. Getting to that place can be difficult as the additional demands on stretched staff can produce an adverse reaction, especially as the approach cuts across traditional attitudes. There is a measure of change. As one contractor put it:

It's not like five years ago – marketing get the work for us.

However, the greater involvement of others has yet to be fully recognised and organised as part of the implementation programme for the marketing plan and therefore as part of sales management.

The *marketing mix* is the traditional approach to marketing. It was developed out of micro-economic theory on price discrimination and is closely related to transaction analysis.[7] One of the ways in which customer or client requirements can be rationalised at the point of sale is to group their needs into categories. With the aid of sales data and market research, this can be carried out in advance. This is a key component of the marketing plan when carried out in a strategic way. The categories become market segments at the strategic level (see Chapter 3). At the point of sale the tactics are to adapt the mix of ingredients in the marketing mix recipe, the aim being to tip the balance in favour of the organisation for whom the sales person works. One example might be to promise delivery ahead of the competitors and another might involve discounting the price in a limited number of situations.

The adjustments at the margins can be made on a client-by-client basis or to groups of customers – perhaps whole segments or particular niches. The nature of the offer would be promoted or advertised in order to target particular groups. There is a fairly high degree of standardisation in such approaches and it fit, the growth markets of mass production in the post-World War II era, when the concept was developed in North America. It most ideally fits those companies and organisations that occupy the 'A' and 'B' type market positions (see Figure 1.5). From the viewpoint of transaction analysis, the internal costs of market exchange were kept at an efficient and low level (see Figure I.1). However, the commitment to customer satisfaction was low too in this era. It really was a case of, *'We sell what we produce, not produce what will sell.'* The key ingredients in the recipe were seen as:

- Product
- Place
- Price
- Promotion.

The variablingredients have come to be known as the 4Ps. What these cover will be considered when they are applied specifically to construction below. The objective was to achieve a mix of these ingredients for the various market segments and to tweak the offer at the point of sale. The role of the sales person is to court the client or customer until the transaction was secured. The overall stages in closing the deal are broadly described:[8]

- Inform
- Persuade
- Remind.

These are largely push factors, rather than pull factors; in other words the selling reflects the 'take it or leave it' approach and does not seek to draw in the customer. The customer is left to react to the selling in one of two ways:

- A wish to satisfy their desire for the product or service
- A beginning to identify with the product or service through the people selling or through association, such as lifestyle or status issues.

The seller is therefore *proactive* at a general level and *reactive* to close the sale. The purchaser counterbalances this by being initially *reactive* and then *proactive* through the purchase.

Application of the marketing mix

In applying this notion to contractors and consultants, the variables take on different levels of importance in general terms, according to the market position. This is shown below in Table 5.1.

The importance of selling is greatest for the routinised organisations. They are the most reliant upon volume of sales in order to generate sufficient turnover secured at lower margins, compared to other market positions (see Figure 1.6). This translates into a particular emphasis on the way that selling is structured (see Figure 3.4). Table 5.2 provides a crude guide on the resultant selling input, which is derived from the structure and should be reflected in the expenditure.

Table 5.1 Ranking of the 4Ps by market position for contractors and consultants

Technology and methods	Market positions	Product ranking	Price ranking	Place ranking	Promotion ranking
Routinised	A and B	1	3	3	3
Analytical	C and D	2	2	2	2
Innovative	E and F	3	1	1	1

Table 5.2 Sales effort derived from ranking

	Sales effort level
Routinised	10
Analytical	8
Innovative	6

General issues

The problem posed by the approach is that not every standard product or service meets the specific needs of everyone. Common-sense tells us that there must be some compromise along the line, particularly on the part of the customer; however, as consumers and organisations have become more demanding and sophisticated in their purchasing and procurement behaviour, suppliers have had to respond.

The initial response was to develop more product lines and variations on each to more closely target smaller segments and niches. The same standard sales approach could then be adopted within each of the lines. This was largely a marketing solution. The second response was to recognise that more needs could be catered for if the number of variables in the marketing mix was increased. Academics and practitioners, recognising this, expanded the 4Ps as shown in Table 5.3.

Adding more ingredients, of course, changes the recipe and therefore is in part recognition that the marketing mix approach had shortcomings. As the mass markets of the 1950s through to the 1970s began to dissolve in some sectors of manufacturing, promotion needed to be diversified within the marketing mix in order to take account of other ingredients. It had been realised too that it

Table 5.3 The marketing mix and multiple Ps

4Ps	5Ps	6Ps	7Ps	15Ps
Product	Product	Product	Product	Product
Price	Price	Price	Price	Price
Promotion	Promotion	Promotion	Promotion	Promotion
Place	Place	Place	Place	Place
	People	Political power	Participants	People
		Public opinion formation	Physical evidence	Politics
			Process	Public relations
				Probe
				Partition
				Prioritise
				Position
				Profit
				Plan
				Performance
				Positive implementations

Sources: adapted from Gummesson (1994).[9]

was less appropriate in other situations, particularly where the product is less tangible, where there was a large service component or where it was a service. In addition, business-to-business markets involved a high degree of selling, usually personal selling. At a practical level, it became impossible for the sales person to juggle the variable or be able to get the appropriate mix in a complex service market. Sales people had been acting more intuitively than the marketing mix concept and were involving others in putting together a service to meet the requirements of their potential client.

This is exaggerated in any situation where the service or product is tailored to the needs of the client. Insurance policies may be considered to be providing a service. The policies are set out in advance and they can be sold as 'products', for the customer knows what they are getting as much as they do when buying a washing machine. Seeking legal advice is a different matter. It is precisely because you do not know what to do that you are seeking advice. You may even have used the services of the firm before, but that does not mean that you know what you will get. At most you may have some idea about how they will conduct business with you. Even that may be different. The expertise needed may demand the use of different personnel, any similarity coming down to the culture and procedures with which the firm conducts its business.

Consulting and construction are intangible services, more akin to legal services than washing machines or insurance policies. For some projects, the requirements may be highly complex and realised over the long term. Relationships with those with whom you are doing business will need to be developed and sustained over the long term. Indeed, relationships can become the key variable. However, this would take the analysis of selling onto the ground of relationship marketing. That will be considered in the next chapter. First it is necessary to look in more detail at the 4Ps and consider how applicable they are to the construction industry in the broadest definition of the sector.

The construction product

Construction produces a *product*. The product is the building or structure. It is frequently the design of someone else, the architect or engineer. This imposes constraints upon the contractor for tailoring the product to meet the needs of clients unless it is a turnkey, BOOT or D&B contract. While many contractors sub-contract the design out to architects and engineers, in-house design or subcontracting, where the contractor is in control, is the norm in most other industries. It inhibits innovation of the technical and production methods that the contractor adopts. It also inhibits the ability of the contractor to serve the client in an efficient way. From the viewpoint of the contractor, the *product* is largely taken out of the marketing mix and the industry is down to '3Ps'. Those selling a contractor service are essentially selling the ability to realise the design of someone else.

For consultants, they control design, but in terms of the marketing mix, it is hard to classify this clearly as a product for two main reasons. First, clients are not receiving anything concrete from the design team. There is visualisation in presentations, drawings, CAD modelling, calculations and so on, but it remains less tangible, even with models and CAD presentation, for the building cannot be experienced in the same way that a manufactured product can be experienced. In addition, the design team is providing a broader range of services in order to represent the client during the contract, so that the building is realised according to client wishes.

This may suggest that the industry needs to be streamlined to be more like others. D&B and other similar contract arrangements are a market recognition of that need. However, in most cases the demands of the client are so complex that the client prefers to break the procurement down into stages so that it can be managed to ensure that they are getting what they want prior to committing themselves to realising the project on the ground. This is to do with the building or structure design, but is also to do with cost, time, quality and risk.

A large part of what is provided by contractors and consultant alike is a *service* rather than *product*. That is a variable that can be adjusted to meet market needs. Previous chapters have considered how that may be strategically steered through the marketing plan. How a service can be set up to satisfy client requirements depends in part upon the organisational culture, market position and the type of clients of the contractor or consultant. Selling a service is heavily influenced by:

- Market position
- Service quality
- Adding value to serve particular client requirements.

These dimensions are key aspects of competitive advantage during the sales process. Market position is the platform from which the selling is undertaken. Dimensions of quality and value are the broad aspects being sold. Under the marketing mix, quality and value are enshrined in the other three variables. That is what is being delivered. How it is delivered to the client is a matter of management. Market and client management is therefore the key to addressing the product dimension of the 4Ps. In the past, this has been viewed as a technical matter – work study process of time management, critical path analysis and all the other tools of project management. There is nothing wrong with this in itself, yet the danger is that the client is put outside the equation. The management is driven by operational criteria and not by client requirements. Sales people, if they intervene, are told that they should concentrate on what they know best and let others get on with their job. This myopic stance has begun to break down with contracts, such as turnkey projects, and continues to do so with partnering and framework agreements. However, there is still a long way to go to achieve a more balanced approach to service management in construction.

Selling the service becomes a fairly standard and non-negotiable process for the sales person under the traditional management approach to contracting. The sales person can raise client expectations by saying, 'We can do that' to everything. They can promise the moon to secure the work, but know they cannot deliver. Not only will they not be able to get the ear of their own management in order to begin to adjust the service to meet the client needs, but, in many cases, they will receive inadequate support during the sales process, due to lack of a strategic and systematic approach. Ironically, it is precisely this relative independence of the sales people that allows them the room to offer the moon without any accountability. After all, their job is simply to hit the target turnover figures set for them.

Worse than this, senior management frequently collude with and become active advocates of this. When it comes to making project presentations at pre-qualification or tender stage, they will say how they do this or that on the assumption that it will happen somehow, either automatically or through the *street-fighting man* approach of bully management tactics. Usually the promises remain as distant as the moon and the expectations of the client fail to be met. Even what may seem like a relative success story from the contractor viewpoint is frequently a case of the contractor having fallen short in the eyes of the client. The disappointment is put down to the low expectations that have been formed of the industry.

Of course, this is a bigger issue than what the sales person does. However, two things are worthy of note. First, the shortcomings of contracting are encapsulated in the sales process under this marketing mix category of *product* or *service*. Second, if contractors wish to break this, it has to be addressed from the start, that is, from the first point of contact with the client. Breaking it concerns what the sales people and others involved say to the client, the support which is given to the sales effort, how it relates back to the strategy and how the promises are delivered once the contract is secured.

Place

Place is irrelevant in construction in the terms of classic marketing. It is about the way in which the product or service reaches the customer. It concerns choosing:

- ❑ Where to sell the product
- ❑ How to get it from the place of production to the place of sale or consumption.

In terms of consumer goods, *place* is therefore about choosing which retail outlets will sell the product and how to supply the retailer. The choice of retailer may be limited in order not to saturate the market or intensify the competition, or in order to create some exclusivity to the item. The outlet opportunities may be restricted by competition, for example Coke tend to sell their products through shops, while Pepsi tend to have negotiated deals to sell through other outlets, such as cinemas and leisure facilities.

A warehousing and distribution system may be necessary in order to service the outlets. How they are operated and who operates them is a marketing issue. The process is different in fashion for mail order, compared to using retail outlets. Retailing may require a large regional warehouse with a fleet of trucks servicing the different geographical locations. Mail order may require a warehouse from which courier or parcel delivery services are required. The decision has implications for selling. In retail outlets, advertising is supported by sales staff who should be trained to serve the customer and who can provide some technical guidance on the range of garments on sale. The sales staff are then required to encourage the customer to make a purchase. Mail order also uses advertising, but here the aim is to get people to ask for the catalogue. Any contact with people will probably be over the phone. These staff need not have any knowledge of fashion, but they do need excellent skills at dealing with people at a distance, where the issues are to do with payment, delivery times and return of goods.

In construction, *place* has very different connotations. At one level, *place* is about where the client wants the building – there is no need to choose a sales outlet. The father of the post-war property development sector in Britain, the late Lord Samuel of Land Securities, coined the dictum that there are three criteria for a successful commercial property development, location, location and location. In this sense, the *place* decisions are taken out of the hands of the contractor and are entirely in the hands of the client.

The exception is where the contractor is also a house-builder and developer, in which case it comes back to decisions about site purchase.

The decision that the contractor and consultant have to make about distribution and outlets comes down to geographical coverage. This has already been raised in terms of market strategy in Chapter 3, while addressing segmentation. It tends to be a strategic decision; however, there are sales issues. It was pointed out that contractors are more likely to be successful while having a 'hands on' management approach for markets close to or adjacent to their headquarters, yet give greater local autonomy in distant markets geographically (see Chapter 4). In the former case, it is better to have sales staff that conform to the procedures and processes set down for them and to provide a system into which all those involved with the sales effort can relate. In distant markets, the self-starter or highly motivated and self-disciplined sales person is needed, especially where the operation is smaller than the region in which the headquarters is located. Where the local operation is large, it may be appropriate to let the management there develop their own system and procedures for their sales effort within some overall guidelines.

There can be other constraints. Selling in international service markets is difficult. The intangible nature of services makes it difficult to deliver locally across a broad geographical market. For example, within the EU the 'four freedoms' – free movement of goods, services, people and capital – have been implemented for goods but not in the service sector.[10] There are barriers to selling construction and consultant services in some European markets. These barriers can be overcome through the acquisition of local companies, but shareholding structures can act as a barrier in some countries. Although these barriers are complained about, one way of considering them is to acknowledge that they are aspects of competitive advantage. In this case, they are *place*-related barriers, many of which have been institutionalised and built into the structure of conducting business.

Across international construction markets the primary issue is who will do the selling. The main avenues for expansion were stated in Chapter 4 (see also Table 5.4). These are repeated below, with the sales implications set out alongside from another angle.

Each sales option has an implication not only for the way in which the service is sold, but also for the way in which the service is delivered to the client. In other sectors, *place* refers as much to the distribution channel as to the place of purchase. Distribution channels normally embrace everything from files delivered through e-mail and courier services to warehouse and truck

Table 5.4 Sales implication of geographical expansion

Option	Sales implication
Follow the client	Sell directly from home market
Franchise	No sales responsibility
Local agent	Indirect control over selling
Alliance with local partner	Indirect and direct control over selling
Joint venture with local partner	Indirect and direct control over selling
Acquisition	Direct control but potential cultural, hence management gap
Set up a regional or local office	Direct control either by sending out sales force from HQ regularly or setting up local sales effort, success being dependent upon HQ support

distribution. In construction, distribution is not by road or the superhighway. The delivery channel is the procurement route and the type of contract. It has already been stated that this is determined in contracting organisations through structural decisions (see Chapter 3). Structure dominating process in this way closes down the option to adjust the marketing mix along this dimension of *place*.

What we are observing under *place* is that there is limited room for manoeuvre. As the specifics of *place* are translated into the particularities of construction, decisions and actions tend to have a one-way outcome; in other words, they become embedded into organisational structures and thus by nature cannot be changed or adjusted to take account of client requirements. The sales person is unable to negotiate around this issue. They are relatively powerless in this context, using the marketing mix approach. The scope is therefore limited at present under this heading.

Promotion

A former managing director of a specialist contractor has argued that marketing is a sub-set of sales.[11] A well-known author[12] on *promotion* says:

> ... *sales promotions are conceived and implemented by people who are not specialists.*

Kotler *et al.*[13] argue that *promotion* covers four things and others[14] have added a fifth:

- ❑ Advertising: non-personal presentation
- ❑ Personal selling: oral presentations
- ❑ Sales promotion: incentives
- ❑ Public relations: creating good relations, publicity and image
- ❑ Direct marketing: mailshots and related techniques.

Co-ordinating the promotions is a complex task in many product and services area. The tone is set by senior management. Senior management have a special role – they are demonstrators that others will emulate and therefore must 'sing from the same hymn sheet'. They are also required to provide personal support. Eldridge and Carvell[15] put it this way for professional practices:

> *So, unless all the partners can agree, you stand little chance of creating a clear corporate persona. To do this, every statement about the firm must add to the agreed picture of it. It only takes one divergent remark to confuse the listener and diffuse the image. If there is dissent it should be sorted out before you spend any money at all trying to raise your profile.*

Eldridge and Carvell allude to a broad picture. Others link *promotion* with personal selling.[16] Certainly in contracting, promotion is an area that has been neglected.[17] It is associated with 'loss leaders' in supermarkets, coupons on packets of cereal and incentive deals to book your holiday early. While house-builders may use some of these ploys, in contracting and in the professions this is seen as irrelevant, even unprofessional.

However, 'let's not throw the baby out with the bathwater!' Architects do *speculative work* – a lost leader. Contractors provide *construction finance* to developers sometimes – an incentive scheme. Contractors sometimes *buy work* in recession or occasionally if they wish to develop long-term work, with a client – another lost leader. Contractors sometimes spend time on feasibility work – yet another lost leader. If these promotional devices are frequently used, perhaps there are other promotional opportunities!

Promotional tools can be used as a second prong of attack in the sales effort, thus complementing personal selling. Pincer movements are effective. Promotions work at three levels:

- *Functional benefits*, for example service quality, delivering added value
- *Image*, for example quality building as part of the corporate branding, boosting staff morale
- *Extra benefits*, for example loss leaders, free 1 year maintenance service as an after-sales service.

There are ten *desired outcomes* to be derived from promotion campaigns (see Table 5.5).

Research has shown that British contractors only use *promotion* to reinforce selling in the following way (see Table 5.6).

Work has also been done in the international marketplace on the use of *promotion*, specifically British and Hong Kong contractors and consultants. This work has focused on the use of media,

Table 5.5 Potential promotional outcomes

Potential desired outcomes	Examples
Increasing volume	Further contracts
Increasing trial	Pilot or phase one contract
Increasing repeat business	Term contracts
Increasing loyalty	Partnering, alliances
Widening usage	Management contracts, total facilities
Creating interest	Innovative site hoardings/billboards, advertisements
Creating awareness	Site visits, logos
Deflecting focus from price	Adding value, negotiating fast track contracts
Gaining intermediary support	Courting and supporting client's representative
Gaining display	TV programme, such as the former UK programme 'Challenge Anneka', sport sponsorship

Table 5.6 Promotional outcomes sought by contractors in Britain

Generic outcomes sought	Specific goals
Increasing volume	Increase number of enquiries
Increasing trial	Increasing number of clients
Increasing repeat business	Further contracts
Creating interest	Improve the company's financial performance

Source: developed from Fellows (1994).[18]

such as brochures and advertising. The comparison is provided in Table 5.7.

What is interesting about this type of *promotion* is that the focus is on pushing the message out, rather than drawing in the client or stakeholder. It is operationally driven, with sales being seen as just one piece in the operational jigsaw. A client orientation is largely missing. In one sense this is consistent with the origins of the marketing mix, in that the approach is a mass market one. On the other hand, the full armoury of *promotion* is not being used. The question is whether that is appropriate in the contracting and consulting context.

It appears that in different geographical locations the emphasis upon *promotion* is different (see Table 5.8). Different factors are relevant in different markets, possibly for cultural reasons. It may also be that the cultural factors have encouraged the development of different *promotion* tools, although each of the tools could

Table 5.7 Comparison of media used in promotion

UK contractors (marketing)	HK contractors (advertising)	UK structural engineers (advertising)
Company brochures	Practice brochures	Headed paper
Targeted mailshots	Headed paper	Company brochures
Construction journals	Site sign boards	
Sponsorship		

Source: adapted from Fellows (1994).[18]

Table 5.8 Promotion factors emphasised for contractors

Factors	Clients	Contractors
UK marketing		
Quality of service	1	1
Experienced personnel	2	2
Competitive price	3	–
Financial standing/record	4=	3=
Team work	4=	3=
HK advertising		
Previous project quality	1	2
Previous project cost	2	6
Previous project time	4	3
Previous experience	3	1
Company reputation	5	7

Source: Fellows (1994).[18]

Table 5.9 Distribution of sales activity among UK contractors

	Contractors		Engineers		Architects		Combined
Activity	Score	Rank	Score	Rank	Score	Rank	ranking
Pre-project proposals	244	1	116	1	100	2	1
Personal selling	197	2	112	2	117	1	2
Additional services offered	196	3	80	3	85	4	3
Alternative proposals	188	4	74	6	91	3	4
Information services	180	5	80	4	73	6	5
Corporate identity	177	6	71	8	70	7	6
Public relations	137	9	79	5	75	5	7
Market research	166	7	67	10	66	9	8
Joint ventures	164	8	73	7	52	13	9

Source: Leung (1994).[19]

be equally effective in several geographical locations. It may be that the scope for *promotion* is yet to be fully explored, an absence of professionalism and strategy being the barriers to its development in the industry. However, that has to remain conjecture unless there is other evidence to support it.

Further evidence is to be found in the mismatch between the emphasis upon promotion and client requirements in the table. This is particularly the case in Hong Kong, a peculiar outcome, as it has been one of the most consistently active markets over recent decades. It emphasises how out of touch contractors can be with their clients, which has been further reinforced in marketing generally.[2] It has been found that the emphasis in selling is very narrow. Promotion is almost excluded, the majority of activity being reactive. The activity concentrates upon the sales processes once the client has already expressed interest.

What can be concluded from this is that *promotion* is underused. There is sufficient evidence to show that a range of promotional tools have been used and with some success. Some have also been used in a misdirected fashion. Clearly, promotion is expensive and that may be one motive for not fully exploring its use. However, organisations that have used promotional tools have reaped some rewards. Barratt Homes in the UK successfully used TV advertising for some years until a documentary exposed some poor quality workmanship, in part a media outcome stimulated by the success of the promotion campaign. Therefore, we have to conclude that *promotion* in contracting and consulting remains relatively underexplored.

What is interesting from the sales perspective is that contractors and consultants have rather tied their hands behind their own backs in respect of *promotion*. In the other two of the 4Ps investigated above, a combination of the irrelevance in a contracting context emerged. What is being seen with *promotion* is that contractors have disqualified themselves from using this variable. Sales people cannot use the profile, image and credibility established through promotions as a launch pad for their sales effort. They have to start largely from scratch or certainly from a very low base in every case. This increases the time and effort any sales person has to make towards a client if the selling is to be proactive.

In this context, is it surprising that sales efforts tend to be reactive? The client is the one to have done the initial hard 'leg work'. Most contractors under-use promotion techniques, so some creativity and imagination could be applied to improve matters:

- ❑ Identify some promotion options that fit the culture and market position of the organisation
- ❑ Take each one and decide to use it in tandem with personal selling – create sales pincer movements.

This can be facilitated by starting with brainstorming ideas:

- ❑ Tools to be used
- ❑ Actions to be taken.

Brainstorming is an unconstrained exercise, all suggestions being noted, regardless of any apparent or intuitive use they have. The next step is therefore to sift the brainstorm ideas, particularly in the light of the promotion techniques selected. This embraces several steps:

- ❑ Analysis
- ❑ Refinement and amendment, as necessary
- ❑ Allocation of a budget
- ❑ Giving the experiment a long run to prove itself
- ❑ Setting some criteria for future evaluation.

In any event, what arises from such an exercise will take some time to be put into place and to have the desired effect. What does that leave the sales process focusing upon in the meantime? *Price*!

Price

Contractors and consultants would love to be able to employ the Rhino principle:

We are thick skinned and charge a lot!

The reality is less positive. If the other variables, *product*, *place* and *promotion* are seriously constrained in construction, then it mostly comes down to *price*. This is the worst possible scenario in a sector where entry barriers are low and in which the market is highly fragmented geographically and by the extent of the competition. In the previous section on promotion it was shown that selling is largely reactive. Whether it is reactive or proactive, the main role of the sales person is to pre-qualify and secure a place on the tender list or a place at the negotiating table (see ... *On the case 18*).[20]

An unscientific sample of 23 contractors and consultants, who (as members of the Centre for Construction Marketing) were looking at ways of improving their approaches to marketing, showed that 57% of the proposed initiatives and actions were aimed at improving *sales performance*. The vast majority of these additional efforts were to be concentrated on the *courting* stage: 26 out of 68. This may be a tacit admission that sales are not handled very well, especially during courting, too much emphasis being put on reactivity and not pro-activity. This is an old cry, but it probably means that there is truth in it! Wilson[21] states:

All price pressure comes from the customer (p. 206).

It is only true, he says, when the sellers sell on price. Is that not what contractors and consultants have been doing more recently? They have even educated and connived with their clients in this. In other industries research has shown that price is not a serious factor, even where customers say it is. Customers will not switch from preferred buyers, even if the price increases by 5%.[21] We have to be careful about this in construction, especially in the public sector. The problem for contractors is that there are few preferred contractors. Bechtel and Bovis had that distinction and still do to some extent. IDC (now part of AMEC) had it in process engineering. In the UK, only 17% of consultants had preferred contractors and 31% of clients had preferred contractors.[19] Those figures were prior to the advent of partnering, so it would be interesting to compare with the current position. Selling, it would

... On the case 18

The CEO of a specialist D&B contractor stated:

> *Most major contractors would argue that they have a marketing strategy, but when you delve into it, it is more likely to be a strategy that sets out to* **'get on a tender list'**, and being successful would be seen as merely being given the chance to submit a quotation or tender.

Once on the list, the issue then tends to revolve around price. He went on to say that if price is everything and your organisation sets out to be cheapest, then you do not need marketing or selling.

In housing development, price is only one facet of the financial equation. A Group Marketing Director of a major volume house-builder, stated in 1994 that there were several financial issues:

- A house as a long-term purchase – mortgage costs as well as purchase costs
- A house as a high-price item – financial risk
- A house must be resold – investment and thus resale value.

This house-builder, on its own admission, had been selling what it built rather than building what would sell. Price and location were the only variables. The product was designed and constructed to play it safe in a conservative market, even to the extent that adding value, which would enhance the other financial criteria, was excluded.

appear, has not created preferences and price competition certainly has not.

How useful is *price* from the viewpoint of the seller? It has not been very useful as a variable to secure sales. Low pricing structures do not help the sales person pre-qualify or get the organisation on the tender short list. The power of discounting is not in their hands, nor is the pricing. That is an estimation function, which acts independently of sales in most cases. The profit margin (or negative margin in too many circumstances) is not in the power of the sales person either. Margins are often left to the Board or senior management.

For the whole team, *price* is problematic. The margin is usually set in relation to competition and profit requirements in most industries. Profit in construction is notional, as is the price around which the competition revolves. Out-turn prices may not be known for several years, when the final account is produced. Even at this stage, profit data is not fed back into the pricing decisions. It has already been stated that construction and consulting is largely about cash generation. Therefore, the *price* decisions have little to do with profit and more or less everything to do with cash-flow in the short term and turnover in the long term. Therefore even price competition in construction is notional because the tender figure is just as fictional as the profit margin in the bidding process. In order for the contractor to make a profit, the adversarial relationships come into play. The contractor does all in their power to convert the tender price into a profitable one through price negotiations over variations and claims.

What would be an ideal balance between the various roles during the sales process? This is set out in Figure 5.3.

If marketing targets a profitable line of work and management concern themselves with turnover, then the input from sales will be advice on how much the client will stand and at what different

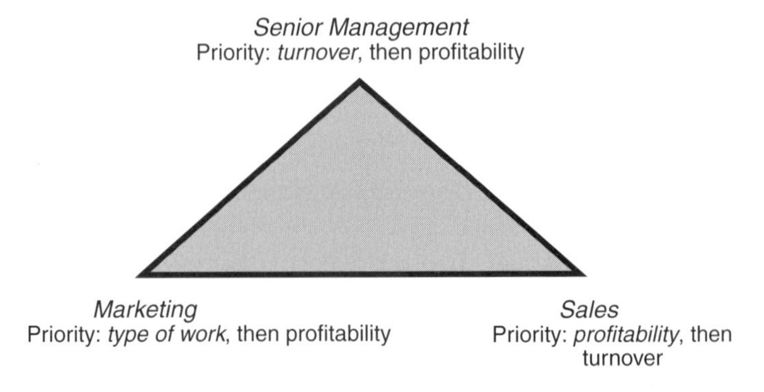

Figure 5.3 Price considerations.

levels of service value. They will therefore help proactively and directly determine profitability. In this way, *price* becomes inextricably linked with quality and service issues. Price will always be a dominant issue in any transaction, but it need not be the only or most dominant one. Those in selling, estimating, those involved in tendering and so on, know that the key variable is *price*. Yet, to raise the importance of the service value, the sales team need to know what is the scope for manoeuvre in negotiations. In other words, the sales person needs flexibility to respond to client needs and to mix the sales pitch accordingly. It is from this level of involvement that trustworthy sales promises are generated. These promises need to be fulfilled to be trustworthy. These promises must be reflected in the way that estimating and other functions process qualification and tendering. For example, in contract documentation the legal terms suggested by contractors often fly in the face of the sales promises,[21] and this may undermine the sales process and ultimately client satisfaction. The estimator also needs to know what *promises* are being sold, the site team should be aware of what promises they have to *deliver*, and so on. What is promoted must be delivered if the client is to consider further work for the contractor. This is important given that it cost fives times more to find a new client than to keep an existing one.

These dimensions have to be taken more seriously because price alone is a competition deathbed. It has been shown that the scope to address wider issues is limited under the marketing mix, yet it can be done. That requires radical action. Failure to do so means that selling comes down to price. While the focus will increasingly be on price the further down the selection process the contractor gets, it does not have to be like that from the outset. Where the competition does come down to price, do not tie the arms of the sales people behind their back. They need to be facilitating and steering the process. A focus on price alone means that the sales person has very little to offer, except the ability to open doors.

How does an organisation address these issues? What has been stated may not seem that big an issue; however, the outcome can require quite radical change. It may involve:

- Restructuring the organisation in order to give support to the sales effort
- Ensuring that restructuring permits the 4Ps to be adjusted during the sales stages, in particularly removing the structurally embedded 'solutions' to marketing

- Training those involved in selling to be more proactive, in order to more closely respond to client needs and adjust the marketing mix within the guidelines of the marketing plan
- Encouraging sales people, who can facilitate the sales process, to be more than just door openers!

Conclusion

The conclusion on the 4Ps is that they do not offer very much scope in contracting to the sales people. The 4Ps therefore do not fit easily into the construction context. Yet contractors have chosen to constrain themselves and close down the options to secure more sales and at profitable levels. *Promotion* probably offers the greatest opportunity for immediate response, although the benefits will take some time to come through. There are, of course, no quick fixes, otherwise everyone would have done it long ago. The point is that, despite the limitations of the 4Ps, they are not being used to full extent. Contractors and consultants alike could invest more time and money in these areas to good effect with a well-worked out programme that has potential to yield a good rate of return, even under constrained market conditions.

Summary

1. This chapter has demonstrated that selling is undertaken in construction:
   - From a narrow perspective
   - With poor management.
2. The chapter has:
   - Set out the scope and constraints of the marketing mix in selling
   - Shown the potential for developing the sales process.
3. Selecting options is part of the aim of securing competitive advantage and the processes themselves provide an added competitive edge when well managed.

References and notes

1. A recent example of this long running saga is to be found in the Egan Report (1998) *Rethinking Construction*, The Report of the Construction Task Force, Department of the Environment, Transport and the Regions, London.
2. CIOB (1993) *Marketing and the Construction Client*, CIOB, Englemere.
   Lim, P.L.H. (1990) An investigation into 'qualifiers' and 'winners' of construction projects, unpublished MSc dissertation, University of Bath.
   Morgan, R.E. and Morgan, N.A. (1991) An exploratory study of market orientation in the UK consulting engineering profession, *International Journal of Advertising*, **10**, 333–347.
3. Harvey, C. (1988) *Secrets of the World's Top Sales Performers*, Century, Random, London.
4. Knutt, E. (1997) The Scouts, *Building*, 21 February.
5. Gummesson, E. (1991) Marketing-orientation revisited: the crucial role of the part-time marketer, *European Journal of Marketing*, **25**(2), 60–75.
6. Good, D.J. and Schultz, R.J. (1997) Technological teaming as a marketing strategy, *Industrial Marketing Management*, **26**, 413-422.
7. Williamson, O.E. (1981) Contract analysis: the transaction cost approach, *The Economic Approach to Law* (eds P. Burrows and C.G. Veljanovski), Butterworths, London.
   Williamson, O.E. (1985) *The Economic Institutions of Capitalism*, Free Press, New York.
8. Kotler, P., Armstrong, G., Saunders, J. and Wong, V. (1996) *Principles of Marketing: the European Edition*, Prentice Hall, London.
9. Gummesson, E. (1994) Making relationship marketing operational, *International Journal of Service Industry Management*, **5**(5), 5–20.
10. Gummesson, E. (1994) Service management: an evaluation and the future, *International Journal of Service Industry Management*, **5**(1), 77–96.
11. Paramor, R.T. (1995) *Marketing and the Bottom Line*, Seminar to the Construction Industry Group Chartered Institute of Marketing, London.
12. Cummins, J. (1989) *Sales Promotion: How to Create and Implement Campaigns that Really Work*, Kogan Page, London, p. 9.
13. Kotler, P., Armstrong, G., Saunders, J. and Wong, V. (1996) *Principles of Marketing*, Prentice Hall, London.
14. For example, see Cummins, J. (1989) *Sales Promotion: How to Create and Implement Campaigns that Really Work*, Kogan Page, London.
15. Eldridge, N. and Carvell, P. (1986) *Promoting the Professions: Which Way Do We Go?*, Surveyor's Publications, London, p. 99.
16. Gummesson, E. (1991) Marketing-orientation revisited: the crucial role of the part-time marketer, *European Journal of Marketing*, **25**(2), 60–75.
    Paramor, R.T. (1995) *Marketing and the Bottom Line*, Seminar to the Construction Industry Group Chartered Institute of Marketing, London.

17. Preece, C.N., Moodley, K. and Smith, A.M. (1998) *Corporate Communications in Construction: Public Relations Strategies for Successful Business and Projects*, Blackwell Science, Oxford.
18. Fellows, R.F. (1994) *Some Perspectives on Marketing in Construction: Person to Person*, paper presented at the Investment Strategies and Management of Construction Conference, 20–24 September, Brijuni, Croatia.
19. Leung (1994) MSc dissertation on construction marketing, Bartlett Graduate School.
20. Paramor, R.T. (1995) *Marketing and the Bottom Line*, Seminar to the Construction Industry Group Chartered Institute of Marketing, London, p. 3.
    Davies, R. (1994) What can housebuilders sell? Paper presented at the *Housing for the 21st Century Conference*, 8–9 June, School for Advanced Urban Studies, University of Bristol.
21. Wilson, A. (1991) *New Directions in Marketing: Business-to-Business Strategies for the 1990s*, Kogan Page, London.

Chapter 6 Relationships and Sales Promotion

Themes

1. The *aim* of this chapter is to investigate the approaches to promoting the services of contractors and consultants using relationship marketing.
2. The *objectives* of this chapter are:
   - ❑ To identify the implications of selling through relationships
   - ❑ To explore the dimensions of client relationships
   - ❑ To investigate the consequences for the management of organisations, especially in client handling.
3. The primary *outcome* of this chapter is to ensure that appropriate selling approaches are adopted, from which client handling models can be evaluated and selected.

Keywords

Relationship marketing, Client management and Client handling

Introduction

The *marketing mix* is one option. Under the marketing mix approach, sales people are to some extent building relationships. The competitive advantage may not just come through the 4Ps or however many Ps the organisation embraces. It may be more intangible than that, but no less strategic and important. The ability to build and sustain relationships may be the key to beating the competition. Such situations are beginning to use *relationship marketing*, the subject of this chapter.

Selling under *relationship marketing* can be seen as a pipeline. The relationship is the pipeline and the selling takes place through it. How can relationship marketing be defined and how can it be used to manage relationships? Selling through relationships is nothing new. I have expressed it this way:[1]

I used to live in Hove near Brighton. The local butcher got to know my flat mates and me. Even though we were students and would only be around for another two years, the staff took a personal interest in us, added in the extra sausage or chop after weighing and writing down the price. The next year, when we moved to the other side of Brighton, we travelled over to Hove just to go to that butcher. He captured our trade through serving us well, building up a relationship which made us want to be loyal and continue to get good service, even if it cost a bit in time and petrol!

When I was a child, our family butcher was the 'bush telegraph', the best local newspaper you could have! He took an interest in customers and knew everybody's business. He would also recommend a good cut or joint for that week. The quality was always good, but that way the butcher persuaded you to spend a little more than you had originally intended. His shop was not cheap but people seemed to delight in saying, 'They're a little on the expensive side, but I wouldn't go anywhere else'.

This is the essence of *relationship marketing*. The first part of the quote is about enhancing the quality of the service through relationship building. The second part of the quote is about service enhancement. Both aspects require an investment by the butcher and both yield a return.

Local builders still work on this principle, having a network of contacts, through which they generate repeat business, recommendations and so on. For the large company, that sales approach breaks down as the organisation becomes large and distant from its customers. Selling through relationships is no longer stimulated by direct contact. Senior management is involved in many diverse activities. An independent or departmental sales effort has to be put together. The effort needs steering and controlling, hence the emergence of marketing and sales management functions. In doing this, the relationship approach has been lost in favour of the more 'rational' or mechanistic approach of the 4Ps: product, promotion, place and price.

Relationship marketing is designed to recapture the relationship approach in a systematic way within the corporate culture. The emphasis is on people and adding value. This is achieved through serving, gaining competitive advantage through more effective working relationships rooted firmly in a systematic context, and through delivering effective value. The outcome of the relationships is profit.

Components and process

Relationship marketing is more theoretical than the marketing mix and therefore application requires a more conceptual approach. While that may mean some struggle with making it operational, it also provides an overall source of competitive advantage, as do the specific options it provides. The options are *soft* compared to the harder edge of the 4Ps, but it offers no 'soft options'! Relationship marketing was developed for business-to-business markets, primarily in the service sector. Pioneers from the 'Nordic School' have provided a model for understanding the process. This is shown in the diagram below.

The model shows the linkage between *perceived value* and resultant *relationship profitability*. The objective of any business is ultimately to be profitable, preferably over the long term, yet the key to long-term profitability is the value embodied into the service. The potential client is offered a service, but it is the nature of the service in which the client is intrinsically interested. How the client views the nature of the service throughout the relationship gives rise to the *perceived value*. The *perceived value* may change over the course of the sales process and during the project; however, at any one point, the *perceived value* is derived from two sources:

- The *embodied value* of the offer. The breadth of the *embodied value* will depend upon the nature of the service, for example D&B or partnering. The depth of the *embodied value* will depend upon the richness of the service offered by the consultant or contractor, which will determine the intensity of the service and hence value for money. There will be a distinctive base line of value that is characteristic of the service of a competitive contractor or consultant, on top of which will be the added value components for that particular client and project.
- The *emotional value* of the offer. The reality of the *emotional value* is where the actual delivery of the *embodied value* meets actual client requirements. For example, the project manager on the client side may wish to have a contract of least hassle or one that will deliver promotion by the contract being seen to be

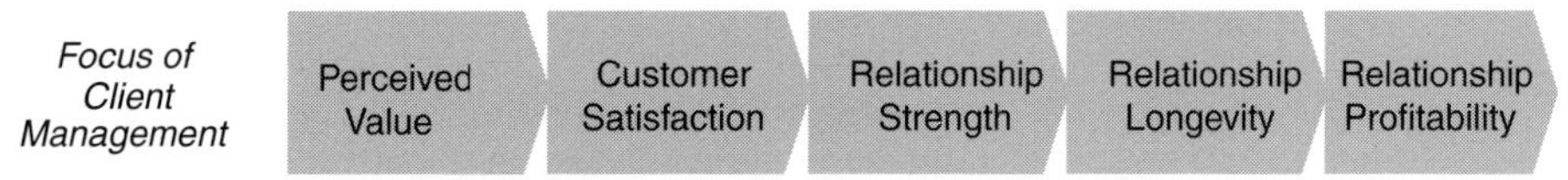

Focus of Client Management

Perceived Value → Customer Satisfaction → Relationship Strength → Relationship Longevity → Relationship Profitability

Figure 6.1 Pipeline of relationship marketing management.

successful in terms of programme, cost and quality. The intensity of the service will help to determine the outcome, providing it is directed to the appropriate areas of the contract. The imaginary element is where the client is psychologically hoping for a particular outcome or need to be met. The imaginary element may be conscious or unconscious. In any case, this type of need is beyond the scope of the contractor to deliver. Meeting this aspect of emotional value will be accidental in the sense that it is beyond the power of the contractor. It is in the client's mind. However, the nearer the contractor is towards understanding the client and the closer he is to satisfying all the other dimensions of the client need, the more likely that the emotional needs of the client will be met.

A successful outcome is where the contractor or consultant has maximised the *perceived value*, that is to say delivered what was offered during the sales process and exceeded this by meeting the most intangible service elements – embodied and emotional value of the service. Such an outcome results in *client satisfaction*. It is the management of the relationship that oils this process.

Client satisfaction goes further than the transaction approach of the marketing mix. If the appropriate balance is achieved across the 4Ps in the most competitive way, then client satisfaction is assumed to have taken place, providing the offer is one of integrity. However, the relationship marketing approach goes several steps further. As it costs five times more to get a new client than to keep an existing one, the aim is to maximise *client satisfaction* in order to get repeat business.

A satisfied client is a good basis for a longer term relationship. If the interests of the client are kept in the mind of the contractor and this is communicated to the client through contact and post-contract service, then *relationship strength* is created. This is an investment for the next sale. It will be the launch pad to put the organisation in the optimum position to secure the next contract, either through negotiation or through open competition. Repeating this pattern over several contracts ensures *relationship longevity*. The client will remain demanding at all times, yet the contractor will neither have to go through a learning curve with the client, nor will they need to invest heavily to achieve continued *client satisfaction*, because the needs of the client will be increasingly understood. The contractor will become efficient at serving the client. The result will be *relationship profitability*.

The marketing mix is fairly mechanical in approach, a type of stimulus–response mechanism. If the supplier gets the mix right then the recipe will produce a successful cake every time! The relationship marketing approach recognises that this is by no means automatic. The linear model set out in Figure 6.1 is crude and therefore only a start. Selling, using relationship marketing, requires considerable *management* input in order that the client is satisfied and the relationship life is maximised. This is shown in Figure 6.2.[2]

In Figure 6.2, the top line deals with the value given to client and the receipts to the contractor or consultant, in other words the transaction. Like any transaction there are costs associated with it and the bottom line therefore has as its focus the investment made in advance and the investment that needs to be made in the sales process, the management of the service delivery and the costs of particular events. Both the value and the costs have to be managed. The contractor or consultant must weigh up the value for money, for there is an expectation that the investment made will yield a proportionally higher return, although not always immediately. Adding the additional value will yield a premium, so the theory goes, but to realise that in practice requires that the scope of management goes beyond project management. It operates at three levels:

- Corporate
- Client
- Project.

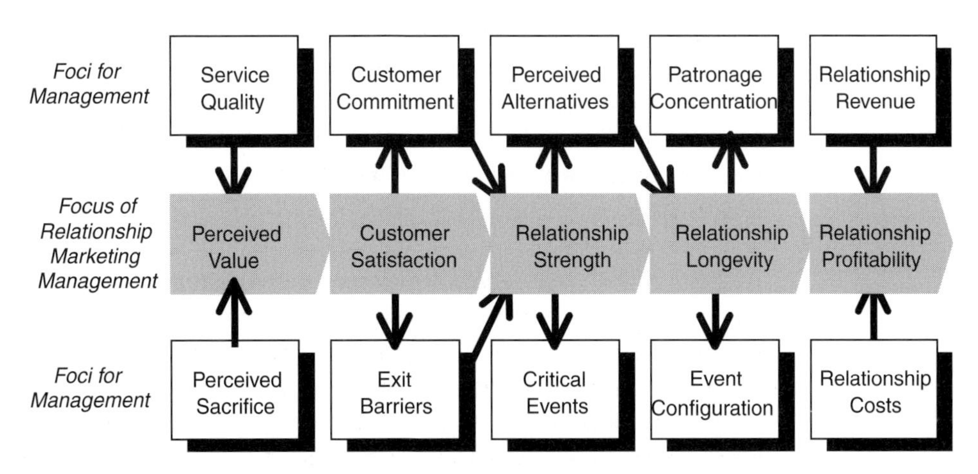

Figure 6.2 Model of relationship marketing management.

Many organisations fail to take this seriously. The structural and cultural context is absent or lacking, investments are not made, the client is seen only in project terms and the project process is not managed in terms of promise delivery. The client becomes cynical. Relationship marketing falls at the second hurdle: *customer satisfaction*. This is very common, for example in the current interest in *partnering*.[3] Contractors and clients alike frequently fail to make the necessary investments. Repeated failure across the industry to do so will ensure that partnering or any other new concept will be consigned to the bin of fickle fads.

Value and profit

Focusing upon the top line of management in more detail, the first step is to ensure that *service quality* is in place. This covers the quality of work on site, the contract management and the wider management support for the client. The responsibility to ensure that this is in place is with management. However, it will start with the sales person, as selling moves from courting towards consideration of a specific contract. It is during this that the sales person must know that they are trusted internally and that they have the support of the management to mobilise the resources to meet the client needs. Some of these resources should already be in place, if the strategy and marketing plan are being implemented consistently. The strengths of the organisation can continue to be developed and the weaknesses will be shored up, as these are identified during selling and subsequently put in place. That is the internal side of the equation. The external side will have been tackled through the market positioning and segmentation process, so that the target clients fit the competitive strengths of the organisation. However, the particular combination of requirements and the balance between those will be particular to each client. There may be some additional 'specials' on top to be resourced; that is (further) *added value*. The sales person needs support to further the negotiations. They will be acting as facilitators. They will therefore need to know when to bring in the experts, for example from estimating, in order to deal with some critical aspects of specification and quality or the management to explore how the client needs can be resourced and supported, hence demonstrating commitment to the client.

Customer commitment is the second factor concerning inputs to the client. The relationship marketing approach may not be used for all clients. Key or core clients may be identified through the segmentation and targeting process. Only to those clients will commitment be developed, others being responded to in the traditional way of opportunism and reaction. The customer or client must know that the organisation is committed to them. Initially, this is in the hands of the sales person. They are developing the relationship through the traditional stages of selling. They are finding out about the structure, the approach, and the hard and soft needs of the client. This will focus at the levels of the organisation as a whole, the key decision-makers, and those responsible for day-to-day management. It will also include any consultants who are retained by the organisation.

As a project comes onto the horizon, the level of investment must go up in order to demonstrate *customer commitment*. The level of commitment at the sales stage is to be carried through to the project stage. The one must be a reflection of the other. It is a mirror image. It is no good just bringing out the 'big guns' to secure the contract if they fail to demonstrate to the client that they are taking an equally pro-active interest throughout. This does not necessarily mean that senior management maintain a daily input in a project. It does mean that when they monitor the contracts manager, they should also be in touch with the client with the same regularity to see how things are going from their perspective (see ... *On the case 19*). It is through maintaining contact in this way before and during the contract that different perspectives can be brought to the problems facing the client and the team. Alternative ideas and solutions can be put forward. These may range from putting a financial package together for the client using connections with banks to a technical solution on site, which can be put into the method statement as part of the pitch for the project.

This is the essence of putting forward *perceived alternatives*. There may even be occasions when the solution is to shelve a project or go for a dramatically scaled-down version. This may seem like a suicide mission for a contractor. Under traditional methods of opportunism, it is understandable that the contractor keeps quiet about such things, the onus really being on the client to seek the necessary range of advice from other parties. In relationship marketing, that is not the case. A sales ploy may be to make a point of being the first to put forward this 'suicide solution'. Why? The

... On the case 19

The importance of turnover from one key client for a medium-sized contractor was such that the marketing director spent two days of the week, ensuring that the relationship was sustained and all the concerns of the client were addressed.

The retail client had used the contractor for new build projects. The programme had come to an end. The contractor appreciated there would be a rolling refurbishment and upgrade programme and continued to build the relationship while competitors began to look opportunistically elsewhere. Through selling the services using relationship marketing, up to 30% of turnover came through the one client.

This became a role model for developing sales with other important clients in order to avoid any over-dependence on the one client derived from the success of the relationship approach.

point about relationship marketing is that the client is satisfied. That will lead to an excellent reputation and increased referral business in the longer run. It can also lead to repeated business, in the case of regular building procurers. The argument is that it is best not to put the first contract on the wrong footing, but to wait for a future opportunity, building trust and commitment in the meantime. Together with successfully satisfying other repeat business clients, a favourable outcome of *patronage concentration* is achieved.

The purpose of putting forward *perceived alternatives*, therefore, is to demonstrate commitment, an understanding of their business. Within that market segment or niche word gets around that your organisation truly does take an interest and has their best interests at heart. In return, you are implicitly asking them to do the same and concentrate their business with you. You are anticipating that the business networks will get the word out, and your own organisation will need to reinforce this message through a variety of other promotion activities. Indirect *patronage concentration* is the result. It is a long game and not for the faint hearted. It reaps rewards, for all the time the contractor is erecting barriers to the competition, entry into the segment becoming more difficult. Even price cutting becomes a difficult option for the competition, for the knowledge, expertise and ability to respond in detail to the client will permit your organisation to achieve price efficiency as well as excellent value for money.

Price cutting is always a strong force during recession. Being able to protect the client base through *patronage concentration* is an important weapon. Such protection may not yield high levels of profitability, yet the starting point is maintaining reasonable turnover levels, crucial for contracting and consulting, which are essentially cash-generating businesses. In recession, *patronage concentration* maintains value for clients and comparatively reasonable turnover for the contractor.

Relationship revenue is that proportion of turnover which is on top of the revenue that is obtained from the traditional opportunism of contractors. Within that *relationship revenue* will be additional profit, much of it at above average profit margins. This works on the principle of added value. If the service is enhanced, then the amount charged can disproportionately increase as the niche market customers are prepared to pay over the odds for the value and convenience. Added value food in supermarkets is a prime example. However, in contracting, the client is also expecting some savings from developing a close relationship and in reward for loyalty. What is at work here is some share out of the financial benefits. Some accrue to the client as savings and some to the contractor as premium profit. This can only happen if investment is made on both sides, especially by the contractor. A return cannot be earned without the investment.

Investment

The investment costs for the contractor is the subject of the bottom line in Figure 6.3. Selling through relationship marketing requires a sacrifice. The *perceived sacrifice* is twofold. First, the contractor has to be more focused in targeting. This is something that should have been dealt with in the marketing strategy. Part of that planning must include an awareness that sales people cannot be as opportunitistic as in the past. More investment must be made in terms of time in the key clients targets. This will inevitably lead

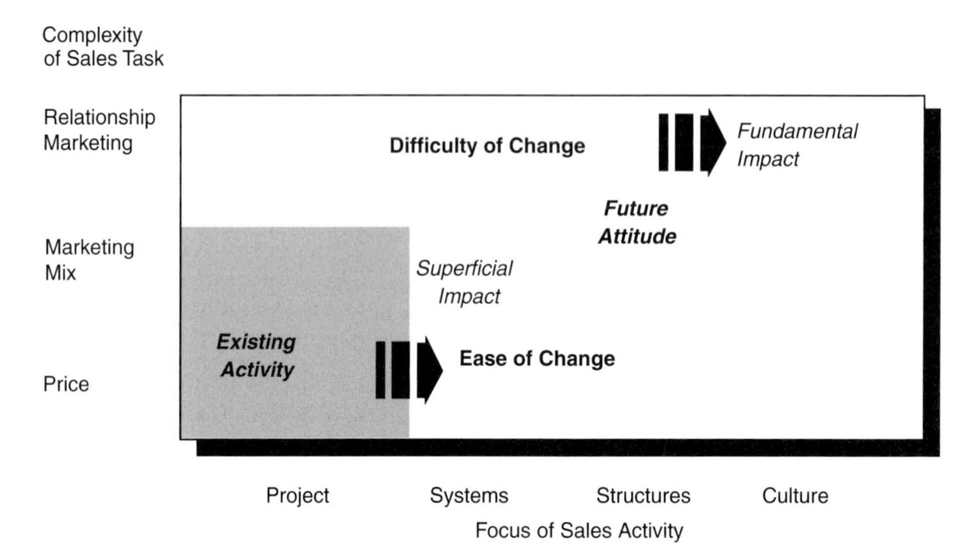

Figure 6.3 Implications of shifting sales activity

to a short term fall in turnover. The sales people must have this built into their time budgets and sales targets. Clearly, the most comfortable time to undertake such a transition is in the growth phase of the construction cycle. It may even be possible to fund additional sales staff. It will pay off during the downturn. It is a paradox of selling that the greatest effort has to be put in during the busiest times. It is too late once the cycle is at or past its peak (see Appendix A, Paradox 12).

Second, the contractor must be prepared to invest resources into the market segments to tailor the services to the general requirements of the segment and to investigate the target clients to a considerable level of detail. On top of this is the investment in time made by senior management in these clients and the time invested by the functional departments to ensure consistency is delivered to the client both in closing a sale and during the contract. There is an opportunity cost. While existing staff are servicing these needs they cannot be attending to another client. The level of investment has to be calculated against the long-term returns in order to get the balance right. This may erode into short-term profits, because once again the benefits take a while to come through. And once again, the best time to make this shift may be on the upturn. Whenever it is done, it is still a sacrifice – it costs!

Having made the commitment within the organisation, what happens if it goes wrong? It may backfire for reasons outside the control of the organisation. There may simply be miscalculations internally. Few organisations examine the costs of exit from a market. These can be higher than the costs of overcoming the entry barriers. The *exit barriers* can be high-cost ones and the costs can be of several types:

- The sales team suited to relationship marketing is to have facilitators, so reverting to traditional selling techniques based around the marketing mix may need a change of personnel or retraining
- Having made the investment in one or more market segments, a strategic switch may be needed and the sales people have to establish new contacts
- The investment in relationship marketing is lost and further investment is made into new markets with no immediate return
- Any success and resultant reputation in the previous markets may themselves act as a barrier
- Special plant and machinery may also have been bought
- Clients for whom work is still being done may notice a cultural change and put no further work in the direction of the organisation
- There could be other costs associated with leaving a market.

Assuming that the organisation remains in the market, there are *critical events* that have to be managed. Contingency and crisis management are the order of the day. A former chief executive of SAS airlines coined the phrase 'moments of truth'. For that airline it was said that every contact with a member of staff was a moment of truth, for the goodwill could be enhanced or eroded with every encounter. That is certainly the case in construction. One claim can be a moment of truth. It can turn client relations sour. That is not to say that claims should not be made. It is the way in which they are managed that is crucial. It is a critical area for relationship marketing. If a sales person promises a harmonious relationship and the client gets hit with insensitively handled claims from day one, customer satisfaction is not going to be the outcome. All the efforts to build a close relationship can be dashed in a trice!

Claims are not the only critical event. Indeed there are many critical events in construction, many moments of truth in client relationships. These frequently come together, often in unpredictable ways, and require steady and sensitive management – *event configuration*. The management of *event configuration* is also of great importance. In some cases, contingency plans can be wheeled out; in other cases crisis management is needed. The relationship objectives with the particular client and with others is important to bear in mind at all stages, especially where the events are dealt with in the public domain (see ... *On the case 20*).[4]

All of these aspects require time and effort. All the time the meter is running. This means that the cost per target and per client is increasing. Unless resources are redirected from the marketing mix approach, overall costs are also going to increase. It has been explained that it costs five times more to find a new client than it does to keep an existing one. The sales approach in relationship marketing is therefore trying to reduce the expenditure on such things as pre-qualification documents, aborted bids and so on. These areas of expenditure are redirected into relationship

... On the case 20

On a major airport project an international contractor began to experience opposition from environmental groups. This was a critical event. The opposition came on site and supporters were able to build considerable support rapidly through the use of the internet, specifically linked websites.

Environmental groups learnt how to use websites as an educational tool, and hence as a propaganda vehicle, in a user-friendly fashion – a lesson contractors have yet to learn.

The contractor had not confronted such opposition hitherto and reacted with crisis management, the strategy being to keep a low profile and to let the client deal with the media. It was the client who mounted a rival website.

The contractor learnt from this crisis management and has subsequently been putting in place a contingency plan to deal with the management and the public relations requirements of this type of event configuration.

building during the pre-contract phase and into client management during the project.

What is the scope for savings? At first these are slender, because of the costs of restructuring operations and the short-term fall in turnover requiring high levels of effort. In construction few clients have regular procurement needs and public bodies are often restricted in placing repeat business with contractors. However, as repeat business does come through, there are considerable savings to be made. Also, the referral market through both clients and consultants will reduce the need for cold calling and selling from opportunism. In other words, the market develops into a network within the segments of operation. Even where generic issues are used to define segments rather then the type of building or the sector in which the client operates, gaps can be plugged with promotion in order to ensure that the message permeates the entire network.

Therefore *relationship costs*, the final category along the bottom line of Figure 6.2, cannot be avoided. They are there and must be addressed. In the short term the costs are very high, as the business is re-orientated. They diminish after a while and in the long term should become equal or less than previous sales costs in the broadest sense of that term. The key to success is achieving a balance in the long term; in other words, the returns outweigh costs or in recession break even.

The financial motive of the *relationship costs* is increased *relationship revenue*. Other factors can be summed up as a relationship investment for the future. The costs are investments in relationship longevity and patronage concentration. The strength of the relationship today is an investment in the trust and loyalty of the client for tomorrow.

30Rs and 5Es

Relationships can take many forms and are usually complex. One author has put together a list of different relationship characteristics to demonstrate the complexity.[5] This is not designed to be comprehensive. It serves to underline that there is no automatic means by which good relationships are translated into profitable streams of business. It all has to be managed and the way in which it is managed will depend upon the relationship. Therefore, no attempt is made here to analyse each of these, but an adapted list is summarised below to show what the author was seeking to show:

R1 *Classic dyad:* relationship of exchange between client and contractor or consultant

R2 *Decision-making unit relationships:* business-to-business transactions require the input of various stakeholders on both the client and contractor or consultant side

R3 *Mega-marketing:* the real client is not in the marketplace, but comes out of a decision made by other organisations, such as government, so it is these organisations which have to be courted

R4 *Classic triad:* competitor relationship can also be important in order to beat the competition, but also to collaborate with them where that makes greater sense

R5 *Alliances:* arising out of collaboration, formal and informal alliances, such as joint ventures and partnering, to curb competition and change market mechanisms

R6 *Profit centres:* creation of internal markets for an organisation, frequently occurring between the development and contracting arm of the same organisation, where relative autonomy is given to profit centres to seek suppliers from outside the organisation as a means to promote keen competition inside

R7 *Service encounter:* the interaction between the client and front-line functional personnel, such as legal advisors, estimators and purchasing, as well as contracts and site managers having direct involvement with the client to ensure that what is offered during the sales process is delivered in practice

R8 *Internal client:* where the front-line functional personnel have the sales team as a surrogate client, for they are representing the client needs within the contracting or consulting organisation

R9 *Full- and part-time marketers:* the front-line functional personnel have a key role in selling services in a fashion that dovetails with the efforts of the full-time sales personnel

R10 *Internal marketing:* ensuring that all other personnel who have any form of contact with clients prior to or during a project carry the same sales message, which may also include key sub-contractors

R11 *Non-commercial relationship:* public and voluntary relationships, for example a contractor as part of the package providing training for a municipal client on a non-profit basis through a charitable trust

R12 *Networking:* where multiple contacts create sales opportunities

R13 *Electronic relationships:* use of internet facilities to promote services and as an instigator of relationship development

R14 *Mega-alliances:* trans-national organisations, governments and aid agencies act as the project catalyst, through which relationships are created to facilitate the sales further down the pipeline

R15 *Quality:* building bridges between service development and technological innovation that can be articulated with client needs

R16 *Personal relationships:* particularly relevant in cultures where friendship or family relationships act as leverage for selling corporate services

R17 *Matrix relationships:* formal mechanisms to facilitate service and client management with sales – a largely absent, yet highly relevant relationship in construction, in order to underpin R7–R9

R18 *External marketing providers:* those organisations involved in market research, public relations, agents. Joint venture sales personnel who are responsible for promoting and referring services

R19 *Client's customer relationships:* understanding the needs of client customers can provide insights into how the client works and what their existing and future needs will be, for example involving the tenants of social housing providers in design or the tenant who has pre-let the developer's building

R20 *Financier relationships:* understanding the client's financial package, or indeed leveraging the package, can change the sales relationship

R21 *Parasocial relationships:* where symbolic images, such as brand names have an impact upon the sales offer

R22 *Law relationships:* clients who are founded upon legal requirements and litigation in their purchasing decisions

R23 *Criminal networks:* these can structure the entire market, whereby premium profits are creamed off as an illegal tax

R24 *Aggregate relationships:* where direct client contact is minimal and clients are served by social, income or segment categorisation, for example volume house-building. This is more common in mass and consumer markets

R25 *Client focus:* where the client draws the contractor or consultant into 'membership' or vice versa, for example through focus groups, as part of the market research and performance monitoring process

R26 *Relationships of dissatisfaction:* unhappy clients feel a relationship as intense, the frequent reaction of the contractor or consultant being one of distancing themselves. Pressing into the client, facing up to any shortcomings and explaining where there may be some misperceptions is also important, followed by remedial action on all sides

R27 *Social responsibility relationships:* environmental, political and social responsibility issues from the policies of the nation state through to health and safety are important issues to handle, especially where interest groups take up shareholdings in the organisation

R28 *Knowledge relationships:* knowledge acquisition is not only a source of advantage against competition; it is also a potential source for collaboration, alliances, innovation and understanding other interest groups

R29 *Media relationship:* support or damage can be inflicted quickly by a fast moving and multi-dimensional media, relationships helping to avoid excesses and misinterpretation through the understanding by the media of the business

R30 *Monopoly relationships:* where any client or supplier has considerable or complete leverage in the market, the importance of not being a 'victim' in the marketplace is important and power can be redressed through mutual, yet assertive relationships.

It is evident from this list that relationships operate on a variety of levels. Some provide a useful reminder of what is subsumed into commonsense thinking. Others can be seen in a new light. Not all are wholesome. R23, the criminal relationship, is a good example. Some organisations are tolerant of such activities. They will pay their dues within the culture, but maintain integrity in all other dealings. This is a difficult tightrope to walk. Many are not prepared to work under any such circumstances. The lesson is to know where the boundaries are (see ... *On the case 21*).

There are other relationships upon which contractors could concentrate considerable effort in order to differentiate their services and gain a competitive edge. R7–R9 are particular cases in point:

- ❑ Service encounter
- ❑ Internal client
- ❑ Full- and part-time marketers.

The role of front-line functional personnel into selling has yet to be fully realised. Implementation of their role in ensuring that promises made to the client are realistic and delivered in practice is necessary, indeed vital, in effective relationship marketing that goes beyond a skin-deep approach. Selling under relationship marketing is very much a team effort. It involves inculcating a process within the organisation that is similar to the facilitative approach to be adopted by sales people. This can be summed up as:

- ❑ Encourage
- ❑ Exhort
- ❑ Enlighten (educate)
- ❑ Engage
- ❑ Empower.

Selling under the marketing mix approach moves quickly to *exhortation*, in other words to directing the client. This only works once the client has already decided to order a building. It does not build relationships. Selling has to first move through

... On the case 21

An international subcontractor for a major terminus had worked in the country concerned before, but not in that location. Setting up their office, they were approached for payment protection. They refused, on the basis of working in the country and being under the umbrella of a reputable international main contractor.

An explosion occurred in the office of the subcontractor over the weekend.

The subcontractor was ill-prepared for the changing circumstances of that location. He had an obligation to complete the contract and thus brown paper envelopes became part of the relationships for the duration.

the stage of encouragement under relationship marketing. This is aimed at showing the potential client that you are not a fly-by-night sales person, but have a genuine interest in their organisation and activity – you are in for the long haul! Listening and understanding can lead to an exchange of information, getting the client to take an equal interest in your own organisation. A basis of mutual trust has been developed and selling can proceed to the next stage of injecting some direction. This may involve putting forward some *perceived alternatives* to problems and issues being faced.

A dialogue and exploration over these is a process of education. However, 'education' is only part of the story. The seller wants to solicit a response. Ideally, this is twofold. There will be an emotional response to the new information – the type of experience where the client goes 'aha', as Germans say. It is revelatory as well as informative. It is hoped that this second emotional response is a closer identification with the contractor or consulting organisation, education and emotional response coming together as *enlightenment* about the service approach and type of organisation. This is a powerful position from which to *engage* in negotiation or biding for a project, exploring more closely a contract, project or agreement to work together in future. *Empowerment* is the final stage. The tables have been turned. They have bought into your organisation rather than you selling them yours. They have an interest in the relationship in the long term. At this stage, the prospect may even be referring your organisation to other clients, even though you have yet to carry out a project for them at this stage. The power lies with them. Just as they now trust you, so you are trusting them to invest in your organisation through future procurement decision.

This process may need to be conducted with a number of individuals in the client organisation. It is a one-to-one process, but the desired outcome is a corporate decision. It also needs to involve the wider team on the contractor or consultant side. Sales can be seen as the front and back end of the project, or the contract can be seen as an extension of the sales process. The choice is a matter of perspective. Integration of the two is what is being sought. How is this to be overseen?

Models of client handling

There are several ways in which the sales process can be set up. In each case, the set up has to work during the contract stage. In other words, a framework or process is needed to manage the client throughout their life as a customer. There are basically two models:

- Relay team
- Account handler.

Relay team

The relay team recognises that there may be different people involved at different stages from the first client contact, through the contract and until the next contract. At each stage it may be best for a different person or small team to take the lead. The person in charge becomes the client manager for that stage.

For a contract, one team may be put forward to secure the work. They start the project, until the contracts manager is taken off for the next job secured. Another person takes over until the contract is about to finish and then an older person is asked to finish the project. The new teams are often introduced to the client or client's representative as a complete surprise, without consultation. This is not the way to do it under relationship marketing. In addition, the new contracts manager usually has no induction and no handover procedure with any real meaning. Contractors and design consultants are generally very poor at managing this. There is usually a mix of good and bad reasons behind such changes. An example of a bad reason is to take off the 'A' team in order to secure the next contract. The relationship marketing priority is to serve the existing client to full potential. An example of a good reason is that the person starting a contract is good at setting things up, motivating staff and mobilising resources, but not very good at completing things and paying attention to detail. Such handovers need to be managed.

The *relay team* model has at its heart the notion of passing the baton. This can be a difficult manoeuvre, requiring understanding and a great deal of practice. The baton provides the continuity of coherent, consistent and detailed briefing. There is even an overlap as the two run side-by-side momentarily at the same speed.

The race is longer than the contract. It starts with the beginning of the sales process and the baton is passed down the stages until it comes back into the hands of the sales person to start the race again towards the next contract. There can be a number of different points where the baton can be passed on:

- *Third-party sellers*: advocacy, referral and recommendation – most suited to innovative market positions (see Figure 1.5) or where agents are used in distant markets. This is the earliest handover in terms of sales stages, but the leg of the race can be the longest as courtship is long term
- *Introducers*: essentially 'door openers', who allow others to close the sale and manage the project – most suited to routinised market positions, that is the traditional contractor approach. This is the most limiting for the seller and for the organisation. Once again the handover is early in the proceedings

... On the case 22

A major international contractor, which is part of a larger conglomerate, involves the contract team at the early stages in selling its major project services. The sales management lead at this stage. The involvement of the contract team grows and formal handover occurs at the tendering stage.

Another international conglomerate has a more formal procedure. At the tender stage, a contract is signed between the sales and contracts team to agree to the handover. The contracts team has the power of veto if they are unhappy with the promises made or the contract terms and the sales team then have to manage the project until they themselves have overcome these obstacles or demonstrated that they are workable. This moderates the sales team to making realistic promises and challenges the contracts team to understand the nature of the offer being put forward.

- *Closers*: responsible for introducing the client and closing the sale – most suited to routinised and analytical market positions, especially relevant for highly standardised products and negotiated work. The baton is handed over in this case as the contract is closed, although this may be at the tender stage or after the post-tender negotiations.

Account handler

The account handler is in contrast to the relay team model. It is borrowed from the world of the advertising agency. It is also used by consultants, where personnel are divided into project teams who are assigned to clients. It is most suited to the analytical and innovative market positions, where high quality and high levels of repeat business are required.

In the account handler model, one person or small team is given responsibility for the client throughout their life with the organisation. This literally goes through from the first to the last contact. The management responsibility stays with the one person. In construction, there are three main options as to who plays the role of account handler:

- Sales person
- Contracts manager
- Director.

The choice may be derived from the corporate culture. It may be a matter of distributing the workload. It may also be dependent upon the type of segment and client for which the organisation is operating. If the objective is to develop the account or penetrate their market, that is to say there is greater potential for repeat business, then the sales person may be the best choice. A contracts manager may be most appropriate where the client is demanding intensive support and input during contracts. Where account maintenance is the key, the director may be best suited, so that the client neither feels overlooked nor imposed upon in an overzealous way.[6]

The construction and consultant markets are highly fragmented. From a client-handling point of view it is vital that a consistent and hands-on approach is adopted, precisely because the market is so fragmented and competition intense. In the selection of the model, thought has to be given to the circumstances in which the organisation operates to select the model with the best fit.

... On the case 23

A leading British and international contractor started an initiative to put the 'customer' first. Part of the initiative was to appoint key *customer account managers*. It is their responsibility to manage the customer, including:

- ❑ Maintaining close relationships
- ❑ Identifying customer's business and corporate objectives
- ❑ Understanding the decision-making processes
- ❑ Looking for selling opportunity across the contractor's business
- ❑ Managing all internal systems *vis-à-vis* the customer.

Conclusion

Selling is alive and well in construction and in the professions. That is the good news. The bad news is that selling is frequently carried out in a narrow way. The self-imposed constraints come from:

- ❑ Attitude and thinking
- ❑ Low levels of creativity
- ❑ Lack of management understanding
- ❑ Limited range of promotion tools used.

The emphasis on new clients has been changing over recent years. The chapter has set out how selling can be approached using relationship marketing. However, the processes needed to enhance selling are complex. The way in which these fit together are shown in Figure 6.3.

Current sales emphasis is towards *fragmented* and *discrete* actions. This is depicted in the bottom left-hand side of Figure 6.3. The potential emphasis is towards a more *integrated approach*, as shown in the top right portion of Figure 6.3. In consumer markets the trend has been towards the top right-hand portion. As the journal *Marketing*[7] suggested as far back as 1989:

Instead of selling a product or a brand image, we had to start selling our integrity.

Summary

1. The purpose of this chapter has been to demonstrate that selling can be undertaken in construction:
   - From a comprehensive perspective
   - With supportive management.
2. The chapter has:
   - Set out the scope and constraints of relationship marketing in selling
   - Shown the potential for supporting management structures and the choice for the client-handling model.

Selecting the management processes is part of the aim of securing competitive advantage and the processes themselves provide an added competitive edge when well managed.

References and notes

1. Branch, R.F. and Smyth, H.J. (1996) *A Client Orientated Service*, Centre for Construction Marketing, Oxford Brookes University, p. 7.
2. Storbacka, K., Strandvik, T. and Gronroos, C. (1994) Managing customer relationships for profit: the dynamics of relationship quality, *International Journal of Service Industry Management*, **5**, 21–38.
3. Partnering will be returned to in Chapter 12.
4. Preece, C.N., Moodley, K. and Smith, A.M. (1998) *Corporate Communications in Construction: Public Relations Strategies for Successful Business and Projects*, Blackwell Science, Oxford.
5. Gummesson, E. (1990) Making relationship marketing operational, *International Journal of Service Industry Management*, **5**(5), 5–20.
6. Cf. Guiltinan, J.P. and Paul, G.W. (1989) *Marketing Management: Strategies and Programs*, McGraw-Hill, London.
7. *Marketing* (1989) Strategy for the 90s, 19 October, 22.

Chapter 7 Selling the Service and Product

Themes

1. The *aim* of this chapter is to investigate in more detail the marketing mix approach to selling for contractors and consultants.
2. The *objectives* of this chapter are:
   - ❑ To distinguish between product and service
   - ❑ To explore selling using examples
   - ❑ To investigate the implications for different procurement routes.
3. The primary *outcome* of this chapter is to enable contractors and consultants to develop a mix in an applied way at the points of sale.

Keywords

Marketing mix, Selling, Services, Products

Introduction

The first ten seconds and the first ten feet (or three metres) into a Warner Bros store is what matters. If you are not sufficiently captivated by then, you are not going to be a good customer for them. Therefore, the key to targeting their customers is to get the right mix of products, advertising and image over to the target customers in the first 'ten by ten'. That is the sales pitch.

Ikea, the Swedish furniture and household store, wished to have a store in downtown New York. The organisation was unable to find suitable premises, so they built an out-of-town one. However, the key to the success of the out-of-town store lay downtown. Ikea took a small corner plot and fitted it out on one theme, for example a kitchen. The whole fit-out was then changed into another room within a matter of weeks. This was literally the shop window for the out-of-town store. Customers are drawn in to find out more and directed to the greater selection in the out-of-town store – an innovative way to tap into the downtown market.

Having examined promotion in the context of the marketing mix and relationship marketing, the same pattern will be used with selling in this and the next chapter. This chapter therefore

Table 7.1 Perception of architectural practices in Britain of the means to secure work

	Important (%)	Unimportant (%)	Don't know (%)
Talking to others in the sector	53	36	11
Monitoring planning system and property markets	50	40	10
Establishing a rapport with potential clients before they need you	48	39	13
Inviting clients out socially	43	45	12
Calling clients informally to talk about non-architectural matters	43	42	15

Source: *RIBA Journal* (1995).[3]

returns to the marketing mix approach. As a reminder, the objective of the sales message is to:

- Inform
- Persuade
- Remind.[1]

From this, potential clients are in a position to assess whether they will wish to consider purchasing the product or service. What are the overall key categories in which clients are interested? Promotion of products and services work at three levels:

- Functional benefits
- Image
- Extra benefits.

In this chapter, it is the *tactical* issues which are the main focus.[2] One way of beginning to understand the tactical issues is to have knowledge of how the sector gets its work. A good, intuitive knowledge is already possessed by most people. Some detailed data will shed further light on the securing sales.

Securing sales

How do organisations secure their work? There are a number of factors at work. One starting point is to see how architects secure their work (see Table 7.1).

Table 7.2 Pre-qualification criteria of contractors

1985 Findings	1993 Findings
Financial position	Experience and reputation
Reputation	Completion on time
Completion on time	Building type track record
Relationship with client	Financial position
	Expertise

Sources: adapted from Baker and Orsaah, 1985 and CIOB (1993).[4]

Contractors pre-qualify for construction along the lines shown in Table 7.2. *Reputation* arises as a most important criterion. It is suggested that this is an amalgam of a number of issues to do with selling. Essentially, all the factors put together infer that this contractor is *capable* of doing the work. The final selection comes down to a mix of real factors of selection and arbitrary ones that are mobilised in order to ensure a selection. Price comes last on the list in terms of chronology and is often the most arbitrary for the tender price seldom comes close to out-turn prices agreed in the final account.

Table 7.2 compares favourably with guidelines for consultant selection. These suggest some general criteria:

- Overall experience and facilities
- Particular services and experience
- Overall approach.

These are the main determinants. Within each there are further important factors:

- Performance
- Financial position
- Staff expertise
- Track record.

There are others.[5] Some types of projects raise issues. In being asked to bid for major projects, critical factors were identified and ranked, as shown in Table 7.3.

The case of design and build

Taking design and build (D&B) procurement as a particular example, the market has been growing across many countries in the

Table 7.3 Success in the major project market

Rank	Invitation to bid	Percent	Effective tactics to win	Percent
1	Perceived quality	68.5	Person-to-person relationships	44.8
2	Relationship building	54.4	Invitation to bid	34.1
3	Market position	52.2	Distinctive image	26.6
4	Corporate image	47.6	Understanding benefits to client	25.6
5	Track record	46.5	Understanding client business	25.0

Source: *Marketing Business* (1996).[6]

world. In Britain, it has grown to constitute between 35% and 40% of the market. That does not mean that clients are content with the services, even at the selling points. It has been found that 80% of one small, but significant, sample of major building procurers were dissatisfied with the poor proposals, poor quality specifications and lack of detailed programming. That experience during the sales stages was repeated in most cases during the project. For example, 80% of clients stated that there was a lack of communication both during and after project completion. Every client in the sample pointed to a lack of after-care service.[7]

This shows how contractors are remiss at having the interests of their clients at heart. It should start with the sales pitch and continue right through the project. It could be that contractors are most intent on survival. So, do contractors have their own interests at heart? One analyst has found that it is the client who takes the initiative to select the D&B procurement route 54% of the time. The quantity surveyor introduces the route 12% of the time. The contractor introduces D&B 13% of the time.[8] This shows a certain amount of complacency. The contractor is relatively reactive, rather than pro-active. This is merely the flip side of the point made previously that sales people are heavily involved in responding to pre-qualification documents, making verbal and written presentation. Therefore, it is not surprising that the service delivery continues in the same vein – reaction to the force of circumstance. It is also a product of the culture of the industry. The paradox of the *street-fighting man* (see Appendix A, Paradox 1) induces an inward looking culture where the main energy is expended defending individual positions in the organisation. Career politics dominate above organisational interests, so the client interests are a very low priority. The overall interests of the contracting organisation are arguably not being best served by the status quo.

Client versus contractor and consultant perceptions

The client is not always knowledgeable about building, yet almost every client is astute. Getting the sales in better shape is a good first step to improvement. Having a client orientation is helpful. Table 7.4 sets out the importance of how selling is perceived within the transaction.

Much of the strategic and structural improvement has been addressed in the preceding chapters. Yet, as Levitt states:

People, with the exception of those who work in sales or marketing, seldom see beyond their company's walls.

In many organisations, this statement can be taken literally. In contracting and consulting it is symbolic, for the issue is that many do not *see* beyond their own walls. Several questions can help in knowing where you sit on this:

Table 7.4 Sales perceptions

Stage	Selling contractor or consultant services	Purchasing from the client perspective
Before	Real hope	Vague need
Courting	Excited and really trying	Testing and hopeful
Sale	Objective achieved	Judgement postponed
After	Selling stops	Purchasing continues – stage payments and next project
	Focus elsewhere	Focus on having contractor/ consultant selection verified or justified
	Tension released	Tension increased
	Onto the next sale	'You don't care'
	Relationship perceived to be secure	Commitment increased, yet insecurity grows with each disappointment
Long after	Indifferent	'Its got to be better than this!'
Next sale	'How about a new contract?'	'You've got to be joking!'

Source: adapted from Levitt.[9]

- ❑ How much do you know about how your organisation sells its services/products? Your answer reflects your culture!
- ❑ What are the advantages and disadvantages of this? Can your culture be used to obtain greater understanding?
- ❑ How could your role be developed and enhanced through relating your expertise to the sales process? Perhaps everyone has a sales contribution to make, directly or indirectly.
- ❑ What do you do that actually benefits the client once they start using the building? The client is seeking benefits, not just service features.
- ❑ How much of your time actually adds value to the service on offer?
- ❑ What would be the corporate benefits?

In answering these questions, it is important to address how people work together. How do the strategic plans, or lack of them, look on the ground? The contractor always claims that their company 'pulls together'. How this actually works out in many cases is shown in the first diagram within Figure 7.1. The first model is self explanatory.

The second model of teamwork in selling is the introverted one, which is driven by complacency or a pre-occupation with operational issues – a production-driven orientation. It is the paradox of the *lighthouse keeper*, who is communicating with someone out in the darkness, hermetically sealed from the outside world, 'getting the job done' in glorious seclusion (see Appendix A, Paradox 13). This approach to teamwork does not enhance selling.

A third approach to co-operation looks good at first sight. People are pulling together, there is rhythm to it, but ultimately it goes around in circles. The focus is inward, even though the levels of co-operation are high and the culture is positive. Essentially, people busy themselves with looking busy. There is no tension in it, yet there is no fruit either. This frequently goes unnoticed in an organisation because the business seldom penetrates beyond the organisational boundaries. Selling fails to hit the targets.

Ideally, selling should have an outward focus with everyone looking and pulling in the same direction (see Figure 7.2).

Tactical direction

Having got an outward focus in selling, and hopefully in undertaking the work too, how should the sales effort be directed tacti-

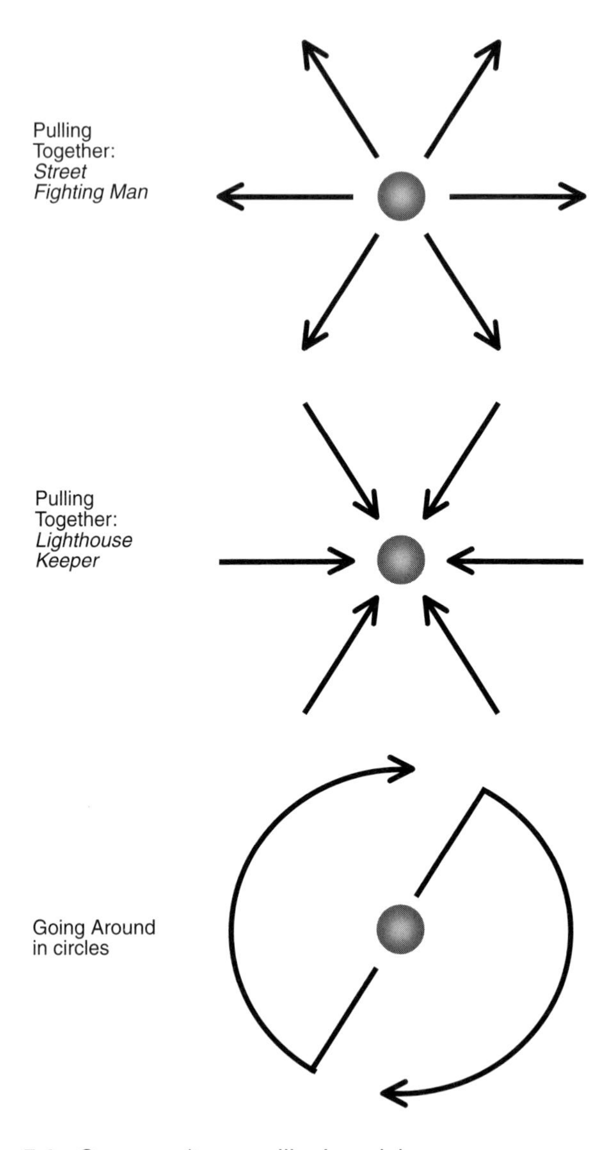

Figure 7.1 Common 'team selling' models.

cally? One sales person per client will have limited impact. A team approach is better, frequently matching the people in hierarchical terms. The sales person, whether a part- or full-time seller, will chose the timing for introducing others and whom to introduce in consultation with others. In this way, a series of person-to-person contacts builds up, each with 'ownership' of the contact. Once this is achieved, it can be legitimately claimed that selling is business-to-business. It has gone beyond the personal.

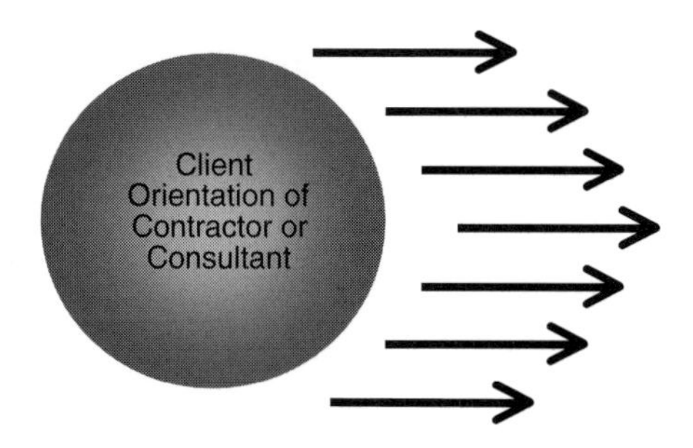

Figure 7.2 Preferred team selling model.

Another model is the network model. Some view this mistakenly as spreading the sales net far and wide. This merely returns to the scatter-gun approach. By definition, nets link, the aim being to come back on the same target clients time and time again and for others within the network to reinforce the message too. The problem within networks is that they are difficult to manage from a corporate perspective. Trust has to be put into the process by the management. Networks also have to be collaborative and not just a matter of contacts.[10] The network is a group of people with an agreed common purpose, membership being based on the ability to further the collaborative goal and trust being the oil that keeps it going.[11] Sales networks are informal, so it is not always clear whether they are proving effective. Some sales environments are clearly defined and here networking is very effective, such as housing regeneration or leading edge architectural design practices.

In most cases, a more simple approach is needed, which is then reinforced in other ways. Initial contact may be established through cold calling or networking, but after that the responsibility should firmly rest within the sales effort. The tactics advocated here involve the *pincer movement*. Essentially, the sales tactics should come from:

- ❑ Two directions
- ❑ Two media
- ❑ And perhaps reach two people in the target organisation, who liase with each other.

Personal selling should be one of those pincer prongs every time. Personal selling embraces both person-to-person and direct business-to-business contacts. The long-term move is from person-to-person to business-to-business contact. The other may be some other form of promotion. An advertising campaign or trade journal article may be another or others. One idea is the billboard or advertising hoarding. For example, it may be that a client has a £400m programme. Half of the key decision-makers live in a suburban sector outside the headquarter sector. All travel to the office by car or train. A billboard campaign may involve occupying just two sites, one at the rail station, the other at a major road interchange. To all intents and purposes it would appear that the advertising is very general. Indeed, a broader audience will view the advertising with a positive spin off, yet the target is a handful of people. Every day these decision-makers travel to work, the personal selling message is being reinforced without them realising. This could be a powerful second pincer prong.

Another example of a second pincer would be to invite clients to visit a series of completed projects and projects under construction. There may be personal contact, yet the pressure is off the relationships, the focus being the quality of work and management.

Most people need to receive the same message several times for it to sink in and make the difference. Personal selling and a second prong can drip feed the message. It has also been shown that recommendation through the referral market is effective. If the targets are beginning to have the same message reinforced from a third direction, then the sales effort improves. It will certainly lead to pre-qualification, and almost certainly lead to being on the tender list if competitors have not got to grips with all this. If there is a strong network, the referral market from third-party recommendations is going to give further impetus. However, the referral market is powerful, yet less easy to manage. The pincer movement can be managed, the third element or media being left to the reinforcing role.

This section has moved from selling on a one-to-one basis to one based upon a number of actors and co-ordinating that into tactical *pincer movements*. As this is done, the sales process moves away from selling products towards selling services. Services can be sold using the *marketing mix* approach. *Relationship marketing* can also be used for products, but it is usually conducted on an aggregate basis, such as reward cards, targeted mailshots and so on, and not on a one-to-one basis, except in very exclusive markets. Selling services is much more intangible and the selling is about 'people'. That truism is frequently used in business contexts in a misguided fashion. Service provision is a people business, but it is not based on the individual. It is not *individualistic*, it is *corporate*!

Selling the corporate service

Contracting and consulting is largely service- rather than product-based. Taking this as the first of the 4Ps, how can the service be sold, given the preceding tactical context?

As Kotler stresses *inform, persuade* and *remind* in the marketing mix context, then the focus for this section is only on the first two of 5Es considered in the last chapter:

- Encourage
- Exhort.

Having directed the client, the onus falls onto them to buy. The information being imparted to the client is about the nature and range of service, in other words the *functional benefits, extra benefits,* that is the added value, and finally the *image* associated with using the company and its services. The *image* embraces the product brand, the goodwill and service reputation.

Deciding what services in which markets and where is strategy. How to sell them is tactical. Case studies are used heavily in this and subsequent sections to illustrate the points. Before investigating the construction sector, financial services offers some interesting insights. It is an interesting contrast because it has been seen as high status, and certainly more sophisticated than construction. It does share some similarities. Financial services has been a sector of rapid growth in marketing communications over the last 15–20 years. While entry into the sector is for the large corporation, it is not impossible for smaller companies, as retailer and building society moves into banking illustrates. Service or product emulation is easy:[12]

> *It is a marketplace where, in theory at least, any new product can be imitated at the flip of a spreadsheet, though working out exactly how best to sell it, and to whom, may take a little longer.*

Like construction, the customer interest in financial services is low most of the time until something is needed, and then it becomes critical and is important to get right. Customers are not experts but they are astute, so they can see through the 'flannel'. The clearing banks spent a great deal of money refurbishing their outlets 15 years ago to make them more user-friendly. It did not take the customer long to see that the real motive was simply to sell more services. The banks had moved into selling insurance, mortgages and a whole package of loans just as if they were any fast-moving consumer good (fmcg). The fmcg approach created scepticism among customers and many believed there was a conflict of interest.

Banks are currently in the business of redefining themselves. Having stressed personal contact, the new effort is in remote banking using the telephone and the internet. An international banking group that was amongst the first into electronic banking found that nearly 90% of its customers were satisfied with the service.

They also found that the main reason why potential customers were not switching to electronic banking was not to do with satisfaction with conventional banking – they were dissatisfied in many cases. The real issue is inertia, in other words the hassle of transferring over to electronic means. A secondary advantage of electronic banking is that it drives banking costs down. The means of selling therefore changes. This in part is a reflection of place, that is the distribution channel, but in a service context this frequently changes the nature of the service, because it is the management that is being bought.

Consultant and contracting business is largely about people. The way that different people in a consultant context, for example, portray the same service is instructive (see ... *On the case* 24).

... On the case 24

A highly successful multi-disciplinary surveying practice, employing around 150 staff across four regional offices has experienced changes in selling. First, they have now found architects selling their services to them, rather than the other way round.

Sixty per cent of work comes direct from the client, less from other consultants.

Partners tend to sell from the level of the *mission statement* using flipcharts, overheads and slides. Others tend to use a more tactical and responsive approach to particular situations. Both are needed, the issue being timing. This is problematic for the practice when selling is driven by personal, rather than client interests. It is hoped that the move towards categorising clients into *existing key clients*, *long-term prospects* and *prospects* will provide further focus for tactical activity.

Strategic solutions can help to frame tactical responses, but do not always provide answers. Categorisation can help. A major international glass manufacturer has achieved this with its product range, hence helping the decisions as to *how* to sell the product (see ... *On the case 25*). The same can be achieved for construction. For example, the service and the various strengths can be set against the type of client and building requirements. The benefits of different types of procurement route can be analysed too. This can provide the tactical sales framework. How to do this is a strategic question and it is hoped that sufficient ways have already been presented to stimulate different solutions, which are part of the process of creating competitive advantage.

Clients want a service that matches their need. Experience and reputation have already been cited as important. One of the

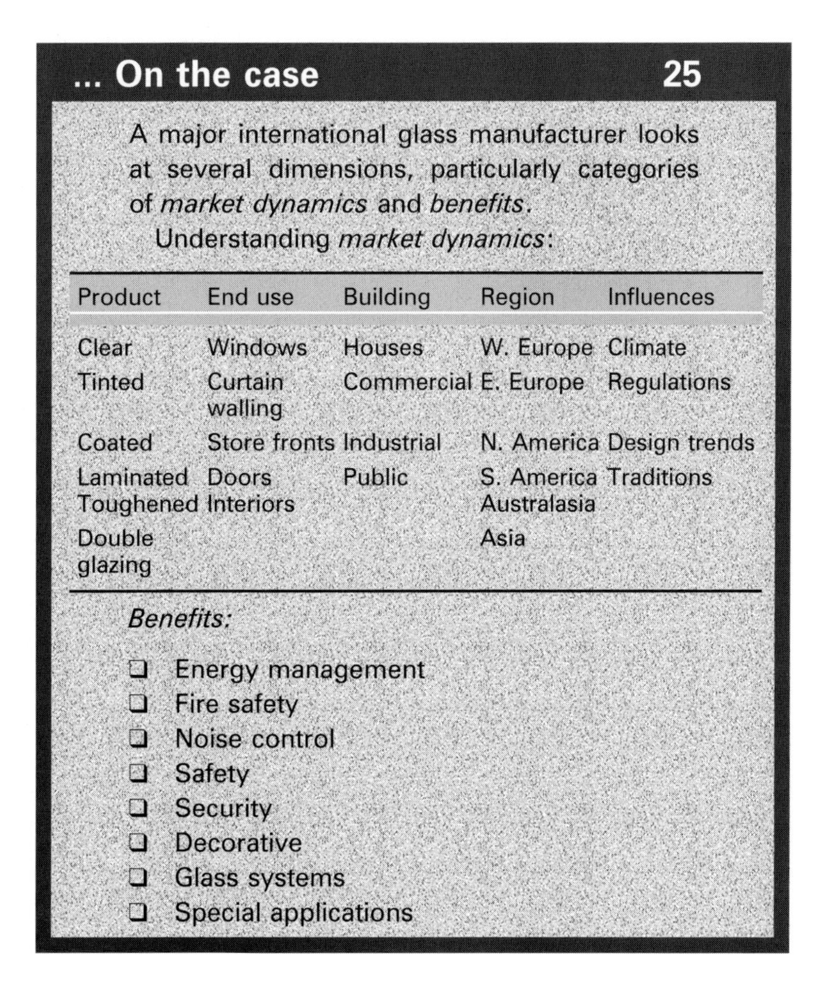

... On the case 25

A major international glass manufacturer looks at several dimensions, particularly categories of *market dynamics* and *benefits*.

Understanding *market dynamics*:

Product	End use	Building	Region	Influences
Clear	Windows	Houses	W. Europe	Climate
Tinted	Curtain walling	Commercial	E. Europe	Regulations
Coated	Store fronts	Industrial	N. America	Design trends
Laminated	Doors	Public	S. America	Traditions
Toughened	Interiors		Australasia	
Double glazing			Asia	

Benefits:

- ❑ Energy management
- ❑ Fire safety
- ❑ Noise control
- ❑ Safety
- ❑ Security
- ❑ Decorative
- ❑ Glass systems
- ❑ Special applications

major leisure groups has specific requirements. A British leisure multiple specifies contractors with track record, for example shop-fitting contractors have consistently proved to be cheaper than main contractors for large leisure fit-outs. The reason is the experience of the responsible management. Competitors have not understood that there is a knock-on effect from this requirement. Specialisation creates a loyalty factor:

> *When I first started I had about ten contractors and then when recession hit I got 'one hundred and ten'. They came from out of the woodwork. They had never spoken to me before, but now because their retailer wasn't doing anything they started putting leisure in their portfolio. These people are not loyal ... . The guys that have done the work, those are the ones that are loyal, so when I demand a good job for a tight budget I can demand that loyalty.*

The loyalty factor feeds back into the service. The client is using it to redefine the quality and price in recessive conditions. This is an intelligent way to use the market to mutual benefit.[13] Educated clients can adopt a more sustainable and beneficial approach. The client uses competition to drive up quality and drive down price. The loyalty factor is not lost. The contractor retains repeat business and is put in a strong position of survival and high market share, ready for the upturn. Profitability, although desirable, is an unrealistic goal for many contractors in a harsh economic environment.

A Finnish consultant, a case in point, has developed a particular expertise in forest management that makes it unique in the international consultant market. It sells its services through a series of branch offices, so that it can both retain its expertise at the leading edge and adapt it to local contexts. It is a highly specialised market and the number of clients is small, so it is easy to manage the selling.[14] The case shows that the service can change under different conditions. A further lesson is that the efforts of the competition are wasted from their viewpoint. This is not always the case, as will be investigated further in Chapter 11. This second example concerns a market that is so specialised that the exposure or dependency is very high. However, the network is restricted and so is more easily managed than more complex and diverse markets. There is a risk of dependency on a limited number of clients. However, the lower costs of selling and the premium help to offset the risks.

... On the case 26

A medium-sized regional contractor has set up a separate partnering division. The group is the result of a series of take-overs of smaller companies, which have proved hard to integrate, thus the need for a separate procurement route for partnering. The main promotion has relied on this packaging, coupled with putting its current repeat-order clients into the partnering procurement route, such as a major pharmaceutical client. The offer consists of workshops with an external facilitator, a visit from the MD and partnering director, plus agreed procedures.

A regional civil and building contractor has established a market niche consistently yielding above average margins. This was achieved by offering a negotiated service and entering discussions at the earliest stage, avoiding open competition as much as possible. From this basis it has offered financial packages upstream, and downstream has entered into partnering agreements, where the clients wishes this. This is a niche approach that permits tailoring the service to meet 'adjacent' niches in the market. The result has been consistent growth – a successful strategy.

Through market research, an international contractor discovered that its reputation and technical expertise were well respected, yet clients were unaware of the service range. It had more subsidiaries than any competitor. Personnel were considered conceited in dealings with client representatives. The organisation was also reluctant to change methods to suit the client.

This led to an initiative to pull together the services and listen to client needs and tailor their services accordingly. It has led to a greater ability to mutually match needs and thus a more targeted sales approach.

The next three examples have all developed 'service products' for the market.[15] They have taken the route of developing a procurement option for clients. However, not all have taken a structural solution, as opposed to the sales driven one. All three are just as targeted in their approach, the implications for selling being markedly different (see ... *On the case 26*). The first case has taken the structural solution. The sales effort becomes concentrated upon a narrow band of potential clients, who have already expressed their preference for the partnering procurement route and those who could be easily persuaded.

The second approach is equally targeted; however, the way the offer has been put together is more open-ended. It permits the sales personnel to be able to listen and respond to client needs. Apart from the obvious advantage that it makes the job of selling far more interesting, it also gives the client the message that the organisation:

- ❑ Listens
- ❑ Understands the needs
- ❑ Responds to the needs.

In addition, it requires that the operational side is responsive to their clients. While the case material does not prove that this occurs, the success of the case study in relation to profitability and to repeat business is a strong pointer that it does precisely that.

Over several years, the organisation has successfully migrated from negotiated work into adjacent markets, such as partnering. The sales team has been successful in responding. Putting this in the context of Figure I.1, it is suggested that the organisation has moved up the curve to a position of high commitment and high investment, hence the sales effort is given sound strategic support. Figure 1.6 shows that the risk for this organisation is manageable as it moves from the square of market penetration to service diversification. Here we see that investment is at a medium level. Combining this with Figure I.1, which indicates high investment, two qualifying comments are necessary. First, the strategy is incremental and thus the investment is spread over time. Second, the additional investment is not great enough to successfully move into an adjacent service market. This is a subtle and important lesson in giving the sales people a realistic set of objectives. Put simply, the organisation has provided the sales team with services that can realistically be sold in the market. Flexibility is combined with targeting.

The third example is even more open-ended. The move is in a sound direction, but it is a more difficult situation. While it may encourage cross-selling and may endear clients to the organisation, it is not necessarily clear what the desired outcomes are. It is potentially leaving a great deal to chance. Furthermore, it may be some years before the benefits come through. The onus is on sales people or on each division to provide the focus. There is no reason why the approach should not work in any company, including the case study concerned. It means that market research, monitoring and implementation management at the tactical level are all key in order to be confident of success.

A further case illustrates how far the service can be changed (see ... *On the case* 27). A major international contractor has essentially switched from the contracting to consultant market – a poacher turned gamekeeper – yet still positions itself in the construction marketplace. The significance of this move was the successful build-up of a consultant culture, supplemented with a series of management techniques derived from engineering and supported by excellent promotion. It put the contractor in a market position, which others have endeavoured to emulate, but failed to fully jettison the contracting mentality. The most successful emulators have come from or entered the market as project managers.

The other case is also about a contractor, where the core of the business is the ability to deliver services to the area and community through carrying out contracting. This too is repositioning construction as a consultant organisation. Contracting is a means, but not the most significant aspect of work. The clients are interested in rebuilding communities as much as building houses and this is the service offered beyond the scope of the competition. Selling in this case is conducted through a tight network of clients, consultants and stakeholders. It is a good example of where something relatively unique, which is also complex, can be conveyed to the decision-makers on a drip-feed basis. Selling is as much about 'being there', because the features can be sold in this drip-feed fashion.

Yet another aspect of service design is to sell the same team management throughout the contract. This is an important feature. In North America, clients sometimes chose the team first and the contractor second. Why? This is a symptom of the absence of a culture and systems approach in contracting. Thus, an experi-

... On the case 27

Client innovation led this international contractor to base its work around a fee system. Essentially, this service takes the contractor out of contracting into consulting. It is a management approach, which permits the contractor to handle the conflict on behalf of the client, indeed the client's representative too.

Traditional model

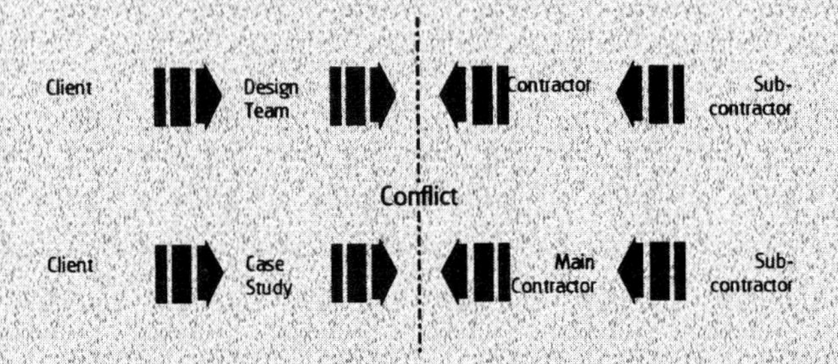

Case study model

The essence is that the contractor acts as the buffer between the client and the construction process, the traditional position of the design team and client's representative.

In another case, the domestic business of a large international contractor negotiated 81% of all its work for one year. They have never put in a claim and subcontract all the work. It is a different culture, working primarily within the community to meet the needs of the various stakeholders involved in urban regeneration. Various senior personnel sum it up as follows:

> *The cheapest marketing is repeat business ... I don't believe we've ever approached it in any other way.*

> *We are now almost more of a consultancy than a housebuilder.*

> *Everyone is a marketing person.*

ence on one contract is not necessarily repeated on another contract, as projects tend to be run on a personality culture. The corporate culture is weak. This is an important feature of which the sales team can inform, persuade and remind the client, although it is the type of feature for which wider management decisions and support are required. This can be given in general or on a project-by-project basis.

The selection of teams is becoming more important elsewhere. The largest software company chose a contractor for its regional headquarters on the basis that they guaranteed to keep the same contracts team throughout. Their bid was not the lowest, but the value for money was considered higher. Again, an international management contractor was chosen for a £12m genetics laboratory complex, on the basis of the team first and the management fee second. The team stage of selection would have been re-evaluated had the fee been considered unreasonable.

All of these cases illustrate the different approaches to service configuration and some of the implications for selling. This is not a comprehensive list, the point being that the options and permutations are numerous. There really is every option for far greater service differentiation than has hitherto been used. Developing different options can be more successful than copying the successful strategies and tactics of others.

Place is one of the 4Ps that have been set out in a construction context in Chapter 5. Place refers to the distribution channel. What is the distribution channel? There is a great of difference between mail order and shopping in the street. Such decisions are about *place*. In construction, the choice of contract and procurement route is about place, yet is inextricably linked with the service (see ... *On the case* 26 and 27). In this section, the focus is upon the contracting 'shop window' – the forum in which the message is sold. Three principle 'shop windows' exist:

- ❑ Potential client market
- ❑ Referral market
- ❑ Repeat business market.

Goodwill should increase as selling shifts from the potential client to the repeat business market. Because of the adversarial nature of traditional contracting, the reverse can be true. The importance in tactical selling is that the sales person should assume that they are building on favourable circumstances. Listening and responding should therefore be the order of the day. The emphasis shifts to *reminding* and from *exhortation* towards *enlightenment*. Any problems of poor performance and any objections should be addressed head on. A virtue can be made out of addressing difficult decisions. This may not make the seller the most popular person within the organisation, but it can be achieved if the senior management give advance support to back them to the hilt. This may mean backing down on some difficult issues, such as set off, retention money in dispute and pride swallowing. It can be worth it if it leads to more business and if the lessons are learnt on the project for next time round.

Selling approaches vary according to the type of construction contract. On the whole, sales people prefer to negotiate rather than tender. There is frequently opportunity to negotiate during the post-tender period. The form of contract is important too. The broader the service, the greater the totality of the sales effort.

For example, design and build involves the architect and other design team members selling their services to the client. The architect is frequently an external consultant chosen by the contractor. Ideally, there should be a strong relationship and understanding here, although this is frequently theory, but not the practice. It involves selling upstream of the site work. BOOT type contracts involve both upstream and downstream sales implications. The access to financial markets is a key point of leverage for BOOT contract sales. While selling may start upstream, the integration of downstream operational services becomes an important selling feature once the project is being determined. Partnering, on the other hand, implies a very close relationship with the client and also sourcing of components and services downstream in an efficient or lean way.

In all circumstances, clients will come up with objections to maintaining contacts or pursuing the project. Objections can always be addressed as a challenge. While there is a client objection to overcome, there is dialogue. A robust challenge is an opportunity, but a balance between persistence and sensitivity is needed. The key principle that when the client stops complaining they are about to switch suppliers[16] holds true in construction and consulting too. The onus is on the sales person to keep the dialogue going, yet they will be supported to a greater or lesser degree by management. The greater the support, the longer the dialogue is likely to be, which in turn maximises the sales opportunity. Some opportunities will be worth pursuing more than others. It is the target markets where the pursuit and promotion should be strongest.

Promotion

Currently, most of the image-making comes from promotional literature of various types.[17,18] Recent work, shown in Table 7.5, produces the following ranking for promotion activities.

On the whole, consultant brochures are far more interesting than those of contractors. They are, however, a true reflection of most organisations – we do everything in pretty much the same way as everybody else. Like the services, they lack imagination. Something can be done about that fairly easily. What is of greater interest is the use of the advertising and promotional media.

In general, advertising media have exploded over the last two decades. TV dominates, yet press and radio advertising has consistently doubled. Using mixed media delivers a greater return,[19] which reinforces the earlier message about using pincer movements. The rise in advertising was 300% for advertising and 700% for PR across UK industry between 1981 and 1991. Across industry about 20% of the marketing budget can be dedicated to PR and lobbying.[20]

Advertising is underused in construction, especially given the symbolic importance of some key buildings and projects. Measuring a campaign is important. Two key variables are recognition (R) and attributes (A). Recognising the organisation as a brand is one aspect, attributing to that brand the type of activity and service is important. An A:R ratio can be produced to measure effectiveness.

Reports have shown that the construction sector could benefit from paying more attention to its image.[21] Setting out budgets according to some scoring system can be helpful. It can be used

Table 7.5 Promotion ranking in construction

Activity	Ranking
Press releases	1
Feature articles	2
Marketing brochures	2
Annual report production	4
One-to-one press contact	5
Company newsletters	5
Corporate hospitality	5

Source: Preece *et al.* (1996).[18]

Table 7.6 Hypothetical promotion budgeting

Elements	Rating for budget allocation (1–10)	Weighting for impact evaluation (%)	
		Expected	Actual
Clients			
Image profile	3	7.5	10
Brand name	3	7.5	5
Advertising	5	12.5	15
News generation	7	17.5	10
Market research	5	12.5	3
Client education	7	17.5	10
Staff			
Senior management education	1	2.5	10
Other staff	4	10	17
Broader community			
Issue PR	5	12.5	9
Sponsorship	3	7.5	3
Lobbying	2	5	2
Total	40	100	97

to allocate resources. It also establishes a benchmark against which to measure the effectiveness of the results.

In the above hypothetical example, two things need to be drawn out:

- ❑ The figures are *subjective*, not objective
- ❑ A *method of measurement* is needed, especially for the results, no matter how approximate they may be.

While too much store must not be put on any one figure, especially without critically appraising the method of measurement and the other factors that may be relevant, an overall guide is produced about the success of a promotion programme. In this case, the greatest success has been on improving internal PR, but more success is needed with clients, especially with the broader constituency of environmental, community groups and government. In one actual example, a leading architecture practice uses a marketing consultant firm. This takes up the majority of the marketing budget, as they see themselves located in an innovative market position (see ... *On the case 28*).

... On the case 28

An innovative architecture practice located in the arts, community and housing project markets, employs a marketing consultant to:

- Look for opportunities
- Targeting and reaching audiences
- Networking, especially in the arena of politics.

This is important for building credibility as, according to one partner, the practice likes people to notice that they *'have the contacts'* and that they are *'dealing with the serious players: that's the message'*.

The aim is to continuously pinprick the market. This media and lobbying activity goes hand-in-hand with partners who network at the level of individuals, rather than of organisations.

One of the curious factors is that clients have begun to see greater potential for promotion than the contractors. One example is the action of a major British food retailer with an annual spend of £500m in recent years, which has filled the promotion gap left by contractors. Guidelines are provided to ensure that sites present the right image of cleanliness and sense of order. The retailer is insisting that contractors wear protective clothing on site bearing the brand name of the retailer, at the contractor's expense. This has been linked in with the broader effort to use the site as a large advertising hoarding or billboard to promote the retailer.

The 'opposition' too have been far more effective users of internet than contractors or consultants. Not only have they been rapid on uptake, they have also appreciated that it is a network. The hard sell is not so appropriate in these cases, so using it as an educational tool is valuable and effective, especially when one person's education is another person's propaganda.

The opposition comes from environmentalists who are against major infrastructure schemes.[22] Opposition groups make the network more effective by providing links to many other relevant websites. This reinforces the educational component and the site managers do not perceive other sites to be competition that may draw interest away from their concerns, but as an opportunity to deepen the message. The tone is user friendly and targets the generation that is currently forming their views and image of construction. The infamous Newbury bypass in Britain was the subject of several websites:[23]

> *[X Burger Giant] have just opened a drive-thru in Newbury. It's on the route of the current A34, on the site of the former [Y] garage, who operate many [X Burger Giant] franchises in the South. Strange that. The [X Burger Giant] was objected to and was actually refused planning permission by the same council that is in favour of the bypass .... [X Burger Giant] won on appeal and decided, weirdly, not to have an opening ceremony. Newbury activists decided to draw attention to the traffic-generating nature of the 'Drive-thru' [X Burger Giant]. About 10 activists, including Daisy the Cow and the 'Smashed Pumpkin man' were joined by about 15 press and 30 police, but none of them was tempted by the Big Whopper.*[23]
>
> *Over 30 anti-Newbury bypass campaigners attended the Annual General Meeting of troubled construction company [Z] today [Friday 6 September]. Critical questioning of the company Board eventually led to a commitment from the Chairman to review the controversial road schemes.*[24]

There could therefore be greater consideration of *promotion*. The reason that promotion is underplayed could be the lack of differentiation of construction services. Strategic marketing is needed to do that and, coupled with carefully targeted selling, is effective. This engenders an approach that will then filter into promotion generally. Carefully targeted promotion will prove effective. Broad-brush advertising and promotion will be a financial black hole. Contractors and consultants are wise to steer clear of that. The right approach will yield positive results and will provide an opportunity to be ahead of the competition.

Promotion, in the form of advertising, could also be used effectively on products and services, which have divided up into structural divisions. Advertising can be used until such time as the organisation decides to restructure to permit a more pro-active approach. For many companies, it could be that structural solutions will continue to work. This is more likely to work with the marketing mix, especially where a routinised market position is adopted. Advertising could provide a strong message about a standard, tried and tested service. In the product area, TV advertising has been used successfully by housebuilders.

Design and build is a good example of a frequently standardised service. Some contractors offer guaranteed maximum price, others specialise in building and project types, and some compete on price alone.

Price

Price competition is a 'mug's game'! It reaches most industries once they reach maturity and their product and service becomes like any other bulk commodity. The *commodity slide* means that selling is on price and price alone.[25] The commodity slide concerns products that mature over time in response to a maturing market. Contracting has yet to mature in the sense of the product life-cycle or it has simply jumped over most of the interim stages. Differentiation is the solution once again. It limits competition and should tailor the services more closely to the needs of the client: a classic win–win situation.

What is there to say about selling and price? Most has already been said. Discounting is the road to ruin. Many contractors and consultants start to invest in new services, become impatient and go back to discounting the price without waiting to reap any of the rewards of the initial investment.

The lack of differentiation means that sales people are unable to sell any particular feature or benefit. It is a situation that tends to lead to a lack of integrity as the sales person, anxious to meet their targets, sells whatever they think the client wants to hear, regardless of the real capabilities of the organisation to deliver the goods. This is a classic lose–lose situation, as client expectations reach sky high with no hope of meeting them.

The sales person has his hands tied behind his back, for most do not even have any input into price. They are mere door keepers – not that there is anything wrong with this, but corporate expectations of those selling is also unrealistically sky high!

The required solutions are strategic, not tactical in these instances. The sales person has a vital role to move selling and negotiations away from price onto the features and benefits. The client will always look after their costs at the end of the day. They have to be given good reasons for not being as price conscious as they are today. The market is highly fragmented and

... On the case 29

One regional contractor of a national group won a contract for a prestigious leisure facility. The contractor did not have a track record with the building type.

The contractor had the opportunity to tender, but did not come out the cheapest. It did, however, offer to complete the project several weeks earlier than the competitors, if the client was willing to pay an additional £10 000. The client accepted this offer, trading off the additional capital cost against the additional income of £60 000 they were projected to earn over the extra weeks they would be open.

price encourages this. Market fragmentation of suppliers all selling the same thing is the worst possible sales environment. Limited suppliers all selling different things is the best possible situation.

On occasions, the sales person can trade off price for a feature, as *... On the case 29* illustrates.

Conclusion

In this chapter, a number of case and hypothetical examples have been given for the 4Ps of the marketing mix. It is viable to use this approach for contracting and in the consultant market. There is more scope in the construction sector as a whole than in many others. This is simply because the sector has yet to prove itself in these ways.

Change in the external environment has been encouraging contractors and consultants to address the markets in new ways. Continuing *differentiation* of services between companies and *concentration* of supply into fewer hands is the foreseeable way forward. Service differentiation does not mean that organisations need to develop an extensive palette of services. That is unrealistic in an industry that tries to keep overheads low. It is also unrealistic in the sense that few organisations are able to sustain a competitive advantage in more than about three areas. It is equally unrealistic to expect sales people and all those involved in selling to continue to sell into a market where there is frequently little to chose between one supplier and another.

One cautionary note is needed. If sales people require something to sell that is genuinely different, that habit of seeing what the competitor is doing and then copying has to be resisted. That zero-sum game will not work in the long run. The sector benefit comes from each organisation differentiating in varied ways. That is what will bring about *differentiation* across the sector and hence reduce fragmentation and market competition in the sector.

The marketing mix offers considerable scope still, yet it does also have limits. The contractor is likely to be fearful that they will focus on a small market area. Not every market of differentiation can be lucrative or substantial in size. That is right when considering just four variables. However, every client has a different profile and each project has different needs. Just as each personality is different and the way we do things is different, so is each client. Servicing the needs of clients in this way offers a greater scope for differentiation. In so doing, the selling moves almost totally away from the product and gearing the services to meet that end. The emphasis shifts to the transitory and intangible service. It shifts to the relationship between client and contractor and other parties, such as the design team. This is where the marketing mix gives way to the *relationship marketing* approach.

Summary

1. The purpose of this chapter has been to demonstrate selling using the marketing mix approach:
   - There are opportunities to use the 4Ps in a creative way
   - Sellers have as many problems in selling stemming from their own organisations as they do from constraints in the market.
2. The key to selling is to get clients to focus on the service features and benefits. The constraints facing contractors, and to some extent consultants, are:
   - Fragmented market
   - Lack of strategic service differentiation
   - Consequential lack of price differentiation.

References and notes

1. Kotler, P., Armstrong, G., Saunders, J. and Wong, V. (1996) *Principles of Marketing*, Prentice Hall, London.
2. Jain, S.C. (1993) *Marketing Planning and Strategy*, South Western, Cincinnati.
3. *RIBA Journal* (1995) The anatomy of the architect, March.
4. Baker, M. and Orsaah, S. (1985) How do customers choose a building contractor? *Building*, **31**, May, 30–31.
   CIOB (1993) *Marketing and the Construction Client*, CIOB, Englemere.
5. Connaughton, J.N. (1994) *Value by Competition: A Guide to the Competitive Procurement of Consultancy Services for Construction*, CIRIA, London.
6. *Marketing Business* (1996) Bidding tricks, December–January, 24–26.
   CIOB (1993) *Marketing and the Construction Client*, CIOB, Englemere.
7. *Construction Manager* (1997) Why are design and build clients unhappy? *September*, **3**(7), 24–25.
8. Akintoye, A. (1994) Design and build: a survey of construction contractor's views, *Construction Management and Economics*, **12**, 155–163.
9. Levitt, T. (1983) After the sale is over, *Havard Business Review*, September–October, 87–93.
10. Cravens, D.W. and Piercy, N.F. (1994) Relationship marketing and collaborative networks in service organizations, *International Journal of Service Industry Management*, **5**(5), 39–53.
11. Taylor, B. (1994) *Successful Change Strategies*, UK Director Books.
12. Saunders, J. (1997) Distribution, innovation and the consumer, *Admap, May*, 22–25.
13. Thompson, N. (forthcoming) Research notes undertaken for PhD on relationship marketing and the conditions of trust, Oxford Brookes University, Oxford.
14. Valikangas, L. and Lehtinen, U. (1994) Strategic types of services and international markets, *International Journal of Service Industry Management*, **5**, 72–84.
15. Watson, K. (1997) No hiding place, *Construction Manager*, February.
   Knutt, E. (1997) The scouts, *Building*, 21 February.
   Preece, C.N., Putsman, A. and Walker, K. (1996) Satisfying the client through a more effective marketing approach in contracting – a case study, *Proceedings of the 1st National Construction Marketing Conference*, 4 July, The Centre for Construction Marketing in association with CIMCIG, Oxford Brookes University, Oxford.
16. Levitt, T. (1983) *The Marketing Imagination*, The Free Press, New York.
17. Fisher, N. (1986) *Marketing for the Construction Industry: A Practical Handbook for Consultants and Other Professionals*, Longman, Harlow.
   Pearce, P. (1992) *Construction Marketing: A Professional Approach*, Thomas Telford, London.
18. Preece, C.N., Moodley, K. and Graham, C. (1996) Public relations to support more effective marketing in UK construction contractors, *Proceedings of the 1st National Construction Marketing Conference*, 4 July,

Centre for Construction Marketing in association with the CIMCIG, Oxford Brookes University, Oxford.
19. Wilkins, J. and Ford, K. (1997) Detecting the effectiveness of integrated marketing, *Admap*, June.
20. Smith, P.R. (1993) *Marketing Communications: An Integrated Approach*, Kogan Page, London.
21. Centre for Strategic Studies in Construction (1989) *Investing in Building 2001*, University of Reading, Reading.
22. One example, for the second runway at Manchester International Airport is Car2 (1997) *Coalition Against Runway* 2, http://www.mfoe.u-net.com/car2.htm.
23. Friends of the Earth (1997) *Stop the Newbury Bypass!*http://www.foe.co.uk/action/newbury/index.htm, 13 June.
24. Third Battle of Newbury (1997) *Welcome to the 3rd Battle Of Newbury*, 13 June.
Friends of the Earth (1996) *Costain Agrees to Review Newbury Bypass Contract*, http://www.foe.co.uk/action/newbury/index.htm, http://www.foe.co.uk/pubsinfo/infoteam/pressrel/1996/19960906184835.htm
25. MacDonald, M. (1995) *Marketing Plans*, Butterworth-Heinemann, Oxford.

Chapter 8 Selling Through Relationships

Themes

1. The *aim* of this chapter is to explore the process of selling contracting and consulting services through relationships.
2. The *objectives* of this chapter are:
   - ❑ To present the dimensions of selling through relationships
   - ❑ To explore the dynamics of selling through relationships
   - ❑ To describe some of the issues and processes involved under:
     - ■ Different client-handling models
     - ■ Various project stages
     - ■ Different procurement routes.
3. The primary *outcome* of this chapter is to show how relationship marketing theory translates into sales practices.

Keywords

Relationship marketing, Decision-making unit, Conditions of trust

Introduction

On entering a men's fashion store, the beady eyes of the eager salesman immediately pounced on me and he started to motor in my direction. I had barely got past the door and there he was, right in my face. I had come to buy a suit for work and I had not as much as a second to glance at any clothes. He was polite and helpful, asking, 'Can I help you?' ... 'What sort of suit did I want?' ... Did I know my size?' ... and so on! He scored ten out of ten for sales efficiency – no question about that, but was he an effective sales person?

I recall another experience, this time in a large department store. The family had been kitting out the kids. We had bought several items for the three children. We had been good customers, as we had spent quite a lot of money with them that day. Suddenly we remembered that we needed to contact someone urgently. We could telephone. We neither had a phone card nor the right change on us. I asked the cashier, who had just served us, if she would change the note I had. I was told very abruptly, although not rudely, that she could not. I pressed her, saying we wanted to use the phone in the department store. She said, 'It is not company policy to give change from the till.' I said that I understood she was doing as she was told, yet reminded her of how much we had spent in the store and from our point of view we were in a bit of a fix: 'Could you please make an exception on this occasion?' At this point, she was very abrupt and informed me that I was holding up business and others were now waiting to pay for their goods. I became quite irritated at this dismissal and started to ask if there was a manager I could see. My wife wisely thought it was better to make a quiet exit and persuaded me to do this.

'The customer is always right' – a frequently quoted cliché. But is the customer always right? Clearly, there are times when the customer is not right. In construction, we have all come across occasions where the client has demanded the architect to make a design change and subsequently protests about picking up the design and construction costs flowing from their instruction.

The cliché that the customer is always right is not about precision, but about attitudes. In the examples, both people were doing their jobs. In the first story, the salesman wanted to emphasise the product as quickly as possible. He wanted to get me into the place where the suits in the right price range could be identified, where the features and benefits could be shown, where the right fit in the right style could be found. In the second case, the selling and exchanging of goods had to be managed as a quick and efficient service to all. Hold ups were to be avoided and the policies in place streamlined the approach.

In both cases, one ingredient was left out – you are dealing with people who have feelings and are sometimes subjective in the way they behave. They have needs and like to conduct their buying in particular ways. One of the great management success stories over the last few decades is the growth of what Drucker calls the pastoral churches.[1] While some mainstream denominations, mainly in the developed world, have been losing people by the score since World War II, the charismatic and evangelical streams have been expanding at exponential rates. They have stuck to Biblical principles, yet have asked, 'Who are our customers?' and, 'What is of value to them?' Perhaps there is something to be learnt from this and other examples. Yet, in society today, we have programmed ourselves to think in terms of functional roles. After being introduced to someone socially, the next question is invariably, 'And what do you do?' This is about people's functional role and not who they are. Yet relationships are founded on *who people are*. This has been stated in a slightly different way earlier in the book, yet it is so important. It is central to selling through relationships.

In my first experience, I was being sold to, despite who I am. I was just another sales target walking through the door, another person to try out the latest range of step-by-step sales techniques. For me it is a real put off. We all like to think that in some way we matter, and so I like sales people to treat me personally. Now, the aggressive approach must clearly yield results or it would not be used. A standard solution that works, using the marketing mix, is highly appropriate for some market positions. I like to browse and then ask for help. I like to begin to form my opinions about the product and then draw in the sales person for advice and comment, at which stage they can use all their techniques, because I am hooked by then. The selling is built around the relationship, however cursory it may be in that case.

In the department store, I had asked for something out of the ordinary. It simply did not fit the selling and exchange process. Yet the approach taken, I felt, discarded me. The result is a loss of goodwill. I have been back to that chain of department stores since, but I use them less than I used to do.

In the case of church growth, the streams stick to their Biblical principles. They are not in the business of compromise or allowing their integrity to be diluted. It has been said in a business context that all around things change but that the organisation must hold onto its essential values and *raison d'être*.[2] This defines who you are or what your organisation is. Then, you see how you get on with people by exploring common ground and mutual interest. If the relationship is developing, the expectation is for a lasting friendship. In essence, this is the process the sales person is initiating with the customer under the theory of relationship marketing. It is the first step in selling through relationships. As in the social context, the hope is that a relationship will last a while, perhaps a long time. As Levitt states:[3]

> *The relationship between a seller and a buyer seldom ends when a sale is made. Increasingly, the relationship intensifies after the sale and helps determine the buyer's choice the next time around. Such dynamics are found particularly with services and products dealt with in a stream of transactions between buyer and seller – financial services, consulting, general contracting, military and space equipment, and capital goods.*

The sale then merely consummates the courtship, at which point the marriage begins. How good the marriage is depends on how well the seller manages the relationship. The quality of the marriage determines whether there will be continued or expanded business, or troubles and divorce. In some cases, divorce is impossible, as when a major construction or installation project is underway. If the marriage that remains is burdened, it tarnishes the seller's reputation.

As the relationship begins, the foundations for the future are built. This does not mean that all will go smoothly in future. But if the foundations are unstable from the outset, it will be sure to go wrong and that is precisely the history of adversarial contracting. It is not quite the same picture in consulting, but that has become more cut-throat in recent years. If organisations sell through relationships, it has to be done well. That task appears difficult. Many extol the value of relationships, as Levitt does above, yet the irony is that in Western culture more and more people are demonstrating their inability to sustain meaningful relationships. Divorce and serial marriage are the norm. It shows our belief in relationships, but shows our shortcomings too. The marriage analogy, used by Levitt and others, is not perfect. In marriage, the relationship is an end in itself. While children may come out of it, neither the initial relationship nor the children are a commodity. In business, the relationship is a means to an exchange. The demands of the relationship are less, but some of the same relationship principles still stand; effective communication, understanding, trust, availability, a willingness to serve and so on, all being crucial. One important difference raised by Levitt is that the client will feel that the contractor or consultant owes them a favour as soon as the contract is signed. The client has bestowed the favour, rather than the contractor seizing the opportunity. Therefore, the service provider owes the client.

Firm foundations

From the outset, the sales effort has to instil into the client the hope of a stable and lasting relationship. There are four foundations to be established beneath the surface of the relationship. These are:

- Vision
- Expectations

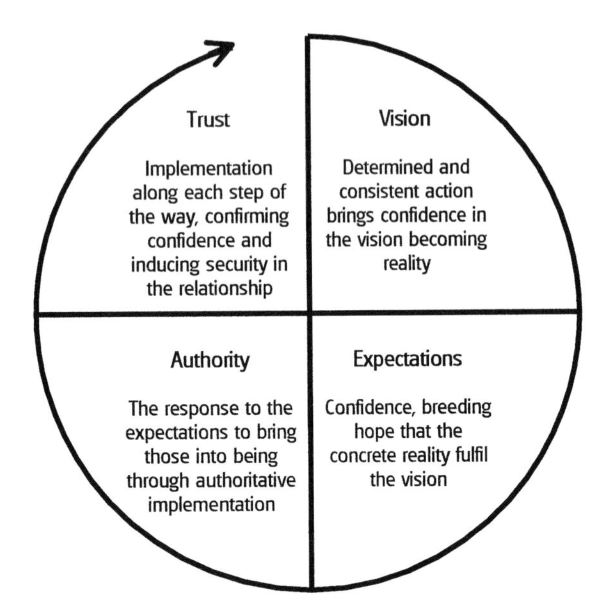

Figure 8.1 Foundations of relationship selling.

- Authority
- Trust.

Selling *through* relationships implies more than selling *to* people. There is an intimacy involved, which involves knowing the client organisation and the decision-makers within the organisation. The human dimension is emphasised. People in the client body need good reasons to consider your services. The *visionary* aspect grabs their attention. This is not easy. The client prospect may not have any urgency to see you or invest in relationship building. Yet some time is needed, coupled with market research in the broadest sense, in order to understand what it is that might envision them. The sales job is made a great deal easier if your own organisation is clearly differentiated, because then you have something distinctive to try to sell. Through the client response, you can gauge what is of interest to them. At this stage, the service selling is a softly, softly one, for the aim is to tease out their needs and see how they work. It is not to secure a sale or get on the pre-qualification list. The objective is to see which service dimensions they require. There are two dimensions:

1. Quantity – giving more for the money
2. Quality – enhancing what is already achieved by doing the same things better or taking out the 'hassle' factors that cost both parties time and money.

In both cases, it adds up to a good service. The client will already be evaluating what they could get out of using the engineering services or construction management capabilities of the organisation. They may have already identified something quite specific, such as a value management dimension to the service or the in-house graphics department of an architectural practice that will enhance the important signage in the building. In other words, *expectations* will begin to form. There are now hopes to fulfil. It is at this stage that most contractors and many consultants fall down. Consultants, especially engineers and architects, complain about the poor briefing from clients. However, it is at the early sales stages that the vision is created and the expectations formed. Implementing that is just as much the consultant responsibility as it is a client responsibility. So when late changes are made, they may arise out of poor briefing, but they also arise out of consultants not continuing to pursue and to get the client to articulate their expectations in relation to needs. It can be seen that these steps in selling through relationships do not have clear starts and ends. They are iterative with aspects overlapping.

Superstructure

So contractors consistently fall short at the *expectation* stage. They help to build up the client expectations during selling, as has been discussed both above and in previous chapters. They then let the client down during the contract. This leads back to the client-handling models. Selecting the right one and implementing the model is vital. Authority is lost without it for the expectations as represented by the 'baton' in the *relay team model* or the person in the *account handler model* are lost. The role of authority is seldom talked about in respect of relationship marketing. There is almost an underlying assumption that it is an anathema. Actually, it is essential. The underlying assumption has its roots in the delayering and flat management structures, coupled with empowerment that has come to the fore in recent years. It has already been discussed that there is a need for personnel to be given the power and to have the responsibility to act in accordance with the needs of clients within realistic and common sense boundaries. Building on that, the contractor or consultant must corporately act at the various management levels to carry authority and develop relationships. This is true in sales, and indeed, for other roles too.

This authority must be supported and reinforced by the actions of senior management. This can be illustrated by looking at sharing information. Relationship marketing needs good communication and issues have to be shared. In some relationships, such as partnering, the sharing can be too widespread. Stable relationships are built as much on confidentiality, and hence respect, as they are on openness and thus sharing. In fact, confidence and confidentiality are closely related. How this works is illustrated in Figure 8.2.

The figure emphasises the respect between organisational boundaries. The greater the level of openness, the greater the scope for sharing. Typically, senior management will tend towards a more secretive approach, and may impose constraints on the openness of more junior staff. Sometimes this is necessary, sometimes not. We are acknowledging the richness and frailties of human relationships and trying to manage this to best possible effect.

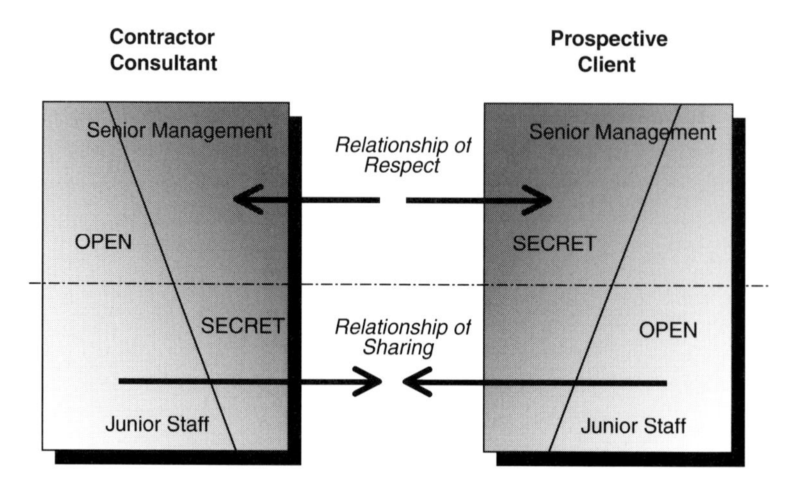

Figure 8.2 Confidence and confidentiality in relationship selling.

Building relationships

In Chapter 6, the types of relationships were looked at – the 30Rs. Also considered were the steps in drawing in the client, namely the 5Es:

- Encourage
- Exhort
- Enlighten (educate)
- Engage
- Empower.

The 5Es steer the sales process through relationships. In the marketing mix only the first two were considered (see Chapter 7). *Encouragement* is about getting alongside so that the sales person can listen and respond. The desired outcome is a befriending – the beginning of relationship. *Exhortation* is the next step, but can only be moved to as the first one is complete. Exhortation seeks to steer the agenda and direct the client. Confidence in the contractor or consultant is necessary for this to work successfully. *Enlightenment* is the third stage. This involves envisioning and educating the client. The 4Ps anticipate closing the sale before enlightenment begins. Demand has been latent until now, so the client knows that what they want is merely matching requirements to the product. In construction in its broadest sense, the sales negotiation is just beginning. Enlightenment helps to build dependence and the client is now buying into the relationship. There is mutual commitment to exploring what will best satisfy the client. As Gummesson[4] states:

> *We accept the ignorance of the customer. We are afraid of educating the customer; that is not the business of business. In my view it is not only an option but a privilege to educate the customer. It opens up the means for improved services and innovations.*

Facilitating the relationship at this stage in order to tailor the service to meet needs as these are becoming fully understood by both parties is engagement. There may be several stages to this – at the sales stage, during briefing and on site – as it is a process. Engagement is more than creating a committed customer. It is about:

- Enthusing them
- Inducing mutual trust

- Their adoption of a few of your ideas, no matter how small or grand these are
- Working together to ensure that their plans are implemented, so interdependency is created.

Both parties, perhaps more accurately those involved directly with each other, need the power to move the relationship forwards – to act on these issues. This is the final stage of *empowerment*. There is compliance to the objectives of both parties. Responsibility for the outcome of the relationship, and projects too, is taken up and initiatives got under way. As selling proceeds:

- Use the 5Es to inform your sales effort
- Do not move from one to the other without a clear understanding that the client is with you
- Moving on too quickly means your message is in danger of being missed – if you close the sale, there may be a mismatch waiting to be exposed
- Monitor the relationship.

Dare to ask directly how things are going – this can be highly informative at times; for example, a client may say they want a close relationship with contractors. If you ask someone how things are going as you prepare to move from enlightenment to engagement, you may discover that they want a relationship as close as the one you have provided, but no further. You would then be into overkill! The question now is whether they are a key client, a secondary client or to be treated opportunistically henceforth.

Gronroos[5] says that implementation requires shifts in the focus of management:

- A client focus, that is 'putting yourself in the client's shoes'
- Long-term and regular contact
- Adding value that is perceived to be quality from the client's perspective, which is going beyond technical managerial competence.

This is a much more targeted approach and therefore more costly per client from the outset. A principle of selling through relationships is the search for loyal clients who will bring long-term business. There are start up or transfer costs in sales. There are also costs across management, which will tend to increase as the relationship builds. Are these costs worth it and are there savings elsewhere to offset the additional expenditure? The savings are the trade-off costs. Savings are made against opportunistic pre-qualification and tendering. The investment costs depend upon perspective. The contractor, consultant or supplier can look at the costs in the following way in a step by step approach to investment:

1. *No cost*: the easiest to implement because the organisation already does this and it may be a question of *extending* sales into current segments of market strength
2. *No cost unless the service is used*: this is easy to implement as it plays on developing or *intensifying* existing strengths, but yields least competitive advantage and is easily copied
3. *Standing costs whether the service is used or not*: this requires comprehensive development and restructuring, not only of the sales effort, but also of the service delivery.

Completing the whole cycle shown in Figure 8.1 involves embracing the restructuring costs in option three above – the only sustainable option of competitive advantage in the long term. However, it has already been stated that it costs five times more to find a new client. It has also been stated that repeat business in construction is around 20%, compared to 60% in other service sectors. So while unit sale cost may go up in the short run and turnover may fall as opportunistic response is constrained, the long-run sales costs fall and turnover and profit increase.

The 5Es mirror what the client is experiencing, as shown in Figure 8.1. How do these relate to each other? Figure 8.3 provides some guidance on this. Reaching a trustworthy relationship is a key. The book will look at conditions that make for trustworthiness at a later stage; however, it has been found that consumers who are able to trust sales people exhibit more integrative bargaining strategies, which lead to benefits for both sides.[6] In other words, they are more likely to understand the whole picture and seek service features that meet their needs. They will be more satisfied consumers. If this is the case for consumers, clients to the construction industry are even more likely to ask for project enhancement and to give repeat business.

Trust is therefore a key ingredient to relationship marketing and hence in the sales process, just as repeat business and increased referral business are key outcomes desired. The Centre for Construction Marketing has found that 15% of people are seeking to build closer relationships with their clients.[7] Trust is an important element and has been so characteristically lacking in the construction process. Trust, then, is a vital condition for relationship selling, in order to achieve the objectives in relationship marketing.

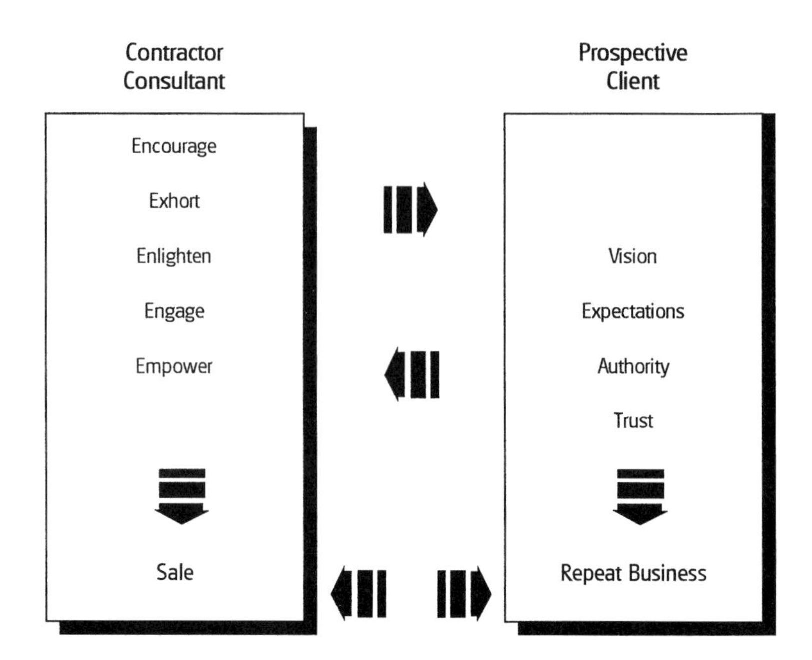

Figure 8.3 Supplier and client perspectives of relationship selling.

Trust is a pre-condition for a successful contract from this perspective. So, trust is an important ingredient to add and hence change the characteristics of some players in construction.

Does *trust* help to secure the sale? There is little evidence to show that it does.[8] Trust may be a central part of selling through relationships, but there is little incentive for sales people to act in a trustworthy way, unless that is built into their job accountability and evaluation. In fact, being trustworthy can be time-consuming and hence costly. It can also be a burden.[9] As was demonstrated in Chapter 6, relationships do not automatically yield results. They have to be managed both in terms of the investments costs and delivery of the service and hence the outcomes (see Figures 6.1 and 6.2).

It is a good management principle to monitor critical events or moments of truth with clients who are potential sources for referral and repeat business (see ... *On the case 30*).[10] This not only provides a valuable resource for improving the service, but is also excellent information for a sales person to use in discussions with clients, perhaps to relate how an issue arose and how it was resolved in practise. Table 8.1 sets out some guidance on this in order to provide a starting point for those wishing to take

... On the case 30

A major consulting organisation holds client meetings every two weeks to discuss the clients. The current work is put to one side and the focus is for each team member to discuss what they have learnt about the client organisation from top to bottom since the last meeting.

Reporting is also carried out to every client staff member.

The agenda is twofold: keep abreast of responding to the client and looking towards the next engagement.

the ideas forward. This can be done while recognising that trust is a key to selling through relationships in order to achieve the objective of increase referral and repeat business. Trust is initiated at the sales stage, yet neither trust in the contractor corporately nor trust in those selling the services secures contracts. Developing trust is dependent upon the client being able to:

- Assess the costs and benefits of trusting the contractor or consultant
- Confidently predict their behaviour
- Confidently predict that the contractor or consultant will deliver their promises
- Assess the motives of the contractor or consultant
- Confirm their assessment from third parties.

Table 8.1 Record of critical event management

Event 1	Event 2	Event 3	Event 4	Event 5	Event *n*
Client 1 Client 2 Client 3 Client 4 Client 5 Client 6 Client *n*	Details by sales or contract stage, type of event, personnel, and (re)solution Analysis of totals will help yield performance results on organisational strengths and weaknesses				

Source: adapted from Storbacka *et al.* (1994).[11]

Being cautious about costs is important, especially if transaction lies between a marketing mix and relationship marketing approach. Some guidance was given earlier on a three-step transition. It is also important to reinforce a further point made prior to this chapter: select key clients and remain opportunistic with the rest. Selection should take account of rewards. Using key clients alone enables sales budgets and time budgets to be established, especially for the client.[10] The initial higher sales costs are to be recouped through:

- Increased turnover
- Higher or premium profit margins
- Lower operating costs through increased effectiveness in working with client
- Less overheads, especially claims and legal fees, plus a reduction in long-term sales costs
- Shallow project learning curve for all clients in the same segment of generic requirements and especially where there is repeat business.

It is helpful to underscore the *radicalism* of selling through relationships. While relationship marketing does seek to recreate what has been lost in the growth of the corporation, it flies in the face of corporate economics. The human dimension to selling through relationships embraces sociology, psychology, anthropology in an intuitive and interpretative way, not in a 'rational' way. Or as Levitt[12] puts it:

> *'This takes more than what comes normally in good marketing. It takes special attention geared to what uniquely characterizes a relationship. That is* **time**. *The economic theory of 'supply and demand' is totally false in this respect. ... To be under these circumstances, a good marketer in the conventional sense is not good enough. When it takes five years of intensive, close work between seller and buyer during which the previous signing of a contract is converted into the 'delivery'... . more will have to be done than in the kind of marketing that landed the contract at the outset. The buyer needs reassurance at the outset that the two parties will live congenially together during the long period in which the purchase is transformed into delivery.'*

Repeat business

Referral markets and repeat business are the best ways to get future business. Referrals occur on a regular basis, but are outside the control of the consultant and contractor unless there is a partnering agreement. Repeat business is in direct control of the client, yet through effective client management there is considerable scope to influence where future business is placed (see Figure 6.2). The importance of repeat business is illustrated by ... *On the case 31*.[13]

Current research shows that considerable effort in investment and time is necessary to achieve high levels of repeat business in construction. On the face of it, the decision of the client to appoint the same contractor for a further project is strongly influenced by past experience and perceptions. Perceptions feature more strongly than experience. Therefore, advocacy in referral markets is more forceful than direct experience.[14] This suggests that either clients have very strong normative beliefs – what *ought* to happen – which will take a great deal to shift, or that they are heavily influenced by the advice of their representatives. Both forces will be at work and the balance will change from client to client. Regular procurers of buildings, and therefore the ones with the greatest levels of potential repeat business, are more likely to be led by their normative beliefs. The investment in selling through relationships may therefore be high, but the rewards are high and the degree of market protection afforded by those beliefs will maintain a stable supply of construction work. Consultants tend to enjoy this situation more frequently and contractors have something to learn from these interdependent relationships.

Decision-making unit

The second example shown in ... *On the case 32* points to the relevance of knowing people in the organisation. In that case, the loyalty was with the individuals and that was the source of repeat business, regardless of their employer. Relationships are always first and foremost with people. In one sense, it always remains like that. At some stage there is the possibility that the relationship with the individuals becomes embedded into the client organisation. It is as if the track record of the contractor or

... On the case 31

A UK subsidiary of an American auxiliary equipment supplier is the largest amongst six major competitors. Over a 4-year period, the following dimensions of repeat business were observed:

	Year 1	Year 2	Year 3	Year 4
Sector growth	1.77%	1.35%	0.68%	−0.98%
Company growth	2.89%	0.97%	−0.26%	−1.92%
Defections	6.66%	7.19%	7.41%	8.09%
New business	–	−12.12%	−11.41%	−14.02%

Customer by age in Year 4:

New business	54 906 units
Customers of 1–2 years	446
Customers of >2 years	49 613
Defections	1120

Long-term business is clearly the most important source of work, but is increasingly being eroded by defections each year. New customers are not plugging the gap and hence the organisation is in decline, and entered into decline before a turndown in the sector.

The case illustrates the importance of existing clients. A company can find it difficult to attract new customers, yet maintain a growth path via its existing client base. It can even grow in markets of decline if it is effective at attracting defected clients from competitors.

consultant becomes so rock solid that the organisation as a whole is prepared to put their trust into that architect or contractor until they are given some reason to do otherwise. The trust is inherited into the formal decision-making and structures of the client, even though the decision-makers may change.

This is an ideal position to be in and is a long-term goal. The important lesson, however, is that it starts with individuals: building relationships with the *decision making unit* (DMU). Up until this point the book has focused upon contractors and consultants in

... On the case 32

An international architectural practice, specialising in public buildings, had neglected its overseas networks of contacts with a consequential fall in the workload. A new partner was appointed, seeing it as his job to resurrect those contacts.

He found it relatively easy to manage this. While the relationships had lapsed, the reputation of the practice was still intact. He was able to use that reputation to establish his own credentials and work began to flow again overseas.

In this case, the relationship hung around the formal arrangements.

A commercial international architectural practice undertook an audit of existing clients. Repeat business was thought to be high. On close scrutiny, it was found that the client personnel brought in the repeat business, not the corporate client body. Therefore, when key client personnel moved jobs, the architectural practice was introduced to work for the new client through the contact.

In this case, the relationships worked most effectively at an informal or personal level.

their search for competitive advantage and for work. The role of consultants in getting work has been underplayed from the stance of the contractor. This is the best place to address that position, for the decision-makers for the client are not always in their own organisation. They include the client's representative and those influential in the design team. This total group of people can be collectively called the DMU.

Each client or stakeholder, therefore, has people who at one time or another make decisions that will have an impact upon the selection of a contractor. Knowing all about the DMU is the raw material source for selling through relationships. The DMU needs to be:

- ❑ Mapped
- ❑ Profiled.

Mapping

Mapping the DMU is simply a matter of determining who the people are. There will be levels of *mapping*, for example from a board member who might rubber stamp a decision to a day-to-day member of the estates department who manages the project on the client side. There will be others in between plus architects, cost consultants and other stakeholders who may have some influence.

Members of the DMU may be allocated to different people in the contracting organisation. It is part of the role of selling through relationships for the sales person to decide who should look after or 'own' which member of the DMU. One rule is to match people at corresponding levels within the organisational hierarchy and this is frequently achieved. What is more difficult is to select the right person. Everyone is a part-time marketer or sales person, so in theory everyone is available. Matching people by character and interests is a further level of screening that can be used. Business lunches and other contacts can be used to help establish where the right 'chemistry' exists and between whom. The sales person may need to facilitate several activities in order to establish the most appropriate matches. Matching areas of expertise is another way, especially if the client needs are technically or managerially precise. Combinations can also be used, but the important thing is to ensure that the optimum pattern is achieved.

This is an area of weakness in contracting. Many contracting and consulting organisations do not do this. Many need to put in place a contact 'ownership' system. Those that have this tend to use the notion of who met them first, or on some occasions there is dispute between partners or directors as to 'who owns whom'. This is an unsatisfactory situation, for it creates tensions internally. It is ineffective too, for the optimum match is the one that will best serve the contracting and client organisation. Nobody would marry the first person they meet, so why match up in this way in business? It simply does not make sense. This is politically sensitive for the power must rest with the sales person with the head of marketing and the account handler if there is one to make executive decisions about this in consultation with all interested parties.

... On the case 33

A civil engineering and building contractor in the south of Britain always allocates its directors to client counterparts. At other levels, clients tend to ask for particular people and that initiative is used to match people. These issues are addressed at the Monday morning business meetings.

The selection of site teams is the decision of management. On one job, the site manager was 'firey' yet methodical. The contracts director was also 'firey' yet a good problem solver. Together they had ideal competencies but their personalities could have led to conflict. The management aim was to serve the client to the best possible advantage, so the management role was to be on top of this source of conflict.

This approach to matching personnel therefore starts at the courting stage and is carried through the contract stage. It is facilitated internally and reinforced on occasions by client breakfasts to monitor and develop the relationships.

The aggregate results are used to assess the overall position with key clients. Each client is assessed in terms of the subjective levels of perceived trust:

❑ Improved levels of trust in the team
❑ Stable position
❑ Traditional mistrust increasing.

Profiling

Profiling the DMU is the second dimension of selling through the DMU. *Profiling* is about understanding what makes them tick. There are two aspects to *profiling*:

❑ Motivations
❑ Team characteristics.

Motivations concern the individual agendas of individuals in the DMU, as observed through their behaviour. Some people have underlying character traits. Some people are performance orientated, in other words they do not have a good feeling about themselves unless they are reaching certain goals. Others are most concerned by what others think of them and their actions tend to be fed by being seen in the best light or seeing themselves in positions of power and influence. For others, being able to accept responsibility is difficult. They tend to blame others. This is common in contracting and should be noted in client organisations too. It's a recipe for adversarial relationships. The victim mentality is another category. Victims tend not to be in positions of power and influence, but when they are they tend not to articulate either their needs or their concerns. You need to invest a great deal of time with victims in order to know what is going on with them.

These approaches to life, or however else belief systems are classified, inform the behaviour of decision-makers. What does each member of the DMU want from the project or relationship? Motivations may exhibit the following behaviour:

- *Bureaucratic or technocratic*, that is being seen to be *accountable* in time and cost terms to the board
- *Career ambitions*, for example seeking to get from the relationship job promotion or job protection within their own organisation
- *Seeking approval*, for example being thought of as a 'good bloke'
- *Least hassle or quiet life*, depending on stress levels, that is wanting the day-to-day project to run smoothly from the start to completion
- *Power*, for example wielding power over contractors or securing power within their own organisation.

Different members of the DMU may have different agendas. By identifying the combinations of motivations, selling tactics for the client can be determined and personnel selected best suited to deal with the motivations identified. Satisfying the client and getting repeat business may depend upon meeting these agendas more than any other factor.

Team characteristics are a further way of profiling, this time the whole team. There are a number of ways in which this can be done. One technique is Belbin Groups,[15] which is very applicable to project type work. The Belbin method is usually applied to groups within one organisation. However, having profiled the

DMU, it is possible to analyse the gaps to select the sales and contract personnel who are complementary and thus help to build a complete team. The process can be extended to the consultant team members too.

What are Belbin Groups? They are like a psychometric profile for a team, rather than an individual. Belbin has identified the key character roles, as opposed to functional roles. The ideal team, that is DMU, contractor and design team, comprises:

- *Plant*: the visionary who is individualistic, serious-minded with a tendency to be unorthodox – characterised as innovative, inventive and creative, yet tending to pay little attention to procedures, systems plus impatience with detailed implementation
- *Shaper*: a motivated person who is dynamic, outgoing and who thrives on a challenge – characterised as over comers of obstacles who thrive under pressure, yet are impatient and tend to be confrontational
- *Resource-investigator*: a good communicator who is extrovert, enthusiastic and curious – characterised as explorers of opportunities, contact developers and adept negotiators; in fact, good in selling in the marketing mix format, yet liable to lose interest after a short while
- *Specialist*: a dedicated individual who can be indispensable as a professional self-starter – characterised as a maintainer of quality and professional standards through the application of technical and specialised knowledge, yet tends to be uninterested in the contributions of others and thus whose role is narrow
- *Co-ordinator*: an able facilitator in getting people to play to their strengths as they work to shared objectives and who is calm, confident and controlled – characterised as mature and trusting with a propensity to delegate and empower others, yet finds analysis and handling complex problems difficult
- *Team worker*: a supportive team player, who is mild, sensitive and has good social skills – characterised as someone who uses their diplomatic and interpersonal skills to draw out contributions from other team members, yet tending to be indecisive and a poor performer under pressure
- *Implementer*: a well-organised individual who is disciplined, practical and likes to know what is expected of them – charac-

terised as systematic, hard working, conservative, dutiful and predictable, yet fears change and finds it difficult to be flexible

- *Completer-finisher*: this methodical person is conscientious and accurate and pays considerable attention to detail – characterised by a determination to see things through to the end, yet tends to worry and finds it difficult to know when enough is enough
- *Monitor-evaluator*: a shrewd judge, who is prudent and considered – characterised as being a safe pair of hands who is discrete and tough when needed, yet lacks inspiration and an ability to motivate others.

While it may be possible to 'best guess' into which team roles members of the DMU fall, the use of a questionnaire is the ideal way to proceed. I have been asked several times, 'How can a contractor be asked to fill in a questionnaire of this sort?' My answer has been that it is a pity that contractors are reticent to take the initiative. Structured in the right way it can be used to:

- Show interest
- Observe more about the client
- Exchange views and learn more about each other
- Tailor services more closely to their needs.

Figure 8.4 shows how the Belbin technique can be used to sell into the client DMU and how that becomes the basis for building a team to serve the client in future contracts. It can be applied to the account handling or relay team model of client management.

Some clients are already asking contractors to be profiled using the Belbin approach. It is especially pertinent in a partnering context. It is certainly relevant in any relationship marketing context. The North Sea oil and gas fields were developed to a large degree by private organisations, each of which invested large sums of money. The British and Norwegian governments encouraged the exploitation of the fields. The oil companies did not nurture the governmental relationships and were taken by surprise when taxes exceeding 90% of market prices were levied. Nurturing the relationships with government, politicians and voters may have reduced the tax burden.[16]

One major UK water utility organisation operates a partnering approach to contracting through the implementation of framework agreements. Contractors are selected from a small panel of six partners. Each partner is required for its staff to undertake a Belbin questionnaire, so that the most appropriate contract team

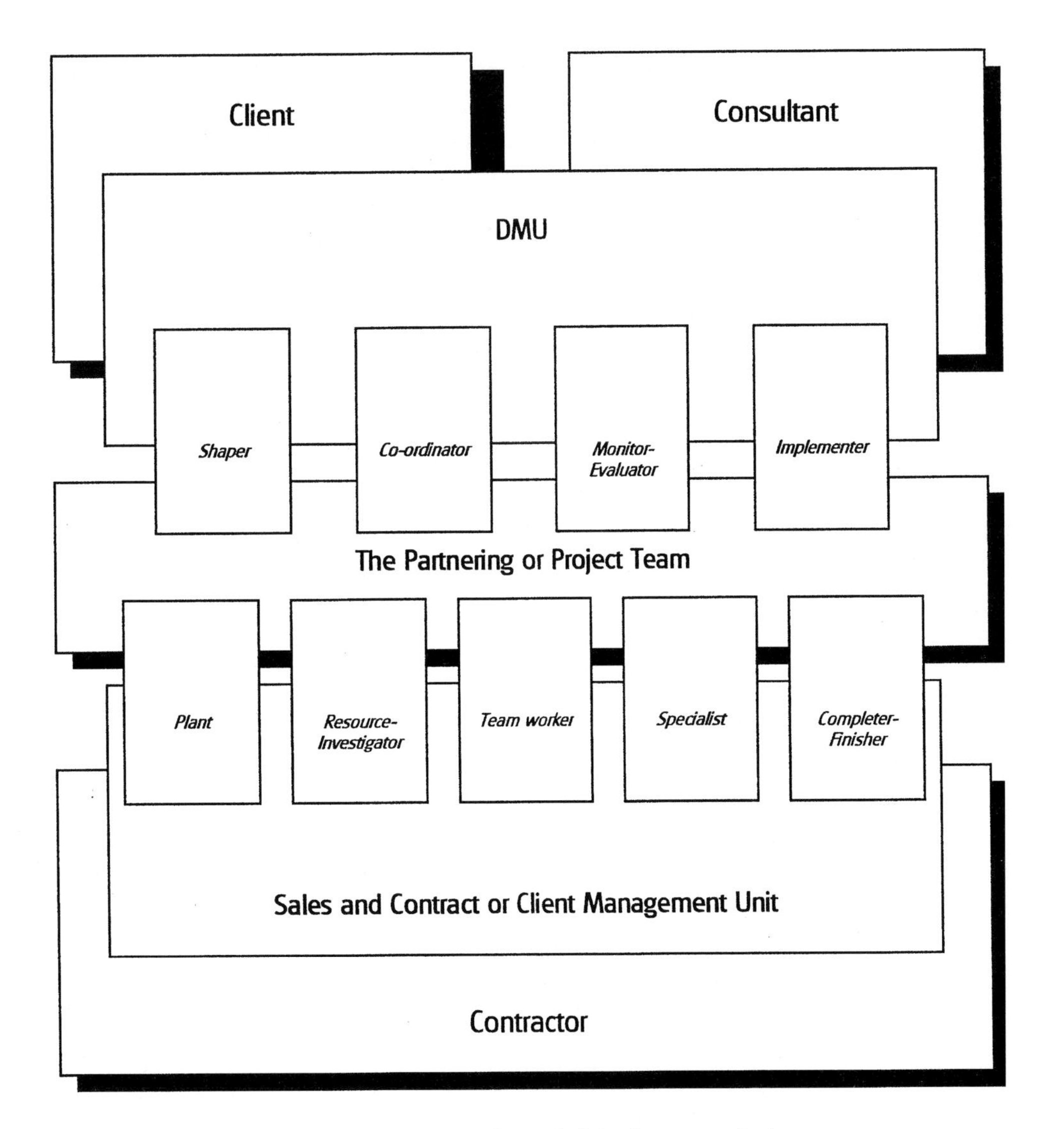

Figure 8.4 A hypothetical team derived from Belbin Group analysis.

can be chosen for all partnering and some D&B contracts. The client demands that the same personnel is retained throughout each contract, in order to preserve the integrity of the team.[17]

Within the DMU, it is important to evaluate where the power lies. Who is it that is responsible for the final decisions? Is contractor selection a team effort; is it made by the board or a senior manager? Who is responsible for day-to-day contact once a project gets underway? This will help in selecting the right people. For example, a resource-investigator would not make a good relationship marketing sales person on their own and would not get around to negotiating a contract or achieving concrete results paired against a plant. Again, it would be a recipe for disaster to have a site manager, who is an implementer, paired with a shaper in the DMU. The two would surely clash, yet they will be continually working together. Of course, there are bound to be conflicts in the team. The Belbin technique is not about harmony. It is about working together, the acceptable creative tensions being used to realise mutual aims:

> *When it takes five years of intensive work between seller and buyer to 'deliver' an operating chemical plant or a telecommunications system, much more is required than the kind of marketing that simply lands the contract. The buyer needs assurance at the outset that the two parties can work well together during the long period in which the purchase gets transformed into delivery.*[16]

Determining who the decision-maker is at the sales stage is vital. They act as a gatekeeper to the organisation. They also filter information to other members of the DMU. Independent approaches to different DMU members are important; however, the gatekeeper should be aware of all that is happening. Indeed, it provides yet another opportunity for contact. It is important to ask for their help to gain wider access if at all possible, but in any case to keep them informed so that they do not feel their role is being eroded. However, they may operate a mirror image of the *little black book syndrome* (see Appendix A, Paradox 9), so ensure that contact opportunities are not squeezed out. What actions can be taken?

- ❑ Map the DMU:
  - ■ Who are they?
  - ■ What is the power structure?
  - ■ How does it work?
- ❑ Profile the DMU:
  - ■ What are the motivations of each member?
  - ■ What are their interests?
  - ■ What are the team characteristics of each member?

How can the sales team and the contracts teams be put together to dovetail the DMU profile? It is important to understand the power relations on the contractor or consultant side. This will affect the ability to co-ordinate the sales effort across the functional roles. In short, the good client relations must be mirrored within the

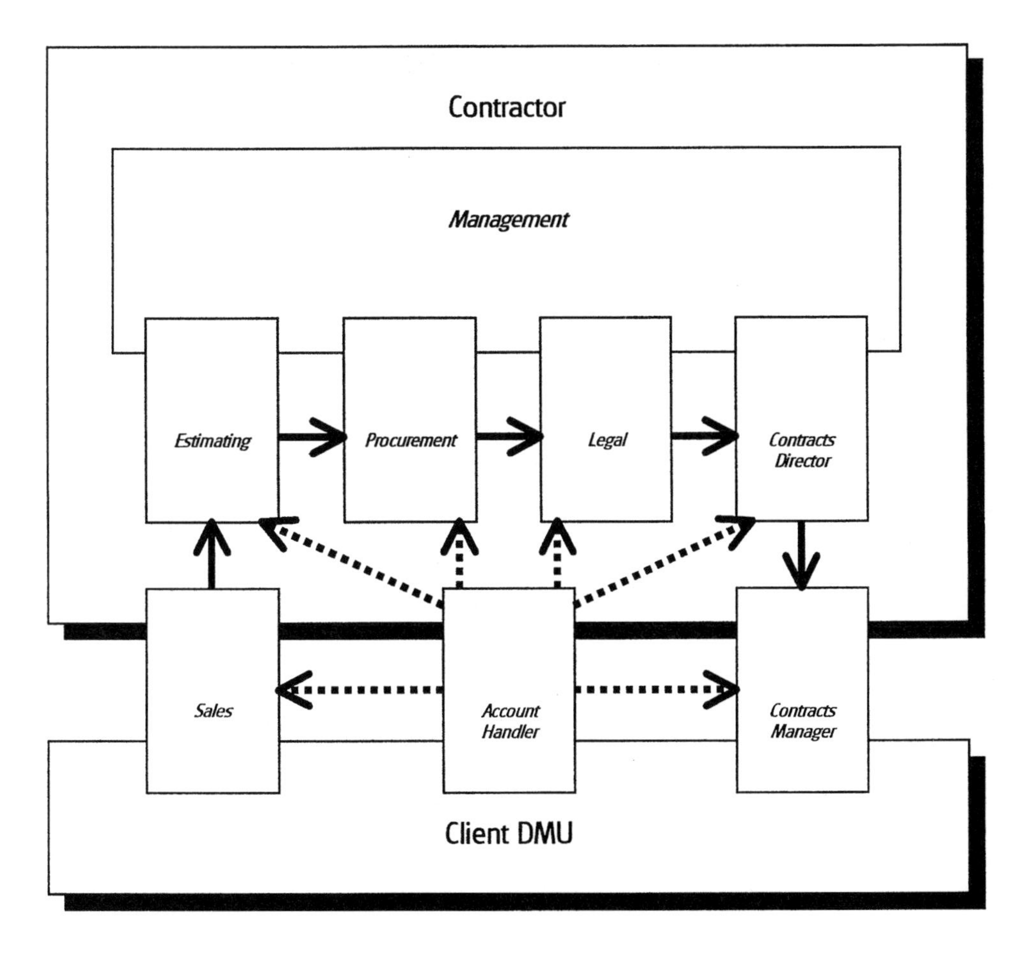

Figure 8.5 Functional role co-ordination for selling through relationships.

contracting organisation. Figure 8.5 gives a cursory indication of how these relationships work and need to be managed to ensure that whatever is promised to the client is actually delivered. The figure shows how the baton should be handed on or how the account handler fits into the picture. It also states that everyone is a part-time marketer or sales person under relationship marketing.

Integration

It was suggested earlier in the chapter that selling through relationships is radical and does not neatly fit into an economic model of exchange. The analysis of the DMU serves to reinforce this. However, Williamson[18] has analysed the *transaction costs*. His analysis suggests that exchange can be analysed along a continuum from markets at one end to hierarchies at the other. Essentially, the hierarchy model exhibits a high degree of integration between the buyer and seller. There are benefits to be derived from that, but the transaction costs are higher too. Selling through relationships in an industry such as construction is not going to replace the market. It does help to manage the market and restrict competition so that the investments can be made on both sides, from which the mutual benefit can be derived. How can this approach be understood in relation to the different procurement routes? How can selling through relationships be located on the continuum? An indicative scheme is presented in Figure 8.6 (see also Figure 0.3).

Where construction practice departs from general market theory is that transactions are extremely costly in construction where adversarial relationships exist. It is managerially and sometimes legally very expensive. Flexible communication derived from long-term relationships can, on the other hand, increase efficiency and reduce costs.[19] This does not mean that clients will not complain. Indeed, when complaints cease, it is likely that the client is looking elsewhere for sourcing design and construction. Complaints will probably increase as a consequence of relationship marketing. For every customer that complains, ten complaints fail to be communicated.[20] Good relationship should encourage

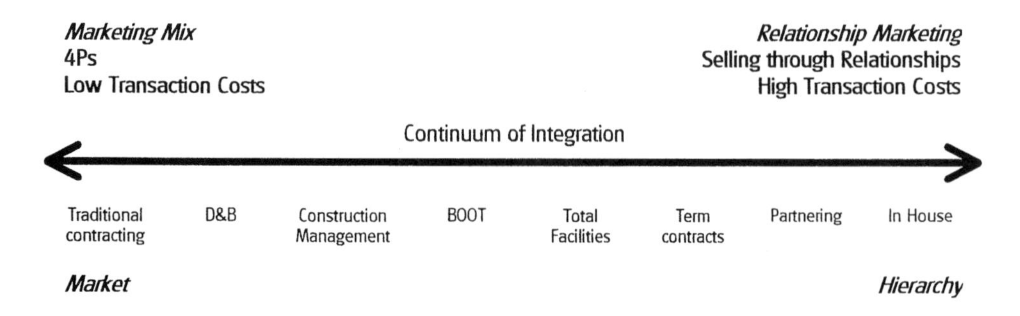

Figure 8.6 Indicative scheme of contracts and transaction costs.

communication of shortcomings and hence the opportunity to put any problems right. This is still a better option than letting goodwill erode away and run the risk of expensive legal action.

Clients have the choice to evaluate their position on their terms. Some will do this more frequently than others, which affects the investment both sides will make in the relationship. There is a balance between complacency and uncertainty. This is easier to observe where a formal agreement such as partnering is in place and review is regular. An electrical retailer with multiple outlets utilises a framework agreement with two construction managers and one project manager. An annual schedule of rates is agreed for tendering during the year, after which the agreements are reviewed. Throughout, performance is monitored, scores being aggregated into a league table. Quality is important, yet reducing construction times and keeping costs down is critical. Other performance indicators are snagging and post-occupancy problems. Agreements will not be renewed if performance falls below the set benchmarks.[21]

Outcomes

The corporate *outcomes* of selling through relationships are mainly judged in direct financial benefits. Turnover and profit margins are the key. The indirect financial benefits, such as shallow project learning curves, are marginal.

What are the profit outcomes of selling through relationships? The main problem is the size of the market. There are too few repeat business clients who have a steady development programme. The referral market is important, but less predictable. Selectivity is therefore the key.

Where selling is through relationship there is a trade-off between investment and return, above average gross margins being achieved with a consequential above average net margin after the additional costs have been deducted. The source of the higher margins is twofold:

- Lower costs in terms of sales, overheads and shallow project learning curve
- Premium paid by client for the higher quality 'least hassle' approach.

Table 8.2 National changes in market share of European contractors

	Percentage change in market share between 1993 and 1996				
	Top 10	Top 50	Top 100	Top 200	Top 300
France	1.1	−0.5	−0.7	−0.4	−0.3
Germany	3.6	0.9	0.7	1.2	1.4
Britain	−18.7	−19.8	−5.2	−3.5	−3.1
Spain	0.0	1.7	1.6	0.2	0.2
Sweden	9.5	2.3	1.8	1.2	1.1
Norway	11.0	5.1	4.1	3.5	3.1
Italy	−6.3	−2.0	−1.9	−2.6	−2.8

Source: Smyth (1998), derived from data in *Building* (1994); (1997).[23]

A contractor may have to prove his worth first before these profits begin to come through, therefore there is an up front investment on the first contract that has to be offset against anticipated profit streams. This requires contractors to be patient. If patience is lost, the contractor gets the worst of both worlds – incurring up front costs and then reverting to discounting in the open market.[22]

Apart from increasing the margin, the other way to raise profits is to increase turnover by gaining a greater proportion of work. Table 8.2 shows how the situation has changed for contractors by nation, rather than by company in a European context.

One can conjecture why British construction groups performed so poorly. While there are statistical anomalies between the data, the loss of share is striking at a time when the overall workload is low. If the work of Madsen[24] is relevant, then high levels of support from head office are needed in adjacent markets. In a sales context, strong market management (see Chapter 4), for which contractors are poor from Britain compared to elsewhere in Europe,[25] is imperative.

Conclusion

The chapter has considered selling through relationships in a way appropriate for consultants and contractors. However, the onus is on the sales people and the organisation to develop particular processes and techniques to implement matters raised here. That will be the source of competitive advantage at this level of operation.

While it is a source of advantage, it is not to be entered into lightly because of the costs, commitments and costs of exit.

Formal arrangements, such as partnering, help to give an impetus to selling through relationships. However, having identified a name for a theoretical tool, contractors tend to go round selling the idea without actually delivering the goods. While that may make short-term inroads for the contractor, the implications for the sector are a continuing poor image. The long-term prognosis for the contractor is unfavourable too. Market management is crucial for sales and delivering the promises made in order to make the best commercial sense of selling through relationship marketing.

Summary

1. The purpose of this chapter has been to demonstrate that selling through relationships is effective yet demanding. It requires commitment and imagination, suggestions and indications being provided about how to proceed.
2. Particular consideration has been given to:
   - Mapping and profiling the DMU
   - Identifying the nature of the market under relationship marketing and relating this to procurement options
   - The need to sell in an economically viable way.

References and notes

1. Drucker, P.F. (1990) *The New Realities*, Mandarin, London.
2. Gardner, J.W. (1978) *Morale*, Norton, New York.
3. Levitt, T. (1983) After the sale is over, *Havard Business Review*, September–October, 87–93.
4. Gummesson, E. (1994) Service management: an evaluation and the future, *International Journal of Service Industry Management*, **5**(1), 77–96.
5. Gronroos, C. (1990) *Service Management and Marketing: Managing the Moments of Truth in Service Competition*, Free Press/Lexington Books, New York.
6. Schurr, P.H. and Ozanne, J.L. (1985) Influences on the exchange process: buyer's preconceptions of a seller's trustworthiness and bargaining toughness, *Journal of Consumer Research*, **11**(4), 939–953.
7. The sample is derived from questionnaire responses among delegates attending courses run by the Centre for Construction Marketing,

Oxford Brookes University, Oxford, which runs short courses and training events, conferences and publishes research.
8. Donay, P.M. and Cannon, J.P. (1997) An examination of the nature of trust in buyer–seller relationships, *Journal of Marketing*, **61**, 35–51.
9. Hakansson, H. and Snehota, I. (1995) The burden of relationships and who is next? *Proceedings of the 11th IMP International Conference*, Manchester, pp. 522–536.
10. Maister, D. (1989) Marketing to existing clients, *Journal of Management Consultancy*, **5**, 25–32.
    Management Today (1998) Keep your sales team sweet, January.
11. Storbacka, K., Strandvik, T. and Gronroos, C. (1994) Managing customer relationships for profit: the dynamics of relationship quality, *International Journal of Service Industry Management*, **5**, 21–38.
12. Levitt, T. (1983) *The Marketing Imagination*, The Free Press, Macmillan, New York, pp. 112–114, emphasis in original.
13. Page, M., Pitt, L., Berthon and Money, A. (1996) Analysing customer defections and their effects on corporate performance: the case of IndCo, *Journal of Marketing Management*, **12**, 617–627.
14. Thompson, N. (forthcoming) Relationship marketing and the conditions of trust, PhD dissertation, Oxford Brookes University, Oxford.
15. Belbin, R.M. (1991) *Building the Perfect Team*, Video Arts, London.
    Belbin, R.M. (1993) *Team Roles at Work*, Butterworth-Heinemann, Oxford.
16. Levitt, T. (1983) After the sale is over, *Havard Business Review*, September–October, pp. 87–93.
17. Standing, N. (1997) VM and tender appraisal, Presentation at *VEA-MAC, Workshop*, 24 September, Oxford Brookes University, Oxford.
18. Williamson, O.E. (1981) Contract analysis: the transaction cost approach, *The Economic Approach to Law* (eds P. Burrows and C.G. Veljanovski), Butterworths, London.
    Williamson, O.E. (1985) *The Economic Institutions of Capitalism*, Free Press, New York.
19. Cf. Mohr, J.J., Fisher, R.J. and Nevin, J.R. (1996) Collaborative communication in interfirm relationships: moderating effects of integration and control, *Journal of Marketing*, **60**, 103–115.
20. Desouza, G. (1992) Designing a Customer Retention Plan, *The Journal of Business Strategy*, **13**(2), 24–28.
21. Cook, A. (1997) Electrical contacts, *Building*, 19 September.
22. See page 222 for a business game that vividly illustrates this point in Senge, P. M. (1992) *The Fifth Discipline: The Art and Practice of the Learning Organisation*, Century Business, London.
23. Smyth, H.J. (1998) The competitive stakes and mistakes: the position of British contractors in Europe, *Proceedings of the 3rd National Construction Marketing Conference*, 9 July, The Centre for Construction Marketing in association with CIMCIG, Oxford Brookes University, Oxford.
    Building (1994) 2 December (1997).
24. Madsen, T.K. (1989) Successful export marketing management: some empirical evidence, *International Marketing Review*, **6**, 41–57.

25. Smyth, H.J. and Stockerl, K. (1998) Strategic marketing planning by UK contractors in an international business environment, *Proceedings of the International Construction Marketing Conference*, 26–27 August, University of Leeds, Leeds.
Stockerl, K.C. (1997) The importance of strategic marketing planning for the UK construction industry in a changing European business environment, *Proceedings of the 2nd National Construction Marketing Conference,* 3 July, The Centre for Construction Marketing in association with CIMCIG, Oxford Brookes University, Oxford.

Chapter 9

Selling Added Value and the Product

Themes

1. The *aim* of this chapter is to investigate some ways of adding value to the product and quality of work in construction and design.
2. The *objectives* of this chapter are to:
   - ❑ Examine added product value
   - ❑ Explore whether added product value yields consultant and contractor benefits.
3. The primary *outcome* of this chapter is to understand the possibilities from added product value in the supply chain.

Keywords

Added value, Product, Unique selling points (USPs)

Introduction

Adding value was a concept that came to the fore in the 1980s. At that time one of the main world automobile manufacturers had made substantial inroads in selling their cars to fleet car and leasing companies. They were looking to make technical improvements. They wished to consolidate, if not improve, their market share. How could they improve the vehicle performance? Market research showed that the majority of users spent most of their time behind the wheel travelling to and from work. And a vast proportion of that time was spent crawling along in queues of traffic. Performance was not really an issue at this time and so the best value that could be added to the fleet cars was a high quality radio and stereo system – one that was better than all the competitors. It proved to be a successful ploy.

This chapter and the next one examine the nature of added value for contractors and consultants. Essentially, the motive for adding value is to beat the competition and achieve a rate of return above the margin. In other words, the rewards should exceed the investment at a profit level above levels achieved in traditional approaches to contracting and consulting. This chapter is concerned with the product in relation to added value, while the next is more focused upon service value. There is more emphasis in this chapter on the marketing mix, switching to a greater emphasis upon relationship marketing in the next. There are, of course, strong connections and linkages in the theory and application of added value. This chapter will therefore also consider service inputs that are experienced in the final product. Pure service will be the focus of the next chapter, yet it is recognised that the division between chapters is to do with the convenience of classification and clarity of the communication to a large extent.

Approaching added value

Adding value to the product can be *direct*. For example, design enhancement in a D&B contract yields better value for money for the client, but at a higher level of value than otherwise would be expected or received. In other words, the value being added is certainly different to and usually greater than that delivered by the competition. It is not a static concept. Contractors and consultants are quick to emulate.

Indirectly adding value looks at the service element, where the benefits filter through to the product. A good example of this phenomenon would be improved site management, which yielded high quality of workmanship. Such improvement in management may be difficult to tie down; however, the results are better. In this sense the inputs are as tangible as the improvements to a product, but the ability to link the inputs to the outcomes is difficult, especially in accountancy terms. Certainly the client may not perceive the added service value easily and so the experience is very indirect, yet no less worthy for that. The ability to copy added service value is more difficult where it is dependent upon culture, organisational structure and in-depth relationships.

How can value be added directly to the *construction industry product*? In mature industries, where the product and technologies have stabilised, any benefits will be short lived.[1] Competitors will quickly copy and the advantage ceases. Therefore the incentive to undertake actions that will improve the construction product is small. The scope is limited too. Contractors do not design the product. It is the engineers and architects who have the best scope to add value through quality design and through design innovation.

Contractors therefore have to find new ways of delivering their services and enhancing quality. An example of tangible benefits comes from the D&B market. Contractors believe that clients benefit from D&B over and above traditional contracting;

- 24% improvement in project time
- 38% improvement in cost
- 14% improvement in quality.[2]

The Managing Director of an international contractor recently stated:

> *Having close ties with people does not win work. ... What wins work is excellence of performance and people doing a good job.*

The company has succeeded in adding value on 80% of its projects. It employs 100 people whose job it is to come up with creative ideas. The result is not only the ability to win work, but a profit margin 1% above the sector average.[3]

Markets change over time, so adding value is a constant process and indeed adding value itself changes the market. The drop in price for clear glass by 50% over a decade has to be seen in the context of growing demands for complex glass. Recently, the major global glass manufacturers have been competing over the development of glass with low emmisivity; for example, 90% of all new windows in Germany use this technology.

An example of added service value through design can be seen in the first illustration within ... *On the case 10*. Another European example is provided in ... *On the case 34*.[4]

These examples are *tangible* aspects of the quality of the building product or the service, as witnessed in the way the service is realised into the final building. It is important to understand three aspects of adding value – product, service built into a product and a pure service. These can be located along a continuum of added value (see Figure 9.1).

This chapter focuses on the value that is built into the final project, the first two categories. However, there are combinations across and grey areas along the continuum, which can give rise to many opportunities to add value. There may be nothing new under the sun, yet the combinations of product and service elements help to differentiate the offer made by the contractor and in that way contractors can create *unique selling propositions* (USPs) compared to rivals (see Figure 9.2, see also Figures 0.3, 2.8 and 6.3). Added value in general and unique selling proposi-

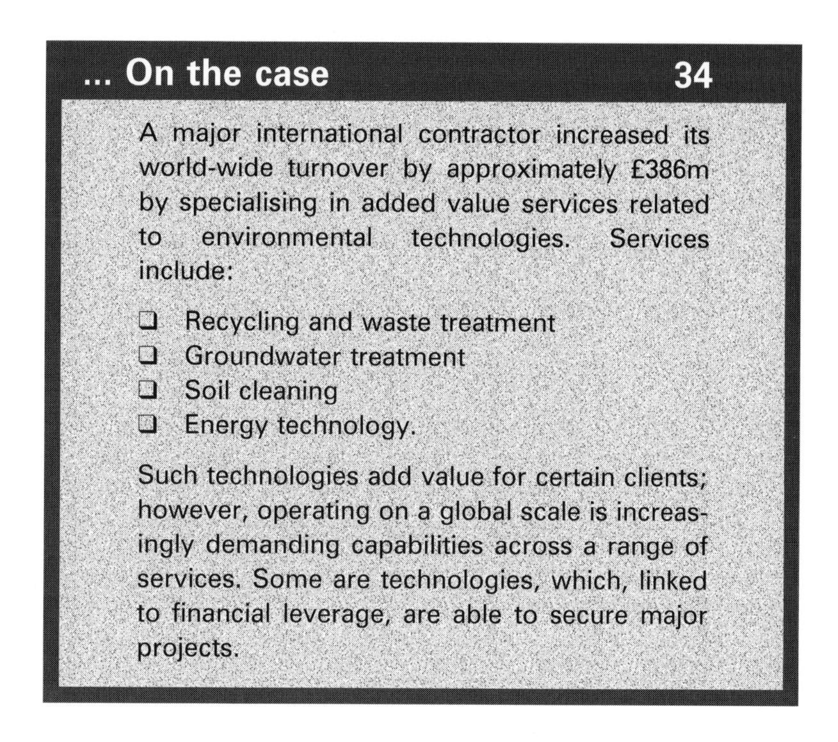

... On the case **34**

A major international contractor increased its world-wide turnover by approximately £386m by specialising in added value services related to environmental technologies. Services include:

- ❑ Recycling and waste treatment
- ❑ Groundwater treatment
- ❑ Soil cleaning
- ❑ Energy technology.

Such technologies add value for certain clients; however, operating on a global scale is increasingly demanding capabilities across a range of services. Some are technologies, which, linked to financial leverage, are able to secure major projects.

tions specifically can range along a continuum of added value, based around how tangible the value is. Depending upon how tangible the value is, the client will receive the added value as:

- ❑ What can be seen and known by the senses
- ❑ What can be known through experience
- ❑ What is experienced, yet is hard to articulate or is an investment that is yet to be experienced.

At one end of the continuum a low-energy building has design features that can be seen, while construction education may not be put to use at the time it is received and thus quite how important it is and the ways in which it will be used have yet to unfold. Ironically, it is often the non-quantifiable that is the most exciting from a human perspective.

Figure 9.1 Product service continuum.

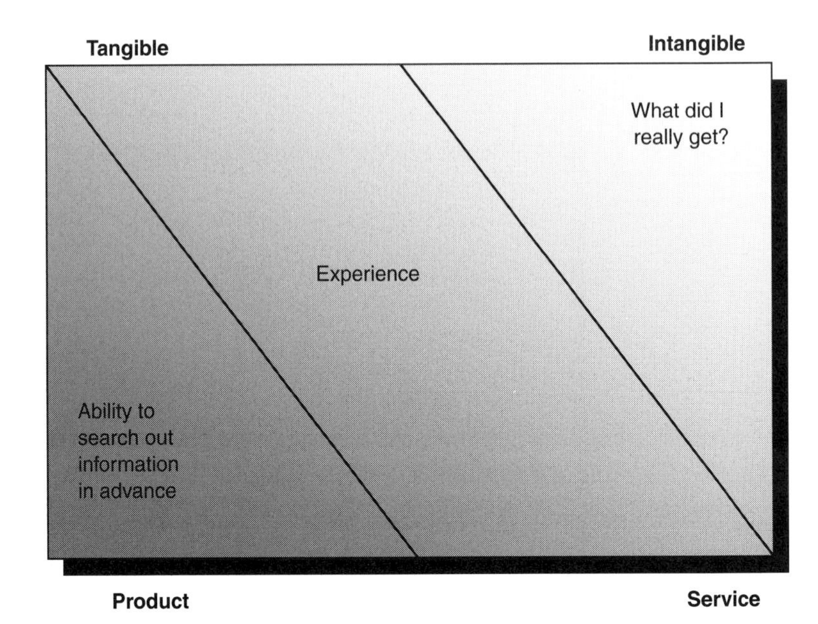

Figure 9.2 Added value continuum.

Zero tolerance and plus factors

It is increasingly the case that clients across the world are less tolerant of problems with cost, programme and quality. They expect these basic requirements to be met. If there is anything surprising about this, it is that this intolerance has taken so long to emerge. The issues increasingly focus on what else is on offer. Performance concerns are therefore becoming more demanding, setting standards across and beyond the traditional dimensions. The response is to add value. Such added value is a response to competition. It is reactive, is with the client driving the process. A more positive response is for the contractor or consultant to sell added product or design value. This yields a competitive edge.

The broadest remit for adding value is to do so for all the prime stakeholders:

- Clients
- Staff
- Investors.[5]

In this framework, adding product, or indeed service, value to the client is intrinsically linked to staffing. It is people who ultimately deliver the value, directly or indirectly. They hold the key to success. Yet construction does not always value people. However, it is the staff who carry the knowledge and experience in order to deliver value. They are also the ones in whom the client has placed its trust. Both knowledge and trust are requirements to add value in this sector. For the contractor or consultant, that implies that there must be employment stability. A system that is unstable, with a high degree of *churn*, either in terms of staff moving between projects or clients, will be unable to sustain value creation. There is some trade-off in flexibility, yet the rewards are theoretically maximised.

To maintain competitive advantage requires a staff development programme. This is likely to have two goals:

- Ensuring that all staff responsible for product or service delivery understand the key elements and dimensions of the added value component, and that the implementation of these dimensions is consistently carried through
- Creating the opportunities to further enhance the added value, building upon the key elements and dimensions.

The above points are clearly illustrated in ... *On the case 35*, which illustrates how value can be delivered in the design product. It also shows the need for carefully managing the added value in a way that keeps the focus internally on the main benefits of the product for the client and the extent of competition externally.

Developing a product or a service that adds value to the final building or structure can provide competitive advantage. In itself that is insufficient, a point seldom appreciated in construction and consulting. Being able to effectively enter and sustain a presence in the quality competition market requires the development of a distinct, authoritative image for the service and for the organisation.[6] In fact, the more specialised the service is, the more prominent that image must be. Indeed, the greater the specialisation, the easier it is to target the promotion and image development. The more international the operation or the global the demand, the more prominent that image must be. Conversely, promoting an international image is more difficult to manage because of the complexity and cost of brand development on a world scale in a market, which is a long way off from becoming truly global.[7]

Joint venture companies and partnership agreements for BOOT type projects can create difficulties to adding product

... On the case 35

A leader in interior design pioneered the use of market segmentation in the retail sector. The practice conceptualised market segmentation in the spatial layout of the store. The aim was to improve circulation flows and present the offers made in each segment to its best advantage to the browsing shopper.

This added value to the stores found resonance in the market for two main reasons:

- The diversity of offers, especially in fashion, was growing exponentially and the so-called designer labels were coming to the fore
- Franchising spaces to multiples within stores had become a dominant mode of operation – maximising the opportunity for each space had the effect of maximising rents.

Business boomed and expansion of the design practice was exponential. As new staff came on board, sight was increasingly lost of the original value of the service. Some staff failed to understand or adopt the key elements and dimensions of the service, while others took the ideas to such extremes that segmentation looked like fragmentation to the consumer. Shoppers were left to browse in retail environments of confusion.

Direction was lost in the marketplace. The practice began to experience a decline in the demand for its services. There was a squeeze from two sides. Externally, competitors were trying to emulate the service and internally competitive distinction was lost as the key elements were diluted and became unclear to clients.

value. Maintaining the corporate culture, the integrity of the value and creating the right image or branding is more difficult. These sorts of deals are best pursued at the level of price at the contract stage, value being leveraged higher up the supply chain, especially in financial packaging and in project design. Project emphasis becomes total quality management (TQM) procedure. The other opportunity for adding value further up the supply chain is innovation. This tends to require high levels of vertical integration, as for example found amongst German, French and Scandinavian construction groups.

In some procurement routes there is a higher demand for value at the outset. For example, the BOOT contract requires the contractor to operate a major bridge or health care facility over a period, frequently a 25-year period. This implies that the contractor will ask the architect to design a building of low maintenance and running cost for the period. In such a case, this may have more value than for a similar project procured through a different contract, but it is *not* added value for those circumstances; it is a minimum reasonable requirement.

Action

A hypothetical example may be useful in illustrating how value can be added to the product in practice. Design and build (D&B) has become a popular procurement option in many countries. In its purest form, the architect is selected by the contractor and works for the contractor. The contract is negotiated or competed for in a tender process. Traditionally, the contractor tenders against the same design as the competition. Under D&B, the design is one of the variables. Price competition will drive down the design quality in order to drive down the price. This may not lead to the client being satisfied in the long term with the building. How can a contractor decide to compete on value, especially where the client is more demanding and is seeking good value?

One way would be for a specialist D&B contractor to select architects that fit the purposes of the clients. What would be the most important factor in delivering 'product' or design value? They can be divided into direct and indirect factors, each of which is giving a *weighting*:

- ❑ Direct factors
  - ■ Generic track record 5
  - ■ Specific building type track record 8
  - ■ Specialist knowledge 3
- ❑ Indirect factors
  - ■ No practice learning curve 3
  - ■ D&B culture 4
  - ■ Same design team personnel 6
  - ■ Buildability of designs 7

Each potential architect could be evaluated according to each weighted factor. They would be *rated* on a Likert scale of −2 to +2. Therefore, if an architect was considered to be excellent in terms of generic track record, then they may receive a *rating* of +2 and this would be multiplied against the *weighting* of 5, giving a *score* of +10. They may be rated −1 for buildability. Multiplying that against a weighting of 7 yields gives a score of −7. This would be carried out against each measure and all the scores added together to give a *total score*. The practice with the highest score would be the one regularly used in that market segment. The quality of the work should be high.[8] Further opportunities could flow from this. The certainty of working together across several projects could permit both organisations to strategically agree what aspects of design, specification and buildability they would work on to deliver further added value in the future.

In this example, a method is established to choose a practice in a way that will add value, yet will bring down the price to a competitive level as the service element feeds into the product. The important management issue is to decide which value benefits the client targets. The value has to be value that can be created and delivered in a competing environment with advantage.

Monitoring the focus and implementation is crucial with periodic re-evaluation of how effectively that is being carried out within the organisation and the extent to which competitors are catching up. Opportunities for further enhancement will be necessary from time to time. Enhancement ahead of the competition is what creates the opportunity for unique selling propositions, even if they are short-lived. Combinations of product and service enhancement can also deliver USPs. It may be found that added value that comes into the product through paying attention to the service on offer will lead to a greater and greater focus upon the service. New added value may become pure service added value over time, the subject of the next chapter.

Conclusion

In any contracting or consulting organisation, there should be a high regard for quality. Sustainable added value requires:

- Obsession with quality
- Quality benchmarking
- Rewarded quality
- Training for quality.[9]

These factors embrace motivation, measurement, developing and maintaining efforts to add product value. Adding value is a dynamic process. There is a requirement to keep reassessing the components of value. The more intense the competition in this market, the faster the rate of change.

The areas of lowest risk are to add value through continuous improvement, next being adding value to existing services and products and finally adding new products or services. Finally, the competitive advantage that is obtained through adding value also carries a risk, especially at the front end of offering and delivering the service.

Summary

1. The purpose of this chapter has been to demonstrate that selling added value requires constant review. Yesterday's added value is tomorrow's norm.
2. Particular attention has been given to:
   - Indirect and direct product value
   - Added value continuum
   - USPs.

References and notes

1. Blois, K. (1997) Are business-to-business relationships inherently unstable? *Journal of Marketing Management*, **13**, 367–382.
2. Akintoye, A. (1994) Design and build: a survey of construction contractor's views, *Construction Management and Economics*, **12**, 155–163.
3. Chevin, D. (1998) The good knight, *Building*, 4 September.

4. Stockerl, K.C. (1997) The importance of strategic marketing planning for the UK construction industry in a changing European business environment, *Proceedings of the 2nd National Construction Marketing Conference*, 3 July, Oxford Brookes University, Oxford.
5. Reichheld, F.A. (1994) Loyalty and the renaissance of marketing, *Marketing Management*, **2**(4), 10–21.
6. Valikangas, L. and Lehtinen, U. (1994) Strategic types of services and international marketing, *International Journal of Service Industry Management*, **5**(2), 72–84.
7. Smyth, H.J. and Stockerl, K. (1998) Strategic marketing planning by UK contractors in an international business environment, *Proceedings of the International Construction Marketing Conference*, 26–27 August, University of Leeds, Leeds.
8. Smyth, H.J. (1996) Design and build marketing: issues and criteria for architecture selection, *Proceedings of the 1st National Construction Marketing Conference*, 4 July, The Centre for Construction Marketing in association with CIMCIG, Oxford Brookes University, Oxford.
9. Smyth, H.J. and Branch, R.F. (1996) *A Client Orientated Service*, Centre for Construction Marketing, Oxford Brookes University, Oxford.

Chapter

10 Selling Added Value and the Service

Themes

1. The *aim* of this chapter is to investigate some ways of adding value to the service and client relationships.
2. The *objectives* of this chapter are to:
   - ❑ Examine added service value
   - ❑ Explore whether added service value yields consultant and contractor benefits.
3. The primary *outcome* of this chapter is to understand the possibilities from added service value in the supply chain.

Keywords

Added value, Service, Loyalty

Introduction

Speaking to a family friend, who was once an air hostess, discussion turned to my work in a former Eastern bloc country. She was surprised to learn that I now flew with the national carrier from that country, rather than a British carrier. 'Why?' The answer was simple: 'Because you get a second cup of coffee.' She was astounded that I should choose an airline on the basis of something as trivial as a cup of coffee. But sometimes decisions do come down to things as simple as this. I like to have at least a couple of cups of coffee, if not more. Getting a second cup was definitely the bit of added value service that made all the difference for me. The flight costs were the same. The airlines flew within ten minutes of each other. On both airlines, the British carrier trains the cabin crew. The difference in practice was the enthusiasm of the British crew to move to in-flight duty-free sales as soon after the meal had been served as possible, whereas the other carrier was a little more relaxed about this and consequently I was more relaxed with my coffee.

The above story illustrates what matters to certain customers or clients. Of course, there may be other customers who really get quite annoyed with passengers who want an extra cup of coffee and they really need that aftershave or perfume for their husband or wife. However, this is a matter of carrying out market research, targeting the customers and advertising accordingly. It is exactly the same in construction and consulting. The emphasis is on face-to-face contact in this market, rather than advertising.

The construction industry is widely accepted as an underachiever. There is little evidence to say that the industry is 'backward'.[1] It may not be at the forefront of technology, but it survives on the stock markets, is flexible enough to ride the roller coaster trade cycles and is becoming increasingly sophisticated. All of which is not to say that it is in great shape. A school report saying, 'Fair, could do better!' would not go amiss.

Purpose

Differentiation of services across the market is a matter of positioning and segmentation.[2] Differentiation against competitors is a matter of targeting services to meet needs.[3] An important aspect of differentiation against competitors is adding service value. The primary focus of this chapter is on service for the client. This is strictly aimed against the competition. However, like every offer made in the marketplace, the goods have to be delivered. Therefore, delivering added service value looks as much to the future as to the current contract. It looks ahead to:

- Referral markets
- Client loyalty and hence repeat business.

While the aim of adding service value from the viewpoint of this chapter is not specifically about seeking to retain existing clients, it cannot be ignored. Client loyalty is so desirable. There is a distinction, because the investment made in adding value to a service for the benefit of a series of clients may not in itself do anything specific to manage a particular client relationship so as to significantly secure improved loyalty levels (cf. Figure 6.2). Adding service value looks ahead to client loyalty and may have a positive effect. To put this in another way, it can be and usually is predetermined what the added service value will be. There is not the same requirement to continually respond to the demands of the client. It is a more standardised approach and can fit well in market positions of routinised service provision (see Figure 1.5).

Table 10.1 Base differences in added service value between contractor and consultant

Client service expectations	
Contractor	Consultant
Corporate trust sought	Personal trust sought
Competence required	Understanding of client
Financial credibility	Common interests
Organised and managed	Reactive and spontaneous
Work more tangible	Work less tangible

A non-random sample of contractors showed that 39% were wishing to take action to improve the value component of the service.[4] There are some subtle and fundamental differences between contractors and consultants adding service value and these are set out in Table 10.1.

Investment

Contractors tend to focus on getting the job done, rather than how to do it and how to enhance the service (cf. Paradox 6 in Appendix A). The differences between contractor and consultant can be summed up in terms of the service delivered by consultants being more reactive and less tangible than that of the contractor. The relationship with the client tends to be more personal and less corporate in style. It is the corporate organisation and culture of the contractor that assures the client: whether they are financially sound, have a trustworthy reputation and a reputation for competence. However, neither party performs well, even on these terms. The poor performance and image of the contractor has been addressed. Several reports have found that clients experienced general dissatisfaction with the services of consulting engineers.[5] Specific problem areas are:

- Lack of understanding of client needs
- Lack of flexibility
- Lack of innovation
- Poor value for money.

Clients, it is suggested, are finding it increasingly difficult to distinguish between the scope and quality of services offered in the marketplace.[6] Clients want to be able to see the value and understand the benefits for them. Everyone promising, 'We can do that!', whatever the client may say, is disingenuous and patronises clients. Lack of knowledge or expertise should not be confused with a lack of discernment. In fact, most consultants prefer to work for discerning clients and the consultants' services could be dispensed with if the client had the expertise. One remedy is the development and marketing of technical expertise,[6] in other words enhancing the service and communicating more clearly in promotion and specific pitches for commissions.

It has been stated that enhancing value helps to differentiate the consultant or contractor from the competition (see Chapter 9). This applies equally to the service as to the product. In service provision, the more value that is added, the more *intangible* the service tends to become (see Figure 9.2). At the extreme range of the added value continuum, the client may not be sure what they are getting. Education is a good service example. The higher the value of the education, the less sure you might be of its value. It may be interesting, but it may not be obvious as to what use it can be put. In fact, some will never be put to good use, but as one of the purposes is to prepare people for their future it may be many years before its full value becomes apparent and can be fully evaluated. Similarly, this means that the consultant or contractor has to clearly communicate the service features and potential benefits. This is why the adding of service value takes the selling beyond the 'take it or leave it' approach of the 4Ps and through enlightenment, towards engagement and empowerment in the 5Es.

What is the sales person selling? In essence, added service value is:

$$\begin{aligned}\text{Added value} &= \text{Knowledge} + \text{Expertise} + \text{Relationship value}\\ &= \text{Service quality}\end{aligned}$$

There are several generic options:

- Adding value with no additional cost to the client
- Adding value with a proportionally smaller contract cost increase, where the contractor's cost is marginal, hence enhancing profit margins
- Adding value with a proportionally equal contract cost increase, where the value for money is enhanced above the marginal cost for the client.

Adding service value in one sense provides nothing new for the client. What is more likely is the combination of service features and their emphasis is unique. That is how to create a *unique selling proposition*. Some elements of added service value have to be considered strategically. That has been dealt with in Section I. The tactical service response is a sales issue and is in the hands of those in face-to-face client contact. Figure 10.1 shows how the scope of added service value can be conceptualised.[7]

How can added service value be developed in practice? Decide to add tactical service value. Set about the process of deciding what value is needed and what is realistic:

- ❑ Listen to clients and research their demands as you court them
- ❑ Use open-ended questions, supported by specific ones as you identify service issues
- ❑ Talk with management and site personal to assess the genuine strengths of organisation
- ❑ Map out between three and five areas of service value
- ❑ Sell clients combinations that meet their needs.

Enter the clusters of service demands into the market plan to inform and improve segmentation and targeting. The tactical developments can therefore feed into strategic decisions on the

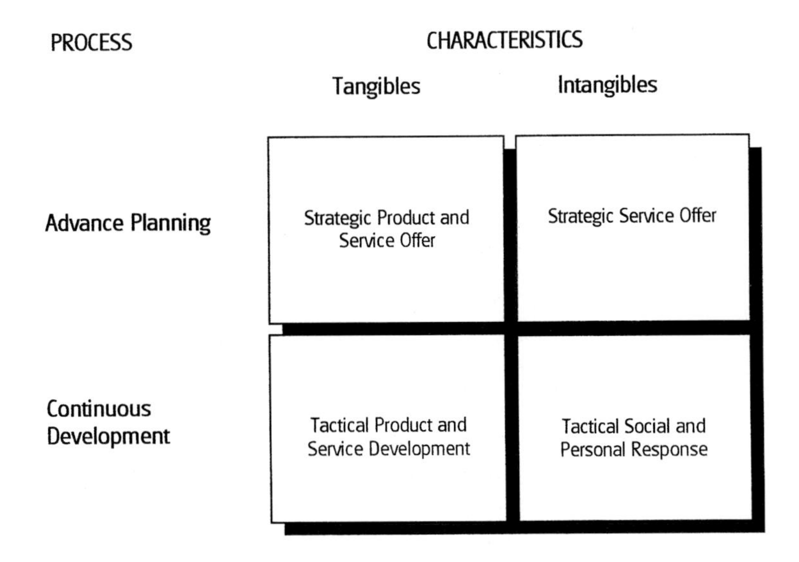

Source: Smyth, Branch and McIlveen, 1995

Figure 10.1 Dynamics of USP creation.

range of added value to be developed and offered within market segments.

In selling service value, it is important to be sensitive to the client. There is a danger of soliciting the client response characterised in the famous McGraw-Hill Magazines advertisement.[8] The client says:

- *I don't know who you are*
- *I don't know your company*
- *I don't know your company's product*
- *I don't know what your company stands for*
- *I don't know your company's customers*
- *I don't know your company's record*
- *I don't know your company's reputation*
- *Now what do you want to sell me?*

In relationships with clients, bonds can be established around several service dimensions:

- Technological bonds
- Bonds of strategic alignment
- Economic and legal bonds
- Knowledge bonds
- Social and cultural bonds
- Bonds of shared belief and perception.[9]

Each of these areas offers an avenue for adding value to the service. The areas become more intangible the further down the list, yet will potentially yield greater levels of client commitment to the service for the ease and cost of exit from the client perspective is greater. The greater the relationship strength, the greater the level of customer or client satisfaction. However, this assumes that the client relationship is being actively managed (see Figure 6.2). Sometimes bonds can be forged, but fail to be capitalised upon and client satisfaction will wane during the project or across several projects. The value is not being injected into the relationship and service in a consistent way. It is a management problem. However, the converse can be true. Value can be added, which strengthens the client relationship, yet the client does not value the highly satisfactory service. The *test* is to see whether the client invests in the relationship. When selling in this situation, the sales person or team must evaluate whether to sell sufficient value to secure that contract or whether to leverage sufficient value to secure loyalty and thus repeat business. The level of contractor or consultant investment is different, being geared to the level of client investment, aided by taking the client through the 5Es.

Is there a moral question? Should not all clients receive the 'best shot'? The answer is, 'No', because the client only gets what he pays for. While the tender price may be similar for both cases in order to secure the project, the investment and hence the value is greater in the case where loyalty is being sought. Here the contractor is investing more as a trade-off against the cost of finding a new client for the next project, which will cost five times more. The client receives the additional value in return for paying a higher margin next time around. They will also be saving money in the costly selection process of finding a new contractor. The monetary exchange is fair. The difference is efficiency, in that the whole process is speeded up. It is also efficient because the client gets better value for money as contractors learn to deliver added service value, which over time becomes the norm. Innovation tends to reduce long-term prices. The contractor is also avoiding the costs of the learning curve and in theory prices drop in this way too. The opportunity cost of buying buildings goes down and the hassle factor is reduced, so the long-term market grows too. Contractors will also have weeded out some of the competition through service differentiation and so projects will be concentrated into marginally fewer hands. The consequence of selling service value at the level of the organisation has a positive effect across the whole sector.

Adding service value can be begun in small ways. It can be an end in itself or an entry point into relationship marketing. Three examples of the types of added service value that can make a difference are set out below:

- ❑ A well-handled problem breeds loyalty
- ❑ Little things mean a lot
- ❑ Under-promise, over-deliver.[10]

Management

Linking specific service requirements of the client to management is important:[11]

> *'What constitutes customer service performance is the **sum** of the value creating **processes** for which staff are the coordinating agents and customers are the participants.'*

... On the case 36

A specialist refurbishment contractor operating in an historic capital city has no marketing strategy. It adopts a very fluid approach. It neither has a marketing nor sales department. Therefore, no targeting is employed. There is not even a sales budget.

Contacts are made by clients coming to them or by 'accident' through their network. The members of the company sell to individuals and not the organisation. What they sell is their track record in general, but specifically they concentrate on selling the technical input and expertise that can be mobilised for that client.

It is the reputation of the contractor, specifically the value of its service above the competition that is important.

The sales people therefore have to be proactive in engaging with management in order to mobilise the services that will add value for each client. It is then the role of the management or account handler to ensure its effective implementation once the project is secured. One international bank decided to formalise this internally by setting up diagnostic review groups and critical change workshops. So, how might a contractor or consultant set this up?

- Set up structure to develop and manage added service value according to the combinations needed by each client
- Identify a person who is an innovator and facilitator, who is given the power to bring people together to create added value
- Analyse existing service provision, breaking it down into its detailed component parts
- Propose alternative ways, re-organising the old ways if necessary
- Identify empathetic people who will champion the new ideas through the organisation
- Reinforce with formal training
- Make the sales person central, not someone working out on a limb.

How is this sales effort related to the rest of the organisation? Figure 10.2 shows where the added service value has its impact, yet realising this in practice requires a wider remit. In other words, the old chestnut of making sure that the organisation delivers what is promised is being raised once more as this is such a poor area of performance amongst all those in construction, especially contractors.

All of this may come as something of a jolt for sales staff. It may seem like marketing; however, the visionary content of this book is to redefine the sales role to make it more facilitative and thus more cerebral in nature (see Paradox 3 in Appendix A). Serving and thus differentiating at a tactical level needs to draw on strategy, draw in management and feed into the refinement of strategy.

The sales person has to be thoughtful and pro-active in understanding and responding to client requirements. The role and the skills are being redefined. This is the competitive edge that will secure the future. How this applies may depend upon market position. One response would be; 'Well, our organisation serves a

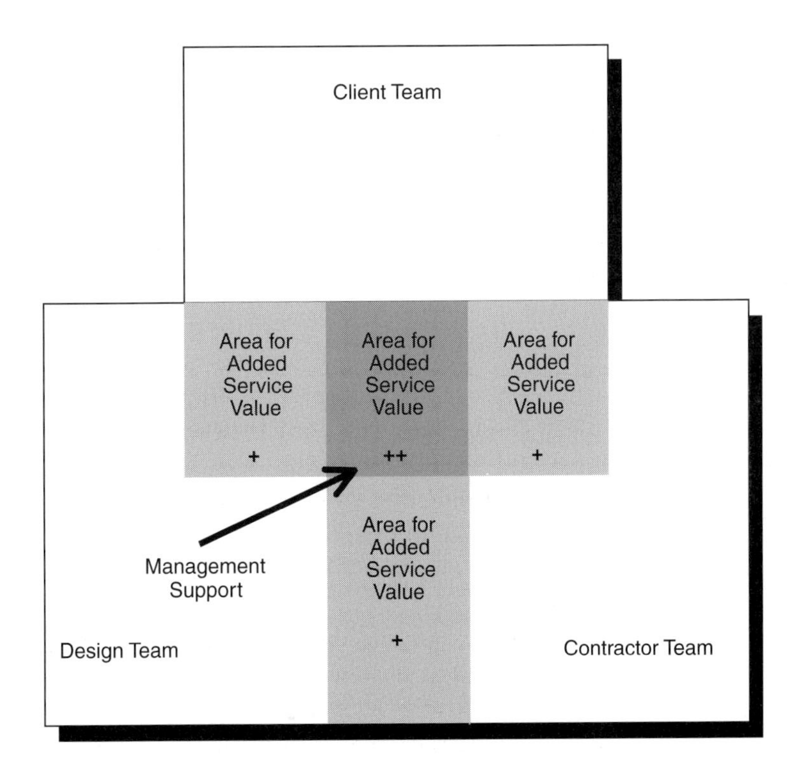

Figure 10.2 Delivering added value service promises.

market with routinised requirements. We have a standard approach and find it is best to keep it simple.' That is an entirely legitimate approach under two conditions:

- A mass market
- Where competition is not intense.

While these may exist to a limited extent among consultants, it is a rare event indeed in contracting. A simple and standard service may make even more demands to add value, as the CARE^{3}S programme in ... *On the case 37* explains.[12]

In the international market, technological supremacy is not the issue any longer.[13] Concerns for US contractors are the intangibles, as shown in Table 10.2. Each of the areas shown in the table offer opportunities for delivering additional service value. Taking such steps shown in ...*On the case 37* and Table 10.2 does not necessarily mean getting a long-term and close relationship with the client base. In the banking sector, it has been suggested that it is not always a good thing. Contractor or consultant dependence on a few sources of work can have an adverse effect during recession or calamitous events.[15] As construction is probably more cyclical than other sectors, this is not likely to be a concern for its clients; however, it does highlight the need to choose clients carefully where extra service elements are to be added, especially where a loyal relationship is being sought.

Table 10.2 Concerns of US international contractors

Areas requiring change	%
Understanding and appreciating ethics and cultures of other countries and acting accordingly	55
Understanding and appreciating political dynamics of other countries and acting accordingly	21
Developing alternative labour and material sources or accommodating situation	18
Expecting and adjusting to differing levels of education and training	6

Source: Yates (1994).[14]

... On the case 37

A civil engineering contractor specialising in simple routinised contracts, realised that the crucial issue was keeping the client happy in order to sustain a high level of necessary repeat business.

The engineering works were highly disruptive to pedestrians and motorists alike, therefore keeping the client happy meant minimising disruption to the public. The company instigated its customer CARE^{3}S programme at street level:

- ❑ Customer focused
- ❑ Awareness of environment
- ❑ Respect for relationships
- ❑ Enterprise, excellence, enthusiasm
- ❑ Solutions to issues.

The programme tries to affect the attitudes and values of those on the ground, so that all of the workforce have values that are parallel with those of the organisation:

- ❑ Purpose: deliver customer benefits + contribute to bottom line
- ❑ Awareness: understand organisational values + take actions in line with those
- ❑ Customer care: not carried out by a separate department – it is the 'cornerstone of our business'.

Trust

Doing his rounds, the owner of one of the famous London department stores was standing in the office when the phone rang. One of the junior staff members answered the phone and told his boss it was for him. 'Tell him I'm out', came the reply. The junior member of staff handed the phone to the owner, saying, 'You tell him you're out!' At the end of the call, the red faced and angry owner turned to the junior employee and asked him what on earth he thought he was doing. The reply was simple: 'If I can lie for you, I can lie to you.' That employee became the most trusted employee in the organisation and still worked there a couple of decades later.

Business ethics have not improved since that incident, more likely the opposite. Right and wrong are frequently defined in terms of what you can get away with; being found out becomes the crime rather than the deed itself. And yet *trust* is being advocated in a great deal of marketing theory and literature as a saviour for gaining a competitive edge. Relationship marketing demands *trust*. Partnering and alliances demand *trust* to be sustainable. Adding value can use *trust* as the most valuable piece of added value in the business climate of today. Trust cannot be created; it evolves.[16]

Establishing trust might as well start with the sales people. They are the very people who are often least trusted by the client – the slick, fast-talking sales person. If a sales person can build trust during the courting process, service value is already being added and goodwill established for the contract. As one developer commented:[17]

> *Courting is more of a time when we are trying to find out more about each other and the amount of trust during this time is quite low because of a degree of uncertainty and reliance is relatively low.*

So the sales person is starting from a low base and is responsible for building trust. Trust is built by showing to the client the competencies in selling in a reliable and consistent way, which can be broken down as follows:

- Demonstrating the expertise of the sales person in:
  - The art of selling
  - Client management
  - Contract management
  - Project technology
- Power
  - To mobilise management support
  - To mobilise specialist expertise
  - To facilitate the sales process
- Likeable character
- Similarity
  - Shared values
  - Shared interests
- Frequent contact
  - Business contact
  - Social contact

- Length of relationship
- Willingness to share confidential information
- Willingness to customise services – adding value.[18]

Building trust in the sales person is a prelude to the client having trust in the organisation. The best situation is where trust is developed between the sales person and buyer, and the trust in the sales person is 'transferred' to the organisation. In many situations, it has been found that buyers are much more likely to adopt relationship marketing at the start and end of the relationship, the actual exchange process being dominated by the marketing mix factors of price and delivery.[18] In construction, this suggests that the marketing mix dominates the tender and post-tender stages. If that casts any doubt on relationship marketing, then added service value may provide a constant thread to focus upon.

Recent research suggests that as a means to building relationships, trust helps to mitigate risk for the client and hence induce confidence in the service (see Figure 10.3).[19]

What are the conditions that give rise to trust? Table 10.3 provides a list of conditions in rank order of importance.

Receptivity is ranked highest. It contrasts with the adversarial tradition in construction. Receptivity is the ability to listen to and understand the ideas and demands of the client. Trust will also be influenced by how receptive the contractor is toward deal-

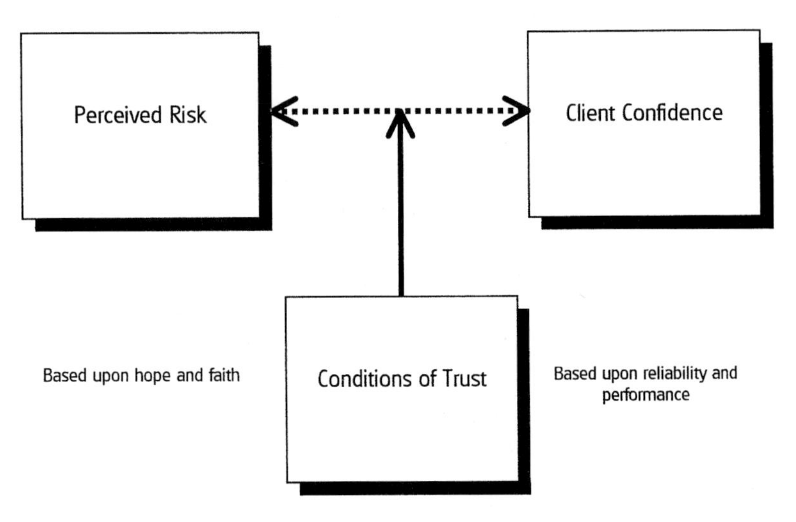

Figure 10.3 Risk-confidence continuum and conditions of trust.

Table 10.3 Rank order of conditions of trust

Rank of trust	Condition of trust	Correlation coefficient (against overall measurement of reported trust by the client)
1	Receptivity	0.84
2	Promise-fulfilment	0.82
3	Consistency	0.82
4	Integrity	0.81
5	Loyalty	0.81
6	Fairness	0.74
7	Openness	0.73
8	Competence	0.73
9	Discretion	0.69
10	Availability	0.52
	Sum of all the above conditions	0.90

Source: Thompson (1997).[19]

ing with requests, challenges and problems. Promise-fulfilment is the second most important factor in the ranking. It starts with the sales person keeping his word, but the proof comes in delivering the added value. The client wants to feel confident that the contractor will do what they have agreed.

Consistency, the next factor, is closely related and thus is also considered an important aspect. In particular, past performance places certain expectations on future performance, so a high standard across jobs is vital. It is felt that consistency between projects is largely down to individuals – a different team results in a different service. There should be systems in place and management support and monitoring to the extent that clients are buying a particular way of working and not a set of very different teams and experiences. What is vitally important for partnering is having the same team throughout. Continuous teams in long-term relationships are a good thing.

Integrity is very important in developing trust. Associated with integrity is honesty and dishonesty by the contractor. Breaches of integrity are defined by clients as inducements, cover-ups, non-disclosure of important issues on-site and during the contract. Breaches also refer to the failure to comply with regulations, misinformation at site meetings and blatant lies as to why something has or has not happened. In terms of developing trust, clients felt that contractors who identified snags and problems before they did showed integrity. A contractor who demonstrates an honest and proactive approach in every aspect of their management and project work would be considered to be a contractor with integrity.

Loyalty is not generally considered to be an issue by clients. Clients have not expected loyalty from contractors. Loyalty from contractors appears inappropriate in determining level of trustworthiness. However, under a partnering arrangement it is assumed and expectations are generally likely to rise on this score amongst clients in future years.

Responses to the condition of fairness were serious. Clients commonly feel fairness does not exist where contractors are concerned. Contractors are seen as always trying to get away with whatever they can. Contractors push the boundaries of constraint under which they have been placed. This starts in the sales process. It is a greater problem under contract.

Openness ties in with points about fairness and integrity. Many contractors are thought to cover up problems, which puts the onus of vigilance upon the client. Openness therefore includes admitting to problems. This can be turned to good effect. Once identified and communicated, there is the opportunity to show the value of the service by rectifying the problems.

Competence is a minimum requirement today. It is or will become insufficient to pre-qualify. The sales effort has to address this alongside the need to meet cost, programme and quality on site. Before awarding a contract, the client needs to know that the contractor has a full knowledge of a project in order to trust them to do it. In addition, before and during the project the client wants the confidence to know that the contractor knows enough about their needs.

Discretion, a further condition of trust, is less central than others higher up the ranking. It can be interpreted in two ways. Before, during and after a project, anyone in the contractor's team should be careful about what they say and to whom. Those not in sales can let the side down too, by being indiscreet. This breaks certain existing confidences and thus undermines trust. Discretion and openness should act in tandem, using each in an appropriate way.

Availability, the last of the conditions of any importance, involves three key issues.

Regular contact as and when required by the client is important. Access to a person able to genuinely initiate action, to answer questions quickly, was considered important. This should be backed up with that same person attending the sales or site meeting whenever the client wanted. It is generally felt that someone should always be available, especially given current technology aiding communications.

What is particularly interesting about all these findings is that the conditions that rank the highest are those that are the most intangible factors. The rankings tend to affirm the proposition that contractors focus upon 80% of the cost element of their work and underplay or ignore the intangible service elements that account for the remaining 20% of costs. However, it is that 20% service element that makes the 80% impact upon client satisfaction.[20]

These are the aspects to sell during the courting stage and to build upon during the contract negotiations and projects. They are key dimensions, especially of partnering agreements.[21] Any long-term contracting or consulting relationship sustained to mutual benefit must have trust as a cornerstone. Research has found that the one profession that tends to undermine trust is the quantity surveyor. In the team context they are controllers rather than servers.[22] This may be a peculiarly British phenomenon. Accountants and financial people used to play the role of servers, but stock market culture has encouraged the switch to a controlling role; that is, organisations driven by short-term financial criteria at the expense of the longer term. However, in sales terms, this brings price-dominated factors to the foreground once more, with a skewed marketing mix approach holding sway.

Can trust work? This can be seen where even competitors are required to trust each other and work together at the behest of the client (see ... *On the case 38*).[23] Trust can operate across normally adversarial relationships, to build co-operation:

- ❑ Client–contractor
- ❑ Contractor–consultant
- ❑ Contractor–subcontractor.[24]

Trust development can be more important than the skills and expertise in the services provided. There is evidence that cleaning and security companies are proving able to secure facilities management and maintenance term contracts on the basis of the existing client relationships.[25] Figure 1.7 shows how such service diversification works. Indeed, the long-term relationship in security had permitted the organisation to invest in infrastructure, which was going to prove competitively advantageous in construction maintenance, thus the re-configuration of the market through the development of just-in-time maintenance practices (see ... *On the case 39*).[25]

... On the case 38

A contracting company within a construction and property group was partnering a large retail client on a series of refurbishment projects.

The client made demands upon the partners to collaborate – *co-opetition* – so that the benefits achieved by one contractor could be shared with other partners so the client benefited across the board. Targets set by the client included:

- 20% savings
- Dust-management demands
- Communication improvements across all parties
- Site-hoarding management.

Dust management was chosen as the area to pilot *co-opetition* in order to establish the trust and see the output. The parties agreed:

- To pool ideas
- Agree best practice
- Develop expertise in benchmarking
- Pilot dust management across all partners
- Demonstrate added value to client and secure partner status for the future.

Using this approach:

> *The client has expressed considerable confidence in the team and delight with performance improvements made so far (p. 60).*

... On the case 39

A cleaning and security organisation diversified its services into construction maintenance. Using the resources and IT technologies applied to security, the organisation developed a pioneering service for major corporate clients.

Some of the features of the service value are:

- ❑ Inclusive service based upon performance, not specification
- ❑ Highly disciplined workforce
- ❑ Client focus and rapid response
- ❑ Service level agreements with performance-related financial structure
- ❑ Contractor input into building specification, especially information technologies
- ❑ Modularised equipment
- ❑ Continuing high levels of capital investment by contractor.

Conclusion

Selling added value is a careful and considered activity. The sales people need to weigh up:

- ❑ What will benefit the client?
- ❑ Can the organisation deliver the required benefits?
- ❑ What will it cost the contractor or consultant?
- ❑ What are the benefits for the contractor or consultant?

This is a far cry from the stereotype 'salesman' and some way off from the personality culture of the *street-fighting man* in construction. However, this is the future. Those embracing it will gain the market lead, although the initial investment can be high and the potential for making mistakes are large. As is sometimes said, it is best not to be first into a market but second, so you can learn from the mistakes of others.

While selling added value services does not necessarily involve embracing relationship marketing, it has been shown that it comes very close and comfortably integrates with the theory. Indeed, relationship marketing can give rise to considerable added service value in its own right.

Summary

1. The purpose of this chapter has been to demonstrate that adding service value:
   - Starts with the sales people
   - Is a thoughtful process
   - Can be built upon trust and developed through relationships
   - Is created in the relationship established at the sales stage and is handed onto the contract team with careful client management.

References and notes

1. Smyth, H.J. (1985) *Property Companies and the Construction Industry in Britain*, Cambridge University Press, Cambridge.
2. Day, G. and Wensley, R. (1988) Assessing advantage: a framework for diagnosing competitive superiority, *Journal of Marketing*, **52**, 1–20.
3. Kotler, P. (1991) *Marketing Management*, Prentice-Hall, New York.
4. The sample is derived from questionnaire responses among delegates attending courses run by the Centre for Construction Marketing, Oxford Brookes University, Oxford, which runs short courses and training events, conferences and publishes research.
5. Association of Consulting Engineers (1994) *Client Perception Study*, The Association of Consulting Engineers.
   Institute of Civil Engineers (1995) *Wither Civil Engineering*, Institute of Civil Engineers, London.
6. Addis, W. and Al-Ghamdi, M. (1998) The International Competitiveness of Consulting Engineers: the role to be played by technical marketing, *Opportunities and Strategies in the Global Marketplace – Proceedings of the 1st International Construction Marketing Conference*, 27–28 August, University of Leeds, Leeds.
7. Smyth, H.J., Branch, R.F. and McIlveen, A. (1995) *Developing Unique Selling Propositions in Fragile Construction Markets*, Centre for Construction Marketing, Oxford Brookes University, Oxford.
8. Adapted from Kotler, P., Armstrong, G., Saunders, J. and Wong, V. (1996) *Principles of Marketing*, Prentice Hall, London.
9. Storbacka, K., Strandvik, T. and Gronroos, C. (1994) Managing customer relationships for profit: the dynamics of relationship quality, *International Journal of Service Industry Management*, **5**, 21–38.

10. Smyth, H.J. and Branch, R.F. (1996) *A Client Orientated Service*, Centre for Construction Marketing, Oxford Brookes University, Oxford.
11. Ballantyne, D. (1997) Internal networks for internal marketing, *Journal of Marketing Management*, **13**, 343–366.
12. Smith, P.B. (1998) CARE^{3}S in action: a practical philosophy for customer and client care in the construction and property industry, *Construction Marketing Conference*, 27–28 August, University of Leeds, Leeds.
13. Hall, M.A., Melaine, Y. and Sheath, D.M. (1998) Operations by British contractors during the procurement process in a global and multicultural environment: some recent experiences, *Opportunities and Strategies in the Global Marketplace – Proceedings of the 1st International Construction Marketing Conference*, 27–28 August, University of Leeds, Leeds.
14. Yates, J.K. (1994) Construction competition and competitive strategies, *Journal of Management in Engineering*, **10**(1), ASCE as quoted in Hall, M.A., Melaine, Y. and Sheath, D.M. (1998) Operations by British contractors during the procurement process in a global and multicultural environment: some recent experiences, *Opportunities and Strategies in the Global Marketplace – Proceedings of the 1st International Construction Marketing Conference*, 27–28 August, University of Leeds, Leeds.
15. McRae, H. (1997) The financial folklore that cannot be banked on, *Independent*, Tuesday 9 December.
16. Blois, K. (1997) Are business-to-business relationships inherently unstable? *Journal of Marketing Management*, **13**, 367–382.
17. Thompson, N. (forthcoming) Research notes undertaken for PhD on relationship marketing and the conditions of trust, Oxford Brookes University, Oxford.
18. Donay, P.M. and Cannon, J.P. (1997) An examination of the nature of trust in buyer–seller relationships, *Journal of Marketing*, **61**, 35–51.
19. Thompson, N.J. (1997) Evidence on Evidence of Trust, *Proceedings of the 2nd National Construction Marketing Conference*, 3 July, The Centre for Construction Marketing in association with CIMCIG, Oxford Brookes University, Oxford.
20. This work draws heavily upon the work of Thompson in Smyth, H.J. and Thompson, N. (1998) *A Client Orientated Service for Partnering*, Centre for Construction Marketing, Oxford Brookes University, Oxford.
21. Smyth, H.J. and Thompson, N. (1999) Partnering and trust, *Proceedings of the CIB Symposium on Customer Satisfaction*, September, Cape Town.
22. Thompson, N. (1998) Can clients trust contractors? Conditional, attitudinal and normative influences on client's behaviour, *Proceedings of the 3rd National Construction Marketing Conference*, 9 July, The Centre for Construction Marketing in association with CIMCIG, Oxford Brookes University, Oxford.
23. Leveson, R. and Pickrell, S. (1998) Partner or competitor? Re-framing relationships in construction, *Proceedings of the 3rd National Construc-*

tion Marketing Conference, 9 July, The Centre for Construction Marketing in association with CIMCIG, Oxford Brookes University, Oxford.
24. See for example Bean, M. (1997) Developing and supporting a trial performance measurement system, *Proceedings of the 2nd National Construction Marketing Conference*, 3 July, Centre for Construction Marketing in association with CIMCIG, Oxford Brookes University, Oxford.
25. Wood, B.R. and Smyth, H.J. (1996) Construction market entry and development: the case of just in time maintenance, *Proceedings of the 1st National Construction Marketing Conference*, 4 July, The Centre for Construction Marketing in association with CIMCIG, Oxford Brookes University, Oxford.

Chapter 11 Selling and the Construction Project Team

Themes

1. The *aim* of this chapter is to consider the effect that diverse people from different organisations have on the sales process.
2. The *objectives* of this chapter are to:
   - ❑ Examine the multi-organisational project team
   - ❑ Explore the temporary nature of that team and implications for client satisfaction.
3. The primary *outcome* of this chapter is to understand the virtual organisation that delivers the construction and consultant service.

Keywords

Temporary multi-organisational team, Virtual organisation, Collaboration

Introduction

Change is the only unchanging thing of certainty. It is endemic to construction. The project team is temporary by nature. The team moves on after each project. The people in each team tend to move on after each project. Yet, it is people that are the source of competitive advantage. They add service value, then move on and so the nature of that advantage or added value becomes threatened.

In previous parts of the book, the need to maintain continuity of purpose, if not staff, has been stressed in order to deliver the promises and build long-term relationships (see, for example, Chapter 8). That focus was on internal issues and it is a theme to which the book returns in the next chapter. However, the focus of this chapter is on the flux created *across* organisations. Teams are flung together, coalesce, undertake massive and complex tasks together and then disband, so that all the experience is dissipated. The learning curve has to be climbed again on the next project. This is what Cherns and Bryant some time ago referred to as the *tempor-*

ary multi-organisational team.[1] Today we would call it a type of *virtual organisation.*

At one level, this temporary multi-organisational project team is all about the matter of co-ordinating information, actions and so on. That is fine and contractors and consultants are proficient at it, or should be. What is more difficult is the marketing and selling implications. What is it a contractor is selling if they are only contributing a portion into the final service? They can control their part, but cannot control the design, perhaps the most important factor in securing client satisfaction. The architect or engineer can oversee, but cannot control the quality of the work. There is a dependency on each other in order to deliver client satisfaction. In other words, there are worst-case scenarios where the majority pull their weight, but one party fouls it all up for everybody. Why build long-term relationships, when so much is outside the control of the organisation? This is important and needs to be addressed. The consultant or contractor can 'get their act together', only to find that all their best efforts melt away in the face of adversity.

Multiple organisations forming a virtual project team will have multiple values.[2] That causes difficulties and conflicts. Yet many people on a project will feel more wedded to the project team than wedded to the employer, especially if it is a project spanning several years. To the team members, it is the headquarters or the offices that is virtual. The project *is* the reality. They will identify more closely with people from other organisations than many in their own. A project culture and value system can therefore grow that pays minimal heed to the employer organisations. The key values revolve around:

- Innovation
- Risk
- Co-operation
- Deference
- Accountability.

Different values will lead to overlaps, conflicts and gaps during project implementation. In terms of project management, tools have been developed to manage these issues, such as programming, project management practices. In the case of marketing, sales and client management the overlaps and gaps have yet to be managed. Tools need to be developed and used so that the goodwill and client management does not spill through the gaps of the team members of this virtual organisation. The issues can be summarised as:

- ❑ What is really being sold?
- ❑ How can service delivery be managed?
- ❑ Who is credited with client satisfaction and in what proportion? Similarly for blame?
- ❑ How can the benefits of working together be transferred into repeat business?

There are also problems at the end of the project, which include how all the valuable client intelligence and project data has been captured to enhance the service for the future. And, of course, the general answer is that it has not. Therefore, the prime problem to address is whether a sales management system can be devised and implemented to solve these problems.

Building blocks

As it generally costs five times more to find a new client than to keep an existing one, the investment is worthwhile if the client has a forward development programme or work in the long term.

Two of the building blocks have already been introduced. These are:

- ❑ Choice of client handling model (see Chapter 8)
- ❑ Relationship marketing management model (see Figure 6.2).

The first ensures that there is continuity throughout the client relationship, while the second provides a framework within which the client can be managed. However, neither address the *temporary multi-organisational team*. Figure 10.2 shows where the *temporary multi-organisational team* is located – the shaded area, precisely the area in which added service value can be delivered. This underscores the need to manage the client from a marketing perspective in order to maximise sales effort with that client and for others. It is here where promises fail to be delivered.

The importance of having a common project management method *in* the corporate environment has been stressed in a number of industries.[3] This means that the personality culture prevalent in construction has to address project management to include the sales aspect of client management. The start is to recognise three main elements to project organisation:

- Steering function
- Project managing function
- Executing function.

The steering function starts at senior management level, the culture and procedures being set down for the contracts directors, who act as the bridge between the head office and site management. This is conceptualised in Figure 11.1. The figure shows how contractors in particular can move from the personality culture towards a more systematic approach (see also Chapter 3).

Depending upon the client-handling model, either the account handler is required to ensure that the process is seamless or the relay team approach must ensure that all material matters are handled – the passing the baton of promises, briefing, legal and all other functions through each functional stage. This should also include a budget for selling. Selling should be built into the project budget. This may include monies for a number of issues from monitoring site performance and client satisfaction to selling across the *virtual organisation*.

Function	*Issue*	*Action*
Steering Function	Generic Issues Located in the hands of Senior Management in HQ	From the *Personality Culture* of *Street Fighting Man* to a Systems Approach for Project and Client Management
Project Managing Function	Specific Issues The HQ-Site interface	Integrating *Relationship Marketing Management* to Project Management
Executing Function	Project Specific Issues Site level	From an operations focus to a *Client Orientation*

Figure 11.1 Project management and sales systems.

The sales team or person should be addressing the *temporary multi-organisational team*. Just as the DMU needs to be profiled (see Chapter 8), so the other parties need to be profiled from a marketing and sales perspective. Networking across the virtual organisation should cover those outside the DMU with a sales role – sales and marketing personnel, public relations and senior management. In this way, the culture and approach to marketing and sales can be gauged. This reveals several important things:

- The values and beliefs of the culture
- How these values are manifested in marketing and sales
- The specific promises that were made to the client
- The actions being taken to deliver those promises
- The scope for synergy on the client project
- The scope for collaboration for repeat business and working together with other clients.

The first three are relevant to any project. Any consultant or contractor benefits from this type of investigative profiling for any sizeable project. The fourth item – deliverables in marketing jargon – is also of relevance. It is important for repeat business, yet even more so in referral markets, where any shortfall amongst the other parties may adversely affect your own reputation. The fourth and last two items are absolutely crucial in relationship marketing.

One of the messages of this book has been how everyone has a part-time marketing and selling function. It has been stressed how project personnel must have one eye on selling. Here it is being advocated that the sales people work *inside* the project team. While many sales people, especially in contracting organisations, come from a project background, they should not leave it totally behind as they develop their sales capabilities. Profiling by the sales and other personnel provides valuable market intelligence for the contracts director and site manager. It also provides vital data beyond the immediate project horizon.

This type of profiling of other members of the virtual organisation can provide valuable information as to which clients to work with in future and on what sort of contracts, for example on BOOT-type projects or D&B. There are many organisations in other sectors that have developed marketing teams in this way,[4] sometimes across organisational boundaries.[5]

One of the functions of the sales people inside the team is to help manage the values as they relate to innovation, risk, co-

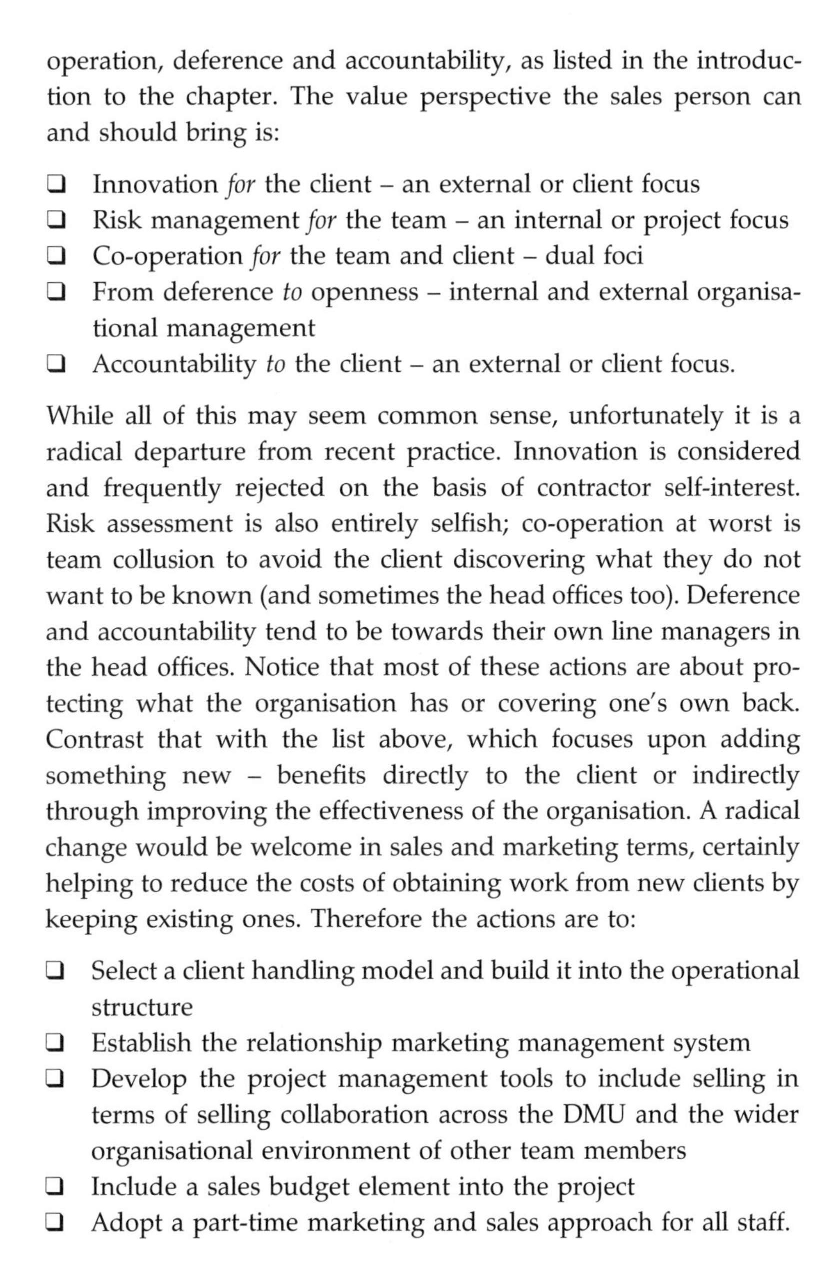

operation, deference and accountability, as listed in the introduction to the chapter. The value perspective the sales person can and should bring is:

- Innovation *for* the client – an external or client focus
- Risk management *for* the team – an internal or project focus
- Co-operation *for* the team and client – dual foci
- From deference *to* openness – internal and external organisational management
- Accountability *to* the client – an external or client focus.

While all of this may seem common sense, unfortunately it is a radical departure from recent practice. Innovation is considered and frequently rejected on the basis of contractor self-interest. Risk assessment is also entirely selfish; co-operation at worst is team collusion to avoid the client discovering what they do not want to be known (and sometimes the head offices too). Deference and accountability tend to be towards their own line managers in the head offices. Notice that most of these actions are about protecting what the organisation has or covering one's own back. Contrast that with the list above, which focuses upon adding something new – benefits directly to the client or indirectly through improving the effectiveness of the organisation. A radical change would be welcome in sales and marketing terms, certainly helping to reduce the costs of obtaining work from new clients by keeping existing ones. Therefore the actions are to:

- Select a client handling model and build it into the operational structure
- Establish the relationship marketing management system
- Develop the project management tools to include selling in terms of selling collaboration across the DMU and the wider organisational environment of other team members
- Include a sales budget element into the project
- Adopt a part-time marketing and sales approach for all staff.

Costs and benefits

All of this may sound expensive and thus inadvisable, yet with the falling costs of project management that can be yielded by using the intranet and majordomos across organisations, this is more

Table 11.1 Impact of information technology on marketing in project teams

Barriers	Impact of information technology
Team size	Improves speed of communication and co-ordination, especially using the intranet
Physical proximity	Brings people together to one place – the computer
Shared values	Key values can be encapsulated for project – the project mission statement – on the intranet
Promises and brief	While not a substitute for face-to-face contact, acts as a reminder of key issues and concerns of the client
Monitoring performance	Can be used for accountability where programme and performance measures are displayed, shortfalls as well as successes being apparent.

realistic than has been possible in the recent past. Table 11.1 lists some of the benefits of using IT for marketing.

The potential for this type of networking, profiling and IT application is greatest in partnering relationships.[5] These types of systems, procedures and actions can be put in place as a way of bringing people together. Trust can be tested and a context put in place for developing the conditions for trust to develop. The criteria by which this can be assessed in the project team are:

1. To what extent is the communication *open and active*?
2. To what extent is there *influence* across the virtual team, whereby people from different organisations are prepared to accept the views and values of others and have their own adopted by negotiation?
3. To what extent is there *control reduction*, whereby co-operation rises above wielding power?
4. To what extent is *forbearance from opportunism* present, whereby people act in a considered way for good of the overall project, client and future business?[5]

This type of activity should be viewed as a low-cost level of investment for developing relationships at the project stage. In this sense, the whole effort has a positive sales push towards relationship marketing for repeat business and in referral markets.

There are strategic issues to address for these activities to be successfully worked out:

- ❑ Work with the project management staff to establish a broader definition of project management procedures, which includes the sales element
- ❑ Seek to develop an open project management system using IT that will facilitate collaboration across the temporary multi-organisational project team
- ❑ Build in a feedback process from the sales side as well as performance measures as far as client satisfaction and quality control are concerned.

Research has shown that the sales benefits of this type of approach are greatest in organisations where there are traditionally low levels of organisational integration. This bodes well for construction and consulting organisations. The following is taken from a manufacturing context:[6]

> *Hence, by aligning the parties' interests and values, collaborative communication may serve as a pseudo-integrating device, thereby making the independent, autonomous dealer feel like more of a partner with the manufacturer. When manufacturers take the time and make the effort to share information, solicit and give feedback, and routinize communication flows, they may see improved coordination of the dealer's efforts with their own. In such independent relationships, collaborative communication may help the dealer work with the manufacturer and protect its interests at the same time.*

The authors state that this type of approach is relatively simple to set up in terms of time and cost. The issue is more to do with willingness than cost or anything else.

A further sales benefit is that such a system can be shared with the client, for example with access to some intranet pages or as a signed up member of the intranet. This is no different to an open book accounting system applied to project management and marketing. Research in construction has shown client perceptions of professional services to be lower than consultants themselves believe them to be (see Table 11.2). Either the clients are correct in their analysis or the project team failed to fully communicate how well they perform.

Apart from *reliability*, the service quality was lower for every dimension in the eyes of the client. In addition, the research also showed a gap between the level of client expectation and the level of service received.[7] Therefore, there is considerable room for improving service quality and the approach outlined above may be one way to do this.

Table 11.2 Comparisons of client and professional perceptions of service quality

	Average perception score	
Dimension	Client	Professional
Reliability	5.17	5.14
Assurance	5.57	5.68
Responsiveness	4.90	5.18
Empathy	5.05	5.26
Tangibles	4.99	5.12

Source: Buttle (1996).[7]

All of this may seem a far cry from selling. This is not the case at either a general level or a specific one. It may simply reflect the gap between existing sales and client management approaches to the potential; in other words, the above is contributing to the envisioning objective of the book.

In concrete terms, it has been found that improving service quality yields higher levels of repeat business and that improving this level by 5% can raise profitability between 25% and 85%.[8] The actual level depends upon the sector and type of organisation. There can be reasonable confidence that the levels of profitability can be higher in construction with the high costs of securing business. This is supported by reported contractor profits, where a focus on partnering relationships yielded a 3.6% profit margin above the sector average.[9] This uplift is achieved prior to taking on board many of the recommendations suggested here.

For the sales personnel, the enhanced collaboration across the team should aid selling and lead to increased client satisfaction. Raising the client expectations is a positive thing at the courting stage. The positive feeling induced appears to leave clients with a more positive perception of the service received at the end of the day (cf. Mohr *et al.*, 1996),[9] even if there is still a gap between expectation and service perception. The sales people also can sell the concept of collaboration across the network of the virtual team organisations. Hence:

- ❑ Sell to the client the greater collaboration, hence enhanced service, through integrating sales into the project
- ❑ Sell across the temporary multi-organisational team the benefits of close collaboration for the project
- ❑ Profile the sales and other management staff outside the DMU

- ❑ Establish and sell, if appropriate, the benefits of long-term collaboration with other temporary teams and of establishing a more permanent relationship.

Conclusion

What works well is to sell a feature *with* corresponding benefits. This is a case in point. Improved satisfaction levels are what the client gets. The contractor or client is seeking greater efficiency and better working relationships on the current project, hopefully for future ones too. The aim has been to inject selling into the virtual organisation of the project team. In doing this, it adds service benefit to both the client organisation and to all other team organisations. It enhances project management and the potential for repeat business. It is achieved by managing the virtual or temporary multi-organisational team from a client orientated perspective.

This is abstract in the sense of, 'Where do we start?' or 'Don't we do some of that anyway?' Yet it is very concrete when looked at in terms of the tools that can be used in the project management to incorporate collaborative mechanisms, such as the intranet, an effective client handling model and a sales presence within the team. Finally, a means of feedback to the head offices needs to be established in order to monitor progress.

Summary

1. The purpose of this chapter has been to demonstrate that managing the virtual project team or temporary multi-organisational project team is an important sales issue in order to:
   - Deliver promises
   - Enhance the service, hence client satisfaction
   - Increase the opportunities for repeat business
   - Establish closer collaborative relations with other team member organisations for the project and for potential business opportunities.

References and notes

1. Cherns, A.B. and Bryant, D.T. (1983) Studying the client's role in construction management, *Construction Management and Economics*, **1**, 177–184.
2. Ahmad, I.U. and Sein, M.K. (1997) Construction project teams for TQM: a factor-element impact model, *Construction Management and Economics*, **15**, 457–467.
3. Mulder, L. (1997) The importance of a common project management method in the corporate environment, *R&D Management*, **27**(3), 189–196.
4. Cf. Good, D.J. and Schultz, R.J. (1997) Technological teaming as a marketing strategy, *Industrial Marketing Management*, **26**, 413–422.
5. Smith, J.B. and Barclay, D.W. (1997) The effects of organisational differences and trust on the effectiveness of selling partner relationships, *Journal of Marketing*, **61**, 3–21.
6. Mohr, J.J., Fisher, R.J. and Nevin, J.R. (1996) Collaborative communication in interfirm relationships: moderating effects of integration and control, *Journal of Marketing*, **60**, 103–115.
7. Buttle, F. (1996) Service quality in the construction industry, *Proceedings of the 1st National Construction Marketing Conference*, 4 July, The Centre for Construction Marketing in association with CIMCIG, Oxford Brookes University, Oxford.
8. Reichheld, F.F. and Sasser, W.E. (1990) Zero defections: quality comes to service, *Havard Business Review*, **69**(5), 105–111.
9. *Financial Times* (1998) Morrison: group places faith in 'partnering', 26 November.

Chapter

12 The Client Perspective

Themes

1. The *aim* of this chapter is to consider the procurement perspective of the client.
2. The *objectives* of this chapter are to consider:
   - ❑ Which criteria a client does and could use to choose a contractor or consultant, and particularly why they should come back for more
   - ❑ Whether there are costs for the client in switching suppliers and under what conditions.
3. The primary *outcome* of this chapter is to understand how marketing can create barriers to competition and hence increase repeat business.

Keywords

Client loyalty, Client satisfaction, Partnering, Repeat business, Switching costs

Introduction

Sixty to eighty per cent of customers say that they are satisfied with their supplier and yet switch to another supplier.[1] Why do customers or clients defect? What can be done to keep a client in the contracting and consulting fields? This is the focus of the chapter. Clients have to have good reasons to stay loyal to their suppliers. There have been few incentives for clients to maintain relations in the long term.

Relationship marketing has become and is increasingly becoming an important way in which contractors and consultants sell their services. Selling over the long term and staying in contact is not quite the same as relationship marketing. It is a start, yet it is possible to use a service repeatedly without having a meaningful relationship.[2] Convenience may be the glue that holds things together.

The primary obstacle to sustaining relationships with clients in the long run is *switching costs*. Put simply, a client moves or can change supplier when the costs of so doing are low. Low switching costs lead to low levels of repeat business. High switching costs will lead to high levels of client retention and hence repeat business. Construction has been characterised by low switching costs and low loyalty levels.

Switching costs are a theoretical component of *transaction analysis* and thus the traditional marketing mix approach. Transaction analysis looks at the costs associated with placing an order or contract from the viewpoint of the client and the supplier. The primary theoretical question concerns increasing the switching costs of the construction transaction through relationship marketing. The practical issue concerns what strategies and actions contractors and consultants can adopt to sustain relationships and obtain repeat business.

In the sector, the adoption of relationship marketing will lead to increasing differentiation of construction services and hence to greater diversity among contractors. Theoretically, there will be a consequential increase of concentration and hence a reduction of competition. Relationship marketing is a source of competitive advantage. The close relationship with a client makes the learning curve steeper for others and thus increases the barrier for the competition. Thus, switching costs are increased.

The benefits for contractor and client of developing relationships over a long period are:

- ❑ Continuing cost reductions – Client
- ❑ Tailored service provision – Client
- ❑ Client satisfaction – Client
- ❑ Repeat business – Contractor or consultant
- ❑ Improved turnover and profitability – Contractor or consultant

The first three benefits accrue to the client and thus increase the value of service. Receiving equal value from another contractor would require greater levels of investment from the client, hence increasing the transaction costs. Switching costs have increased. The latter two yield benefits for the contractor and directly affect the 'bottom line'. Where this results in lower tendering or negotiation costs for the contractor in bidding for the next project for the client, then the transaction costs of the contractor are also reduced. The effect on profitability can be seen in Figure 12.1.

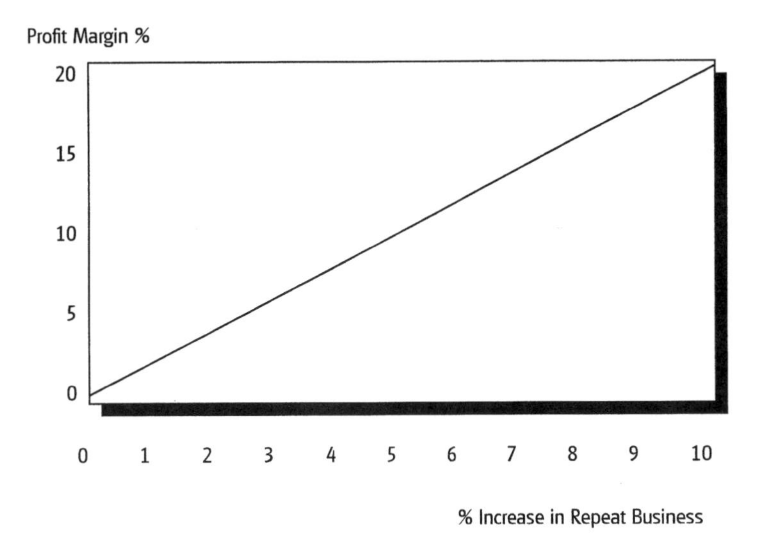

Figure 12.1 Effect of repeat business on gross profit.

Towards mutual co-operation

Repeat business is the most significant contractor and consultant benefit because it holds the key to realising increased profit, both in terms of the mass of profit accumulated and the margins achieved. Compared to other service sectors where repeat business is typically 60%, in construction it is 20–25%.[3] The ability to obtain repeat business is dependent upon the strength of the contractor or consultant in the market and the buying power of the client. There are three main strategies for buying and selling. These are:

- Competitive
- Co-operative
- Command.[4]

The *competitive strategy* is the classical market approach. If the client and contractor behave in a competitive way, then both are acting *independently* of each other and a *perfect market* results (see Figure 12.2). This is a traditional position of consulting and contracting. The construction market is highly fragmented in terms of the number of companies offering their services. It is also homogenous in the sense that services offered by contractors are undifferentiated, inducing an absence of competitive advantage beyond price competition at the tender stage and post-tender negotiations. In these circumstances, the client can adopt a *command strategy*, the contractor having to maintain a *competitive strategy* – the bottom left-hand position in the matrix in Figure 12.2. Under these circumstances, both parties operate *independently* in a *buyer's market*. The roles are seldom reversed. Contractors seldom operate in a *seller's market*. Contractors only operate a *command strategy* with clients being *co-operative* in very exceptional circumstances, such as in war time.[5]

Client retention is about curbing competition and changing the market mechanisms. It is about moving away from open competition to mutual advantage – a more *co-operative strategy*. However, this does not mean that the co-operation is balanced or equal. This is the state of play in partnering within the UK at the time of writing. Experts and clients advocate mutual advantage;[6] however, most clients wish the contractor to be co-operative, while they retain a commanding position.[7]

Where there is an imbalance in favour of the client, the contractor becomes *dependent* upon the client, acting in essence in a *subcontract* role. It would be understandable under such circumstances for many contractors to pay 'lip service' to new working practices, such as partnering as a quick fix to secure work.

Co-operation creates a climate for mutual objectives to be met. Co-operation integrates marketing and procurement through close relationships. It can be formalised through procurement routes, such as partnering, but it is not necessary to pursue formal

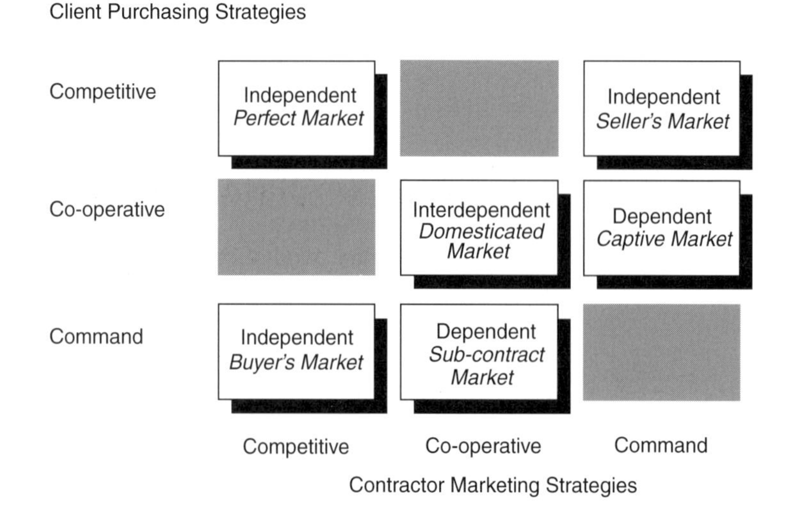

Figure 12.2 Classification of buyer–seller relationships.

solutions. Under these circumstances interdependence emerges between the parties, the contractor operating in a domesticated market. In other words, the clients receive a service that contains many of the benefits of an in-house service.

In summary, the most favourable or equally balanced position is a mutually ***co-operative*** one. There is a high degree of *interdependence*. It is not a relationship of complete dependence. It may take a number of forms:

- A close working relationship – an informal solution induced through effective marketing and sales
- A term contract may be one form – a formal outcome of an informal marketing and sales approach
- A strategic partnering or framework agreement – a formal outcome from a formal marketing and sales approach.

The more satisfied the client, the greater the likelihood that they will be *loyal*. Loyalty bodes well for repeat business. However, there is nothing automatic about this (see Figure 6.2). Many satisfied clients defect, unless they are managed. Client management as opposed to project management is the order of the day in relationship marketing. However well managed a client is, periodic evaluation will take place to assess whether to switch to a different contractor or consultant. From the client perspective there must be good reasons for staying with the same supplier. This is where economics and marketing meet head on. Indeed, the role of marketing at this point is to raise the economic barriers. This gives the client a higher hurdle to overcome in order to justify defecting. In other words, the objective is to use marketing to increase the switching costs.

Economics and marketing

The economic forces at play are the costs of doing business – the *transaction costs*.[8] Developing the understanding of transaction costs further, favourable supply and demand factors meet in the marketplace, under neo-classical economic theory when an exchange takes place. That exchange is not 'free'. It carries its own costs for both parties. In construction, the exchange process is particularly complicated. From pre-qualification, tendering, post-tender negotiations, through stage payments, until the final account is settled are a whole host of administrative costs for the contractor and client which are incurred as overheads. Prior to the tender or contract negotiations there are costs for selecting a contractor, which starts with pre-qualification under traditional contracting. These are the direct *transaction costs* incurred by the client.

There are also indirect transaction costs, which arise from learning to work together – the so-called *learning curve*. Relationship marketing can be used to increase the steepness of the curve in understanding at considerable depth the needs of the client and responding accordingly. Because relationship marketing is a source of becoming close to a client, making the steep learning curve for the competition becomes a barrier. Thus, switching costs are increased.

Relationship marketing reduces some of these costs for the client. Serving the client, going the extra mile, using the techniques described in this book, give to the client both tangible and intangible reasons for the client to stick with the same contractor or consultant. Consultants tend to be better at this by virtue of usually being higher up the supply chain; however, contractors have failed to capitalise on their key position.

Therefore, the client cost of *switching* from one contractor to another is traditionally very low, even where partnering arrangements are in place. In order to increase the switching costs and retain clients as much as possible, investment is required in the relationship (see, for example, ... *On the case 40*). This must be on both sides. Indeed, monitoring this is an important way for both parties to examine how committed they are to a co-operative approach. Therefore the initial transaction costs are high, and the downstream benefits in terms of time, cost, quality and risks are increased. After the initial high transaction costs, subsequent transaction costs are lower. Consequently, the switching costs are low at the outset, yet grow as the relationship matures. In other words, the interdependency takes time to be established.

Risks are therefore high in the early stages of a relationship, for the switching costs take some time to increase to a significant level. It takes time for the benefits of a client–contractor relationship to work through to yield high switching costs.

... On the case 40

An international contractor specialising in leisure had proved to be highly effective in securing repeat business from a high-profile client. Repeat business reached 50% of annual turnover in 1 year. They invested in the service relationship. This was considered to be moving towards dependency for the client in an otherwise co-operative relationship and other core clients were being added to the portfolio of repeat business clients.

In contrast, a medium-sized British contractor entered a lucrative facilities management and maintenance market of term contracts. Being one of five principal players, they believed they were in a strong position to renew their contract. Investment in the service relationship had not been made a high priority. A term contract and investment in equipment was deemed sufficient. However, the client took a different view. The client decided to 'flex its muscles' and failed to renew one of the contracts. This contractor lost its work, yet the client kept a high level of continuity across all its suppliers.

Switching costs and low loyalty levels

Getting close to the client can raise switching costs. That is adding value of an intangible nature. The offer can be enhanced through more tangible aspects of the service. There are constraints operating in traditional contracting:

- Fragmented industry inducing intensive price competition, which size, geographical coverage and differentiation offer the respite and solution
- Design, hence product differentiation, is conducted by others, hence the advantage of D&B and BOOT-type contracts from the contracting perspective, although clearly there is no corre-

sponding benefit for consultants unless a contractor can supply a steady stream of work to a specialist D&B consultant
- Absence of a marketing culture where staff are all part-time marketers.[9]

Where switching costs are low, it is said, transaction marketing is appropriate and relationship marketing becomes inappropriate.[10] This poses a constraint for contracting. Given the above, relationship marketing is untenable. The message is revert back to the price domination of the marketing mix approach. However, clients are saying they need the investment in new approaches to contracting from the industry.[11] The climate is changing, investment from the contractor is being demanded and the client must also make corresponding investments to ensure mutual co-operation. Transaction costs may initially rise on both sides, but will fall over time.

One way to introduce such measures is to adopt a structural solution to a marketing problem. Partnering can be seen as a tool of relationship marketing in this regard. However, in order to reap the benefits of partnering, contractors are advised not to set up separate partnering departments or divisions, as has been the case for construction management, but to permit the marketing and sales aspects to filter across the organisation. This will yield positive benefits to the contractor in changing the culture.

There are four main ways of increasing switching costs through relationship marketing management. These are:

- Commitment to a formal or informal relationship that is already in existence
- Reduced costs
- Rewarding personal working relationships, which may yield a high quality experience for the client
- The high cost of switching to another contractor.

All four are valid. Not every reason carries equal weight. Commitments, especially informal ones, can be reneged upon. There may be several reasons for reneging on a relationship. If cost reductions are not being met, then there may be an incentive to renege on the relationship. The price factor in the 4Ps of the marketing mix comes to the fore. Transaction costs are lowered. Switching costs are lowered as the benefits are eroded. Experiencing rewarding personal working relationships focuses upon a narrow band of people – the client decision-making unit and the project team. This can be vulnerable to other influences within the company and is also dependent upon some stability within the client personnel, even if the contractor successfully manages continuity in their own organisation.

High switching costs is the most compulsive reason for retaining client loyalty. For relationships to be sustainable in the long run, there must be higher switching costs created through supportive management actions. The client project team is usually accountable to a board of directors. The directors operate at arm's length. The obligations towards honouring a relationship may be weaker where price competition is intensive. The client board will not have direct experience of the day-to-day working benefits of a high-quality working relationship. The client project team need to justify their decisions to the directors, especially on cost grounds. Their account of quality working relationships may not hold a great deal of weight, especially in a highly cost competitive industry. It is hard to quantify the benefits of good working relationships and the board may not have to live with the direct consequences in any case. How can the contractor help to raise client awareness at board level of the increase in switching costs under these circumstances? The contractor can help to show that the savings and reductions have occurred on contracts compared to previous traditional contracts. Cost improvements over successive partnering contracts can also be shown. This can be communicated at every level of decision making, including client board level.

If the comparative costs are high to switch contractors, then a client will not undertake that move. This is the tangible measurement. It has suggested that customers will only seek long-term relationships where the product or service is strategic to their business.[12] A new building may be strategic, but it is not of on-going strategic importance whereby there is an incentive to maintain close working relationships. Unless a total facilities package is in place, that structural strategic link is broken after contract completion and the final account. How can this be tackled? If the value supply chain is broken in project terms, then the human investment in the relationships has to be increased. It is increased so that the contractor is making continual input into the client's estate and facilitates problem solving and decision making.[13]

What *tactical* management responses can be put in place to overcome typical problems that arise? Typical problems have been grouped under broad areas of risk, client priorities, client expectations and anxieties. Taking risk as one factor, the first problem is project performance. Client fear of immediate disruption on current contracts may lead to reluctance to place repeat orders while a contract is still outstanding. The client may also fear unsatisfactory performance on the next project because the contractor may become overstretched, hence leading to a reluctance to place repeat orders. This immediately opens the door to clients having several suppliers or partners. This in turn may lead to a pause between contracts. Competitors who subsequently come along can cast doubts in the client's mind, which can lead to switching, even where the original client intention was to stay loyal.

The remedy against this situation is to keep in close contact with the client and to track competitors in a systematic way.[3] This occurs both during a project and continually at a strategic level (see ... *On the case 41*). It requires a reversal of the traditional senior management input. Contact becomes greatest at the end of a con-

... On the case 41

A large architectural practice, employing around 100 people believed its level of repeat business to be high. On analysis it found that corporate repeat business was not particularly high, except for a very few clients.

What it did discover was that its repeat business with individuals was very high. Directors had the capacity to build and maintain long-term relationships based on personal empathy and friendship.

Working for a new client often came about when personnel on the client side switched employer and introduced this practice to the client body.

This proved to be very useful in knowing what was being tracked. It was also valuable in the recession of the early 1990s. As people were made redundant, every effort was made to keep track of the individuals until such time as they were placed into new employment. The practice wished to be seen as the one that remembered them during harder times.

tract and before the decision to commence the design of the next building.[14] It also requires monitoring how much investment the client is making in the relationship in order to assess the strength of the relationship or partnering process.[3,10]

The client corporate priorities may change. This gives rise to the second set of problems. There may be changes in policy from the top of the organisation. Under these circumstances the remedy is both to track change and amend client profiles of the decision-making unit. Rebuilding relationships must then take place quickly, building on the personal and performance goodwill as much as is permitted.

The key staff may change in the client project team. The task is to amend the client profile and adjust the team approach. This recognises the virtual organisation or the temporary multi-organisational team (see Chapter 11),[15] and to rebuild the contractor team between projects in order to closely reflect the new profile in the client project team.[16]

Another area of concern is client expectations and anxieties. One issue is that the more complex the marketing and procurement systems become, the greater the levels of expectations over the service. Overcoming this obstacle requires building the relationship, especially establishing conditions of trust.[17]

The act of buying changes the buyer, who expects the seller to remember the purchase as having been a favour.[18] As has been pointed out, in general 60–80% of customers that switch are satisfied with what they have received. A satisfied client in terms of a close relationship may not automatically yield a (premium) profit[19] and may not automatically yield repeat business.[1,20] The remedial objective is to serve the client and add value to the service.[21]

When the client stops complaining, their interest is waning.[18,22] To overcome this, the contractor must increase dependency and reasons for co-operation. Looking at this from a procurement angle, the purchasing behaviour of the client must be shifted away from a command approach towards one of mutual co-operation – a shift from independence to interdependence.[4,23] A differentiated service, linked to close contact, will yield opportunities to identify where the strengths of the contractor can be mobilised to address particular issues and problems on the client side. This can occur during and between projects. It requires a facilitative approach among sales personnel and a part-time sales approach across all other functions.

The client decision-maker can become motivated by hedonistic variety seeking when letting a contract.[24] A change of personnel on the client decision-making side may lead to the new person wishing to establish a power base. It may also occur against the best interests of the client where career politics overrides other considerations, where one decision-maker wishes to extend their power over that of another colleague or interested party. This hedonistic behaviour contrasts with utilitarian reasons for switching. However, the sales effort of the contractor can reduce hedonism by encouraging multi-level relationships between the key personnel in the client hierarchy.[25]

Creating a scapegoat of the contractor within the politics of client organisation is another issue. Creating a scapegoat is a potent strategy during periods of client structural change, for example invoked as part of a relocation or expansion plan.[26] Unfortunately, there may be little a contractor can do to obviate this, apart from doing their best and endeavouring to explain to the client what is going on.

The contractor may also fail in being an effective partner in the relationship. This may be an error concerning one aspect of the project, a critical event or a so-called 'moment of truth'. It may be the overall performance. These provide legitimate reasons for switching.[26] Overcoming them requires a systematic marketing and management approach, rather than one dominated by profit motives and turnover concerns. If the client is a highly valuable one then considerable investment may be necessary to manage this set of circumstances. However, the contractor must also be assured that such investment will make a difference; in other words, the client will recognise and honour such investment and effort in practice, provided the shortcoming is pointed out or identified in time, before considerable damage to the relationship has been done.

Addressing these issues and the corresponding *tactical* remedies, highlight ways in which the *general* management approach needs to change. While change is bound to be substantial, it can be carried out from the bottom up, that is in response to issues met on the ground. This can be more appropriate than a top-down approach, where many important operational nuances can be missed. In both cases, considerable attention and support are needed from senior management.

A word of caution is needed at this stage. Just as a satisfied customer may switch suppliers, unless actively and positively managed, so a dissatisfied client may remain a service buyer. It should not be assumed that obtaining repeat business is always a sign of relationship strength. There may be other factors:

- Need to commence quickly
- High inherited switching costs, followed by poor performance may lead the client to give the contractor or supplier a 'hard time' on the next project in an effort to rectify falling standards of service
- Lack of confidence in the competition
- Poor procurement approach in the client organisation.

Obtaining repeat business may also mean that the contractor is not getting a good deal. In other words, the client is in a very strong commanding position and the contractor is taking a co-operative stance. Communication has an interesting role in these and similar circumstances:

- When relationships are strong and integrated – interdependent domestic market – collaborative communication has a weak effect upon client satisfaction
- When relationships are weak and lack integration – dependent buyer's market – collaborative communication has a positive effect upon client satisfaction.[27]

Just as open-book accounting policies, once thought to be outrageous, are common practise, so open communication is effective. It is especially valuable where a relationship appears to be weakening or is not as strong as first thought.

The impact of increasing switching costs on market structure

Undertaking these actions will change the operations and management of a number of companies. This will begin to differentiate the services of leading contractors in the sector. Competition is therefore being reduced:

- In the co-operative market
- Through highly differentiated services.

Contractors who are pursuing relationship marketing policies and are successful in sustaining market share will experience improved turnover and profitability. This will provide further opportunities to expand or take over other companies. This in turn will lead to:

- Ownership being concentrated in fewer hands
- Higher switching costs and thus a reduction of competition on price.

Therefore, the knock-on effect of structural changes within companies can begin to change the structure of the entire market. This will not be to the detriment of clients. Indeed, the reverse will prevail. Clients will benefit from:

- Continuous cost reductions through close relationships
- Construction services that are more closely tailored to individual client needs
- Potentially a more responsive and dynamic sector of greater stability and financial rewards, with potential for technical innovation and further service improvements.

Conclusion

Clients may seek to switch contractors. Remedies are required to increase client dependency upon the client. Remedies that are based around relationship marketing have an economic consequence: switching costs are increased, thus these remedies provide incentives to remain loyal. In other words, the economics of repeat business is currently dominated by low switching costs from the client perspective. Relationship marketing offers a means to increase business if policies are strategically and tactically implemented in order to raise the barriers, and hence increase switching costs. This requires investment and changes in structure and approach of contracting organisations. The starting point is to complement the current procurement driven approach with a marketing approach to such opportunities, especially where alliances and partnering feature. Yet it has been emphasised that formal procedures, such as partnering, are in essence tools of relationship marketing and are neither an end in themselves nor can they be divorced from the wider body of marketing and procurement theory and practice. Such divorce will only lead to switching costs remaining low.

It is from this recognition that the management systems should flow. Indeed, management systems for relationship marketing are necessary for relationships to survive otherwise the low transaction costs of switching contractors will overpower the approach. The chapter has therefore provided a useful way of reinforcing some of the issues raised in previous chapters, especially Chapters 8, 10 and 11.

A further important point is that structural change in contracting organisations will induce structural change in the market. In other words, permanent barriers to switching could begin to be erected. This will occur as:

- Sufficient contractors successfully follow the relationship marketing route
- Contractors differentiate their services
- Competition is reduced through fewer contractors operating in the market.

The increasing of switching costs can be undertaken on a project-by-project and on a contractor-by-contractor basis. The relationship marketing approach offers the means to do this and create a firm linkage between client satisfaction and repeat business.

Summary

1. The purpose of this chapter has been to show how relationship marketing can build loyalty and so overcome some of the economic reasons why clients may choose to switch consultant or contractor.
2. A clear link has been established between the multi-disciplinary approach of relationship marketing and the economics of business transactions.

References and notes

1. Reichheld, F.F. (1994) Loyalty and the renaissance of marketing, *Marketing Management*, **2**, 10–21.
2. Blois, K. (1997) Are business-to-business relationships inherently unstable? *Journal of Marketing Management*, **13**, 367–382.
3. Branch, R. F. and Smyth, H. J. (1996) *A Client Orientated Service*, Centre for Construction Marketing, Oxford Brookes University.
   Smyth, H. J. (1997) Partnering and the problems of low client loyalty incentives, *Proceedings of the 2nd National Construction Marketing Conference*, 3 July, Centre for Construction Marketing in association with CIMCIG, Oxford Brookes University.
4. Campbell, N. (1995) An interaction approach to organisational buying behaviour, *Relationship Marketing for Competitive Advantage* (eds A.

Payne, M. Christopher, M. Clark and H. Peck), Butterworth Heinemann, Oxford.

5. See for example Smyth, H.J. (1985) *Property Companies and the Construction Industry in Britain*, Cambridge University Press, Cambridge.
6. See for example Baden Hellard, R. (1995) *Project Partnering: Principle and Practice*, Thomas Telford, London.
   Bennett, J. and Jayes, S. (1995) *Trusting the Team: The Best Practice Guide to Partnering in Construction*, Centre for Strategic Studies in Construction, University of Reading.
   Stephenson, R.J. (1996) *Project Partnering for the Design and Construction Industry*, John Wiley, New York.
7. Smyth, H.J and Thompson, N. A client orientated service for partnering, unpublished paper, Centre for Construction Marketing, Oxford Brookes University, Oxford.
8. Williamson, O.E. (1981) Contract analysis: the transaction cost approach, *The Economic Approach to Law* (eds P. Burrows and C.G. Veljanovski), Butterworths, London.
   Williamson, O.E. (1985) *The Economic Institutions of Capitalism*, Free Press, New York.
9. Gummesson, E. (1990) Making relationship marketing operational, *International Journal of Service Industry Management*, **5**(5), 5–20.
   Smyth, H.J. and Thompson, N. (1997) *Developing Loyal Clients*, Centre for Construction Marketing, Oxford Brookes University, Oxford.
10. Jackson, B.B. (1985) Build customer relationships that last, *Havard Business Review*, November/December, 120–128.
11. Latham Report (1994) *Constructing the Team*, HMSO, London.
    Egan Report (1998) *Rethinking Construction*, The Report of the Construction Task Force, Department of the Environment, Transport and the Regions, London.
12. Donaldson, W. (1996) Industrial marketing relationships and open-to-tender contracts: co-operation or competition? *Journal of Marketing Practice and Applied Marketing Science*, **2**(2), 23–34.
13. Juttner, U. and Wehrli, H.P. (1994) Relationship marketing from a value system perspective, *International Journal of Service Industry Management*, **5**(5), 54–73.
14. McIlveen, A. and Smyth, H.J. (1997) *Adding Value*, Centre for Construction Marketing, Oxford Brookes University.
15. Cherns, A.B. and Bryant, D.T. (1983) Studying the client's role in construction management, *Construction Management and Economics*, **1**, 177–184.
16. Belbin, R.M. (1993) *Team Roles at Work*, Butterworth-Heinemann, Oxford.
    McIlveen, A. and Smyth, H. J. (1997) *Adding Value*, Centre for Construction Marketing, Oxford Brookes University.
17. Thompson, N. (1997) Evidence on conditions of trust, *Proceedings of the 2nd National Construction Marketing Conference*, 3 July, The Centre for Construction Marketing in association with CIMCIG, Oxford Brookes University, Oxford.
    Thompson, N. (1998) Can clients trust contractors? Conditional, attitudinal and normative influences on client's behaviour, *Proceedings*

of the 3rd National Construction Marketing Conference, 9 July, The Centre for Construction Marketing in association with CIMCIG, Oxford Brookes University, Oxford.
Smyth, H.J. and Thompson, N. (1999) Partnering and trust, *Customer Satisfaction: A Focus for Research and Practice in Construction, Proceedings of W55 and W65 – Joint Triennial Symposium*, CIB, 5–10 September, University of Cape Town, Cape Town.

18. Levitt, T. (1983) After the sale is over, *Havard Business Review*, September–October, 87–93.
19. Storbacka, K., Strandvik, T. and Gronroos, C. (1994) Managing customer relationships for profit: the dynamics of relationship quality, *International Journal of Service Industry Management*, **5**, 21–38.
20. Fellows, R.F. (1994) *Some Perspectives on Marketing in Construction: Person to Person*, paper presented at the *Investment Strategies and Management of Construction Conference*, 20–24 September, Brijuni, Croatia.
    Leading Edge (1994) *Capturing Clients in the 90's: a Benchmark Study of Client Preferences and Procurement Routes*, Leading Edge Management Consultancy, Welwyn.
21. Branch, R.F., McIlveen, A. and Smyth, H.J. (1995) *Developing Unique Selling Propositions in Fragile Construction Markets*, Centre for Construction Marketing, Oxford Brookes University.
    McIlveen, A. and Smyth, H.J. (1997) *Adding Value*, Centre for Construction Marketing, Oxford Brookes University.
22. Smyth, H.J. and Thompson, N. (1997) *Developing Loyal Clients*, Centre for Construction Marketing, Oxford Brookes University, Oxford.
23. Smyth, H.J. (1996b) Effective selling of design and build: architecture as a sales strategy, paper presented at *COBRA '96, RICS Construction and Building Research Conference*, 19–20 September, University of the West of England, Bristol.
24. van Trijp, H.C.M., Hoyer, W.D. and Inman, J.J. (1996) Why switch? Product category – level explanations for true variety-seeking behavior, *Journal of Marketing Research*, **33**, August, 281–292.
25. Gummesson, E. (1991) Marketing-orientation revisited: the crucial role of the part-time marketer, *European Journal of Marketing*, **25**(2), 60–75.
26. Judge, P. (1992) Unpublished paper, WPP, London.
27. Mohr, J.J., Fisher, R.J. and Nevin, J.R. (1996) Collaborative communication in interfirm relationships: moderating effects of integration and control, *Journal of Marketing*, **60**, 103–115.

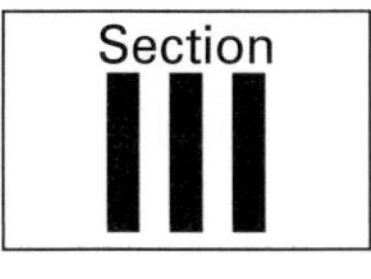

Which Sales Messages and Evaluation Techniques?

This section contains the remaining three chapters. The previous chapter brought together some of the issues of selling and the strategic underpinnings. It considered how the contractor in particular and the consultant can respond from the client perspective to the services on offer. In one sense, it is the culmination of the prime thrust of addressing the external environment. The external focus becomes secondary in this final section.

The emphasis here is on generating the messages internally for external consumption. The first chapter in this section deals with the corporate approach as to what those selling are prepared to and will say to clients. The last two chapters focus on evaluating what is happening in the outside world and what that means internally. In particular, the sales effort has to be effectively monitored and adequate research conducted into the market; these are then combined to refine the strategy. This completes the loop and feedback into the topics discussed at the beginning of the book and especially the chapters on strategic planning and sales systems.

The focus for the section is therefore upon the implementation of the sales plan and feedback from this into the marketing plan.

The sales person needs to be equipped with key messages to direct their sales effort. The *soft images* of communication are dealt with in Chapter 13. Just as advertising performs this function in mass consumer markets, so an equivalent is needed in business-to-business services marketing. This is a neglected area for both selling and public relations.

In contrast, *hard data* is needed for the evaluation processes outlined in Chapters 14 and 15. Hard data can be both quantitative and qualitative. These are the facts from the outside world, which are collected into the internal corporate environment. Facts are useful. The competitive issue is how to interpret them. This occurs at two levels: deciding what information to collect and evaluating it. Both deciding and evaluating are also matters of interpretation as to what is considered important. The decisions, just like all decisions of this sort in science, are subjective, because they are based upon what people believe to be important questions to ask. What one organisation asks will differ from that of its competitors. This in turn depends upon the approach and hence competitive advantage through service differentiation. The second level is interpretation of the data received. The same subjectivity comes into play here. The point being made is not so much a wish to eliminate subjectivity and value judgement, which is frankly a 'straw man', but to recognise the importance of interpretation and manage the process to best serve your organisation. That does not mean 'running' with the issues that suit your viewpoint, but to consider the relationship between what the organisation is trying to achieve and how closely that matches the external environment. This is then implemented in accordance with that environment in relation to the organisation's strategic and tactical objectives. Here the focus is on realising the competitive advantage and hence concerns *marketing effectiveness.*

Once again, therefore, the message that *managing these issues is a potential source of competitive advantage* comes to the fore.

What has been built up over the previous section is a complex picture. Relationship marketing means managing complex sales situations prior to, during and post-projects. The situations involved are numerous:

- ❑ According to the client-handling model employed
- ❑ Acknowledging the virtual organisation of the temporary multi-organisational team
- ❑ Because of the client DMU size, structure and operations.

The results are complex networks endeavouring to respond to the diverse client needs and managing the client experience of all contacts with the consultant or contractor. This is about *sales effectiveness.*

Cravens and Piercy[1] expressed effectiveness in this way:

> *... the issue of effectiveness clearly extends beyond the objective of minimizing transaction costs proposed ... Effectiveness logically includes more than financial performance, market position and customer satisfaction. Flexibility, risk reduction, and other strategic objectives may be important aspects of network effectiveness. Effectiveness needs to be examined in terms of the type of marketing relationship appropriate for a particular network form.*
>
> *Management evaluation and control systems in network structures may call for softer intelligence and analysis (p. 50).*

Systems were addressed in the first section of the book, as were some aspects of evaluation. These aspects are being drawn together in this section in order to complete the functional feedback loop for both the market and personnel.

Reference

1. Cravens, D.W. and Piercy, N.F. (1994) Relationship marketing and collaborative networks in service organizations, *International Journal of Service Industry Management*, **5**(5), 39–53.

Chapter 13 Sales Messages

Themes

1. The *aim* of this chapter is to consider the way in which organisations communicate with the external environment.
2. The *objectives* of this chapter are to analyse:
   - ❑ The nature of sales messages
   - ❑ How messages can be generated for sales and public relations purposes.
3. The primary *outcome* of this chapter is to understand the importance of managing the inputs into corporate communication to its external target audiences.

Keywords

Benefits, Sales messages, Images of the organisation

Introduction

Selling services involves several processes, which can be summarised as follows:

- ❑ Describe features, yet sell benefits
- ❑ Overcome objections
- ❑ Acknowledge problems, sell solutions
- ❑ Turn the person to whom you are selling into an advocate, so that they do the selling for you
- ❑ Ask for a regular update from the client on how you are doing and how long before the sales decision is to be made.

Can everyone be effective in this role? Many of the most effective sales people have the outgoing, positive and friendly approach as a gift. Good sales people are frequently resource-investigators and team workers in terms of Belbin Group analysis (see Chapter 8). Learning enhances that role. It is enhancing the innate ability that gives a competitive edge. Processes need to be learnt, techniques enhanced through training, technical expertise obtained and experience enrich the capabilities of the sales person.

In terms of *outputs*, sales people are often motivated by being appreciated and applauded by others. This means that they are prone to exaggerate how well things are going from a management point of view. Managers are usually aware of this, make allowances accordingly and probe for further information. We shall return to the outputs in subsequent chapters.

The neglected area is *inputs*. What will the sales person say, as they go through the stages above prior to closing the sale? What messages will they convey as they build relationships with the target clients? How consistent will it be in comparison to what colleagues say? Whom should they draw in, or worse still, will there be reluctance to draw in others for fear that their efforts may be undermined? These uncertainties arise because inputs are not developed and controlled.

It is necessary to be clear what these *inputs* are. They are not standard phrases or sales lines. These would be false, inhuman and can be sent by direct mail! No need for sales people – just people to stuff envelops. The inputs are *generic messages*, upon which the sales people can draw to create and form their own sales lines and responses.

Getting the message

How can organisations generate *generic messages*? As a preface, it is necessary to be able to categorise messages. Figure 13.1 provides a framework of categorisation. Specific messages can be divided into two: those that communicate benefits and those that communicate features. As indicated, the benefit message is preferred and more effective. It is surprising how many advertisers still major on features.

What is a service benefit? The benefit is what the client is looking for, so in the case of added service value the client might want least hassle as the benefit of a service, which tries to ameliorate all problems before reaching the client. Examples of benefits might include those listed below, with the aspects from which they are derived set alongside in order to relate back to previous sections of the book:

- Low maintenance – financial spin off
- No surprises – understanding and responding to client needs
- Reliability – relationship benefit
- Receptivity – condition of trust
- Value for money – product and price in the 4Ps

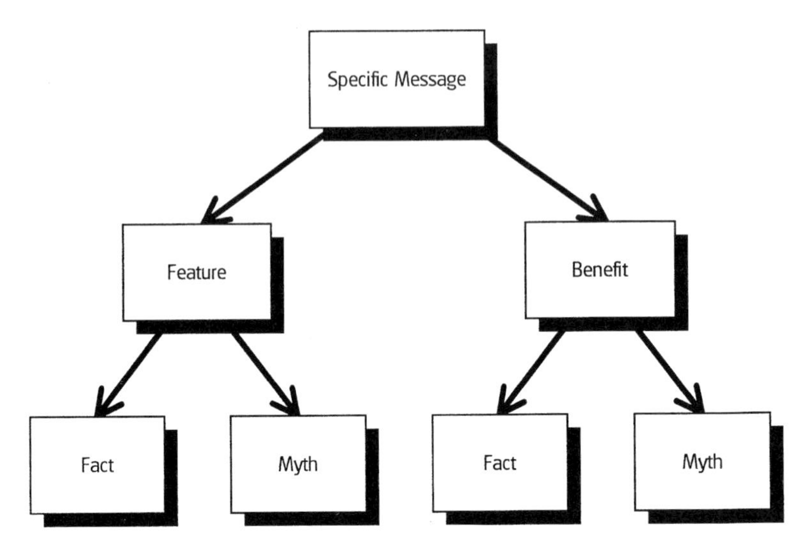

Figure 13.1 Framework for message categorisation.

- ❑ Minimal perceived risk – outcome of high trust levels
- ❑ Hassle-free service – satisfying motivation within DMU
- ❑ Smooth operation – derived from team characteristics.

Each of these can be sub-divided into messages of fact and myth. Myth centres on messages of association, for example in advertising selling a product on the basis of status, lifestyle or sex. There are creative ways of using association, as will be seen below, that preserve integrity. There are also those mythical messages, lacking integrity, that set out to manipulative and deceive. These are to be avoided. The client will soon discern that the people and the service cannot deliver. The preferred route of categorisation is therefore message–benefit–fact. This can be supported by mythical images, especially visual images. Visual images work at an emotional level and have the power to shift attitudes. It is attitudes that inform decisions.

How can this be taken forwards? It is worth running a session in the office with senior management and all those involved in selling. Divide the people into teams, give each a different advertisement taken from a colour magazine and get them to analyse every message in the advertisement, written and visual, in terms of its categorisation.

Get each team to present their findings and analyse which of the adverts works best and why at the end. Finally, take a critical look at a sample of your own literature in this light.

Creating the message

The question about analysing messages has been answered, yet it does not help us in deciding upon the content. It is here that some general guidelines are needed. To be more precise, it is necessary to determine some key ideas that sum up the overall communication process and will inform the nature of individual statements. Some *generic messages* or *images* are needed.

Many organisations have generated mission statements. These can be quite helpful, although they tend to be good slogans, which help to attract attention rather than tell the recipient much about the nature of the organisation or service. A few mission statements-cum-slogans of European contractors are provided as examples in ... *On the Case 42*.

One way to start is with the market position. The market position can be used to generate key messages (see Chapter 1 and Figure 1.5). A few examples are provided as a starting point:

- *A type market position* – routinised-paternal/practice orientated
  - Tried and tested solutions
  - We are builders through and through
- *D type market position* – analytical-corporate orientated
  - A problem-solving approach
  - Seekers of commercial solutions
  - Drawing upon our wide experience and technical expertise
- *E type market position* – innovative-paternal/practice orientated
  - Creating pioneering designs
  - Using cutting-edge technologies.

Even the flavour provided above demonstrates that the output is likely to be a little bland, limited in scope and more difficult to differentiate once competitors have caught up. This can be supplemented from the findings of a SWOT analysis. SWOT analysis strengths are useful.

There are other ways of generating images and messages of a general and generic nature. One way is to use images of the organisation. Gareth Morgan in his book *Images of Organization* has generated a series of 'metaphors' or analogies in order to understand the nature and actions of management.[1] These analogies provide excellent raw material for the types of messages and imagery needed to convey the organisation approach and style. They are reflections of culture and structure and characterise the types of processes a client will encounter when dealing with that organisation.

... On the case 42

Examples of slogans of endorsement or mission statements of leading European contractors are listed below:

- From building to communication
- Built on teamwork
- To be a truly global engineering and construction company
- Building confidence
- From master builder to system leader
- Right from the start
- Building on strength
- Working together.

They give a hint of emphasis. Mainland European contractors tend towards content, whereas the UK contractors tend towards pure slogan. It should be appreciated that more dynamic statements may also be available, although these are not always available for broad public consumption. The slogan type of emphasis is rather vague in content and the recipient is left without clarity as to how this is applied to the business. The applications may be drawn from the following:

- Informing the entire business strategy
- Emphasising the major service
- Reflecting a structural solution to a marketing issue
- Public promotion
- A sales pitch.

The list used should not be viewed as exclusive. Others may be appropriate for different organisations. The metaphors used are:

- Machines
- Organisms
- Brains
- Political systems
- Psychic prisons

- ❑ Flux and transformation
- ❑ Instruments of domination.

Taking three of these images, the type of messages that can be created are sketched out, relating them back to relevant issues raised in the book.

The *mechanistic metaphor* is the traditional hierarchical image of the organisation. It reflects the origins of corporate management, being derived from military command, which, along with the railway companies of the last century with their bureaucratic models, provided the basis for the modern organisation. The military metaphor is top down, and thus the client can be assured that such an organisation will have senior management that will control the type of service delivered. It will be precise, yet probably not flexible; efficient, yet not always effective. It is most suited to routinised types of services. The sales and PR messages can take on a military flavour.

The *brain analogy* is at the other end of the spectrum. It is highly flexible and ideal at reflecting the holistic service. It embraces aspects of cybernetics, information processing as well as aspects of holographic systems. For example, if a piece of a hologram is broken off, what is left will reproduce the whole picture of the hologram. The brain metaphor is ideal with reproducing messages for the learning organisation. The hologram is one analogy for relationship marketing. It is three dimensional, yet frequently very intangible to the recipient. It attracts the client attention and is tailored to them, yet its applications may be limited. The hologram has tremendous potential for science and the arts, yet its potential is still unrealised. There is a sense of excitement about the potential, which can be turned to advantage.

The *political analogy* is close to the experience of many people working within contracting organisations. In this book, I have developed imagery to convey something of the way things currently are, namely:

- ❑ The *personality culture* of contracting, often based upon blame, reflecting the lack of corporate systems and personal support
- ❑ The *street-fighting man* of gang politics, which is about power at the expense of others, is essentially based upon fear.

These are essentially negative images to create the perception of need and possibility of improving. There are positive political images. Being chosen in a democratic sense is useful in communicating how clients chose your organisation. Another positive message is to do with being sensitive to situations, and successfully responding to these. Sensitivity to need is countered by action sometimes being froth rather than substance. Being seen to act rather than to be taking appropriate action is the downside. Empowerment policies can be usefully linked to the political analogy, as can stakeholder concepts.

There are other analogies outside those provided by Morgan that can be used. The *orchestra* is an interesting one, in that it has a leader, but each section is trying to perform its own function. It is both top down in that the conductor engenders activity, yet bottom up in that the conductor tries to bring out the best in each individual section.[2] It is facilitative, encouraging, celebrating individualism as an integral part of teamwork.

This process of image generation can be conducted within your organisation. The advantage of this approach is that it:

- ❑ Promotes greater accuracy of organisational images
- ❑ Creates variety between organisations and aids the differentiation process
- ❑ Permits those involved in sales to be guided and not constrained in their sales efforts.

Generating images and messages is a strategic issue for subsequent implementation through PR and sales. There are several steps:

1. Explore two or three potential images of the organisation and evaluate the degree of fit in terms of where the organisation currently is and where it is heading
2. Review your market position, SWOT and other key strategic issues to realistically assess the current position
3. Generate an image for your organisation using an organisational analogy and use this as a basis for developing *generic messages* for sales and PR purposes
4. Generate several key generic messages derived from the image or analogy, which operate at a general level and are memorable, so that management and all those involved in sales can picture the overall image and remember the key generic messages.
5. Feed into training and staff induction programmes.

The messages should not be used in a constantly repetitive way, but as examples for staff to express in a way with which they feel comfortable and preserve the thrust and integrity of the message. These types of messages can be used not only in direct selling and PR, but also in all promotions:

- ❑ Corporate identity
- ❑ Branding
- ❑ Advertising
- ❑ Promotional literature
- ❑ Newsletters
- ❑ Journal articles
- ❑ Presentation documents.

In Chapter 5, it was shown that the full armoury of sales promotion was not being used and that there is greater scope to do so effectively. It was also shown that there are differences in the media used by those involved in construction (see Table 5.7). There are disparities between contractor and client perceptions concerning the importance of various media.

Some caution is needed in interpreting the findings in Tables 5.7 and 13.1 below into other situations. First, Hong Kong may have its own cultural factors. Second, the ranking is partly a reflection of what is already done. For example, if you do not expect to hear an advert for a contractor on the radio, then you may rank it low in the list, which is not the same as saying that radio could not be influential. What the above permits is the ability of organisations to review their promotion strategy in a twofold way:

- ❑ More closely matching promotion to client ranking where appropriate
- ❑ Selecting the media that most closely matches the image being conveyed.

This is a balancing act and some compromises may be required. It may be decided, in the light of this analysis and the findings in

Table 13.1 Rankings of media used in promotion in Hong Kong

Media	Client's ranking (HK)	Contractor's ranking (HK)	Contractor's ranking (UK)
Professional journal	1	1	1 (with Magazine)
Magazine	2	4	
Direct mail	3	5	–
Newspaper	4	3	2
Directory entries	5	7	–
Hoardings and outdoor signs	6	2	–
TV	7	6	3 (with Radio)
Radio	8	8	

Source: Yung and Fellows (1996); Preece *et al.* (1997).[3]

Chapter 5, that there is a real need to step up the media presence. There are two good reasons for so doing and a realistic assessment must be made as to whether the promotion will work in the required timescale. The two reasons are:

- ❑ To increase market share or seek more profitable work
- ❑ To prepare the organisation for a take-over or to be taken over by a competitor.

One note of final caution when generating images and messages. Discussion and decisions can easily slide towards an internal focus with internal power coming to the fore. Remember the following:

1. Focus upon the target audience
2. Be certain the targets are interested in your services
3. Understand the benefits they need
4. Use testimonies
5. Gain their attention
6. Motivate the targets to respond or act
7. Assure the targets you deliver your promises; assured satisfaction!

All this is not possible in every communication; however, each target should pick up those seven aspects in communication to them – the pincer movement is useful in this context (see Figure 13.2).[4]

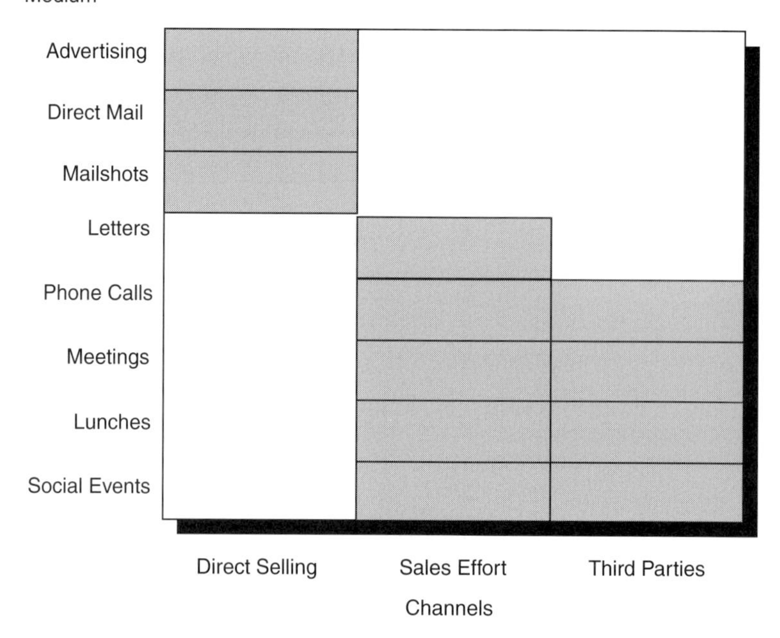

Figure 13.2 Direct communication recipe framework.

Gaining attention is an important issue. There can be a large gap between gaining attention and motivating the audience to respond or act. This aspect can be further broken down:

- Gain attention – decide what to say
  - Make an appeal
  - Develop a theme
  - Present an idea
- Hold interest – decide how to say it logically
  - Appeal to motivations of DMU
  - Appeal to quality, economy and value
- Arouse desire – say it symbolically
  - Performance satisfaction
  - Appearance satisfaction
  - Overcoming shame or victim mentality
  - Avoiding blame
  - Creating joy
- Elicit action – decide who should respond in the DMU.[4]

This breakdown shows the value of using the symbolism of key generic sales messages. They need not only be used to inform communication, but can be the heart of it as well. Most analogies have attractive imagery that can be harnessed to good effect. Avoid being emotional in your tone, although you are aiming for an emotional response in the recipient. Testimonials and project examples that tell a story are always very effective in this respect. This is pictured in Figure 13.3.

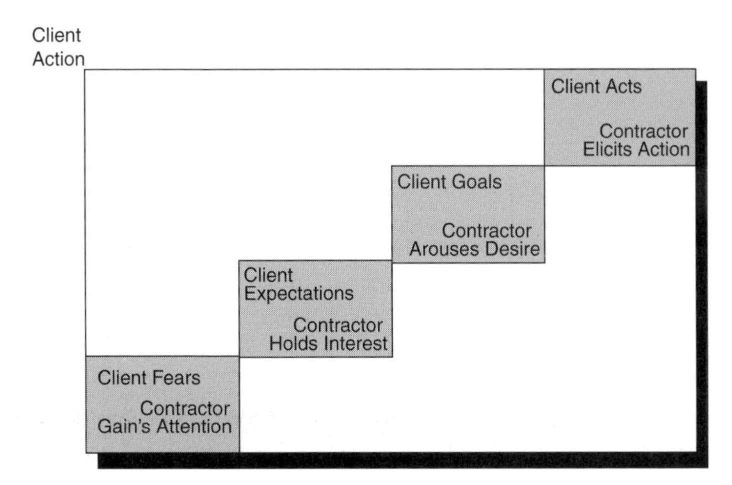

Figure 13.3 Eliciting client action.

Gaining a positive response

How far should an organisation go in structuring its key generic messages and specific statements? It has been suggested that the more an organisation structures its communication, the more compressed the information becomes and the easier it is to communicate through diverse media quickly and effectively.[5] It also makes the information more and more difficult for the recipient to interpret. The messages become more *codified* or jargon-like. This is fine if targeting a niche market, where jargon can be used to build barriers of entry through a veil of expertise.[6] The professions have practised this for generations. If the target audience is more diverse, jargon is to be avoided. However, avoid the bland approach to promotion, such as is found in much brochure production.[7]

Organisational change is often a good time to redress the communications issue. Indeed, it can help people in the organisation to focus on what change means to their public, rather than focus on what may go wrong internally. The law of sowing and reaping is important here – what you sow towards others is what you get from others!

Conclusion

Developing positive generic messages, derived from analogies for organisations opens up exciting possibilities for further differentiating services in the marketplace and communicating the nature of that differentiation effectively to target audiences, whether they are the stock market, clients or in referral markets.

It is important to realise that effective communication to the external environment supports the face-to-face contact of those involved in selling and helps to establish credibility. It also actively sells the organisational services and can be capitalised by the sales people as they build upon the generic messages in the most effective way. This is an area where more attention can be focused in the future and the chapter has endeavoured to provide pointers for such effort.

Summary

1. The purpose of this chapter has been to demonstrate that selling is:

- Informed by key generic sales messages
- More effective using such sales messages.

2. The chapter has:
   - Set out ways of creating generic sales images and messages, using analogies
   - Shown how these can be used to generate sales statements and specific content of communication to the external environment.
3. It is important to gear the content so that an action or response is elicited from the target audiences. This response may initially be purely emotional or subjective until precise action is required to put the contractor or consultant in the frame for a proposed project.

References and notes

1. Morgan, G. (1986) *Images of Organization*, Sage, London.
2. Drucker, P.F. (1986) *The Frontiers of Management*, Heinenmann, London.
3. Yung, F. and Fellows, R. F. (1996) Construction firms' and clients' attitudes towards advertising in the Hong Kong construction industry, *Asia Pacific Building and Construction Management Journal*, **2**(1), 25–32.
   Preece, C.N., Moodley, K. and Humphrey, J. (1997) Effective media relations strategies for civil engineering and building contractors, *Proceedings of the 2nd National Construction Marketing Conference*, 3 July, Centre for Construction Marketing in association with the CIMCIG, Oxford Brookes University, Oxford.
4. Smyth, H.J., Branch, R.F. and McIlveen, A. (1995) *Developing Unique Selling Propositions in Fragile Construction Markets*, Centre for Construction Marketing, Oxford Brookes University, Oxford.
5. Boisot, M. (1986) Markets and hierarchies in a cultural perspective, *Organisation Studies*, **7**, 135–158.
6. Blois, K. (1997) Are business-to-business relationships inherently unstable? *Journal of Marketing Management*, **13**, 367–382.
7. Preece, C.N., Putsman, A. and Walker, K. (1996) Satisfying the client through a more effective marketing approach in contracting – a case study, *Proceedings of the 1st National Construction Marketing Conference*, 4 July, The Centre for Construction Marketing in association with CIMCIG, Oxford Brookes University, Oxford.
   Preece, C.N., Moodley, K. and Smith, A.M. (1998) *Corporate Communications in Construction: Public Relations Strategies for Successful Business and Projects*, Blackwell Science, Oxford.

Chapter 14 Sales Monitoring

Themes

1. The *aim* of this chapter is to address the methods and means of sales monitoring.
2. The *objectives* of this chapter are to consider:
   - ❑ The purpose of evaluation
   - ❑ The process of evaluation
   - ❑ How evaluation relates to sales personnel policy.
3. The primary *outcome* of this chapter is to link sales evaluation with personal support for and evaluation of all staff involved in selling.

Keywords

Sales targets, Sales evaluation, Rewarding sales effort

Introduction

A US comedian said there are three phases to North American history: the passing of the buffalo, the passing of the Indian and the passing of the buck. To those in sales, it often seems that way: when the company is doing well then it's a reflection on the organisation, when turnover and profits increase then it is the product of good management and when turnover falls it must be the fault of the sales person. Somehow the *market* does not feature in this analysis.

Incentive schemes and pay structures are frequently rewarding sales output, when those in sales may only contribute a portion, although a significant one, to the outcome. The need is to select a sales evaluation system that meets the marketing plan and sales objectives and motivates the staff to contribute to the objectives in accordance with their role.

In Chapter 2, it was seen that there were a series of key interfaces:

- ❑ The market and marketing
- ❑ Sales and the market
- ❑ Sales and marketing
- ❑ Sales and the project
- ❑ The project and marketing.

Figure 3.1 shows how these relate to each other. The evaluation system must encompass those elements. In overall terms, the evaluation of sales will differ according to the market position adopted by the organisation. This is set out in Figure 14.1.

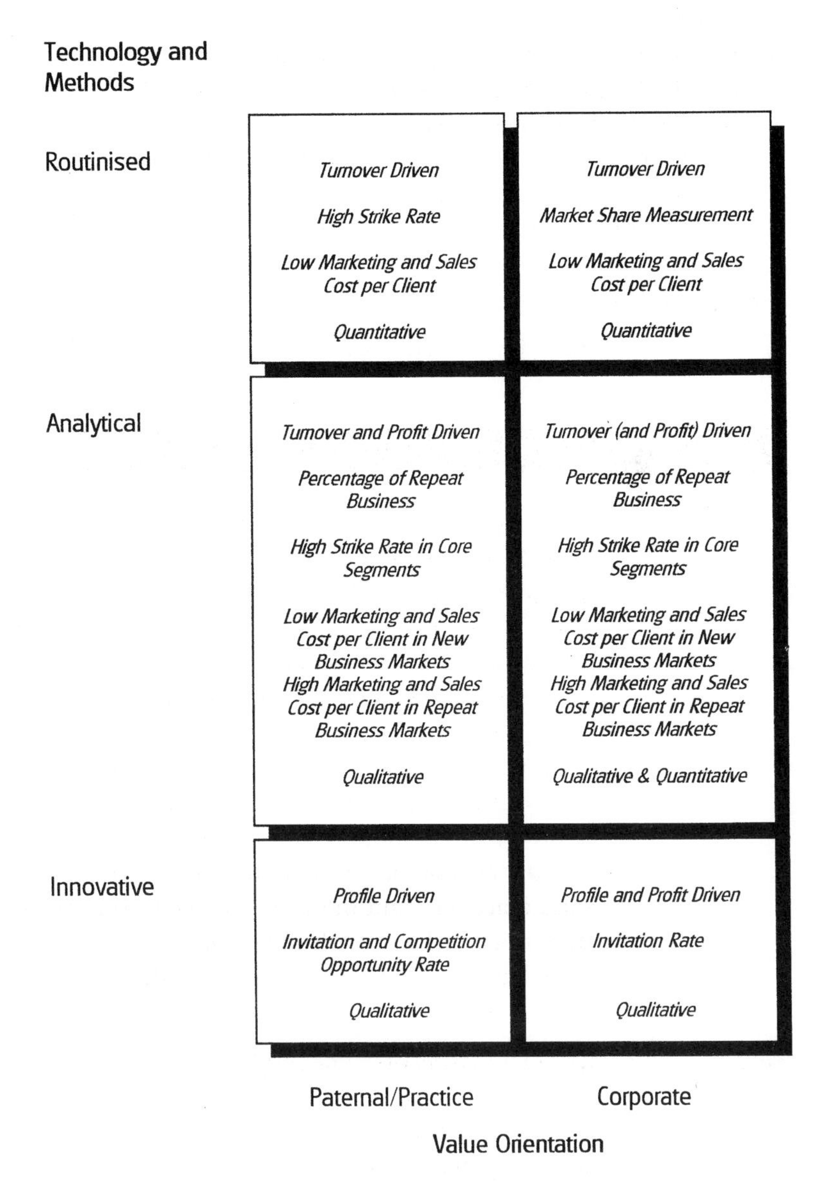

Figure 14.1 Sales evaluation features of market positions.

At every level, the method of evaluation must be appropriate – it must be SMART:

- Specific
- Measurable
- Actionable
- Realistic
- Timed.

Consistency and regularity are important. In contracting, this is the key to being specific. Patterns over time are often better at illustrating the specifics than snapshots, which may not always be clear and their significance uncertain. Selecting the best measure is important, yet secondary. The measures may not be as precise as are needed ideally and hence may be unspecific. Measurement is not always quantitative. Targets are interpreted as both quantitative and qualitative. Evaluation for sales will take place at several levels. These will be driven by the factors shown in Figure 14.1 and can be summarised as:

- Organisation targets (business and marketing plan)
- Sales targets for segments and niches
- Account-handler targets
- Satisfaction targets.

The first two levels are normal in the sense that it is expected practice in well-managed consultant and contracting organisations. The account handler level may seem more of an anomaly to some organisations. There is a tradition in sales generally for high rewards to be given for pulling in *new* business. This makes little sense when it tends to underreward the retention of existing clients.[1] As noted several times in this book, it costs five times more to find a new client than it does to keep an existing one. It is keeping an existing client that might be better rewarded. Monitoring the scope for repeat business and evaluating how close a relationship has become is essential.

Organisation targets

What are key measures at this level? They will vary according to market position (see Figure 14.1 for an overview). Some of the key measurements are:

- Turnover generated
  - Value of projects won
  - Number of tendered or negotiated bids
  - Number of projects won
- Sales effort
  - Number of courted clients
  - Number of pre-qualifications of courted clients
  - Intensity of competition
  - Named competition
  - Named competitor's market advantage
- Overall marketing and sales cost per sale
  - Direct marketing and sales personnel costs
  - Estimated and time sheet costs of other personnel
  - Other indirect costs, such as pre-qualification documentation and speculative work
- Working capital requirements
- Profit generation:
  - Anticipated profit derived from tender or negotiated contracts
  - Actual project profit (related back to anticipated profit).

How may an organisation use these measures to understand the dynamics of the marketplace, especially from the senior management perspective (see Figure 1.3)?

Most organisations use the following as part of the staple diet. They should be seen in sales and marketing terms, as much as in share terms.

$$\text{Anticipated profit margin} = \frac{\text{Anticipated profit}}{\text{Value of projects won}} \times 100\%$$

$$\text{Actual project profit margin} = \frac{\text{Actual project profit}}{\text{Project turnover}} \times 100\%$$

The second measure is used to meet the requirements of shareholders, statutory requirements as well as internal concerns. The first is a useful measure as well. It is a useful predictor for the latter and if it is not, then there is either a problem with the management, especially project management, or the external environment has been misjudged and costs have risen. These measures help to map market dynamics.

The measure below also helps to map market dynamics. This and other measures, when plotted as a time series in graph or histogram format, are valuable indicators of market change. Comparing across the time series aids management in taking policy decisions, for example market improvement, while average tender margins are static, may lead to a directive to increase margins, other things being equal.

$$\text{Average tender profit margin} = \frac{\text{Tender profit margin}}{\text{Number of tender or negotiated bids}} \times 100\%$$

Management also needs to be aware of how this current situation has been reached. Mapping from the stage of door opening, through courting, pre-qualification and tendering is important. Senior management frequently lacks an appreciation of the expectations they have of sales people. The first few measures try to capture the relation of sales effort – full and part-time sales people – to the turnover requirements of the organisation.

$$\begin{array}{c}\text{Turnover generation}\\ \text{per full-time equivalent}\\ \text{sales staff (actual or targets)}\end{array} = \frac{\begin{array}{c}\text{Number of full-time}\\ \text{equivalent sales staff}\end{array}}{\text{Turnover (actual or targets)}}$$

$$\begin{array}{c}\text{Number of clients per}\\ \text{full-time equivalent sales staff}\end{array} = \frac{\begin{array}{c}\text{Number of full-time}\\ \text{equivalent sales staff}\end{array}}{\begin{array}{c}\text{Number of clients}\\ \text{being courted}\end{array}}$$

From this it is useful to calculate the amount of time allocated and spent on selling by all those involved in this from senior management to the most junior sales person.

$$\begin{array}{c}\text{Number of clients per}\\ \text{full-time equivalent sales staff}\end{array} = \frac{\begin{array}{c}\text{Number of clients per}\\ \text{full-time equivalent}\\ \text{sales staff}\end{array}}{\begin{array}{c}\text{Time (spent or allocated}\\ \text{in time budget)}\end{array}}$$

It is important that the part-time sales people include senior management, so that they can judge the investment in effort, time and resources for all levels of selling. The next stage is assessing what the current or projected sales effort is or will realistically be expected to yield.

$$\text{Pre-qualifying success} = \frac{\text{Number of pre-qualifications}}{\text{Number of courted clients}} \times 100\%$$

$$\text{Conversion rate} = \frac{\text{Number of tender opportunities}}{\text{Number of pre-qualifications of courted clients}} \times 100\%$$

$$\text{Overall strike rate} = \frac{\text{Number of leads and enquiries}}{\text{Number of projects won}} \times 100\%$$

$$\text{Project strike rate} = \frac{\text{Number of projects won}}{\text{Number of tender or negotiated bids}} \times 100\%$$

The level of repeat business has been cited as extremely important. It is important that this is not calculated by job number, otherwise the dividing of additions to existing projects will grow further than is already regular practice. It should be a reflection of substantial work, won on the basis of performance.

$$\text{Repeat business} = \frac{\text{Number or value of repeat orders}}{\text{Number or value of projects}} \times 100\%$$

Marketing and sales costs are important too. Given the importance of repeat business in this equation, the calculation for this is included at this point. Total marketing and sales costs should be included, not just departmental, that is part-time marketing and sales costs. It is accounting policy as to whether pre-qualification and other presentation costs are allocated against projects or marketing and sales.

$$\text{Budget and expenditure levels} = \frac{\text{Total marketing and sales costs}}{\text{Turnover}} \times 100\%$$

$$\begin{array}{c}\text{Budget and expenditure levels}\\ \text{per client or project}\end{array} = \frac{\text{Total marketing and sales costs}}{\text{Number of clients or projects}} \times 100\%$$

$$\begin{array}{c}\text{Budget and expenditure}\\ \text{levels on repeat business}\end{array} = \frac{\begin{array}{c}\text{Repeat business}\\ \text{marketing and sales costs}\end{array}}{\text{Value of repeat business}} \times 100\%$$

The last figure can be compared with the residual figure, which, calculated in the same way, will set the costs for new business and produce a ratio for repeat to new business.

$$\begin{array}{c}\text{Budget and expenditure}\\ \text{levels on new business}\end{array} = \frac{\begin{array}{c}\text{New marketing}\\ \text{and sales costs}\end{array}}{\text{Value of new business}} \times 100\%$$

$$\text{Ratio of repeat business to new business costs} = \frac{\text{Budget and expenditure levels on repeat business}}{\text{Budget and expenditure levels on new business}}$$

A neglected area is monitoring what the competition is doing, especially at a specific level. Sympathy exists here, for the market is highly fragmented and therefore exacerbated by a lack of data. However, these obstacles can be overcome. Large contractors and consultant can focus on their markets and others can define their markets in terms of segments and geographical areas. Telephoning around clients, architects and engineers is not hard and can be seen as part of keeping the organisation in the frame for future work. It shows interest.

$$\text{Intensity of competition} = \frac{\text{Total number on tender lists}}{\text{Total number of tenders}}$$

$$\text{Named competition} = \frac{\text{Number of contracts won by named competitor}}{\text{Total number of tenders bid by your organisation}}$$

$$\text{Named competitor's market advantage} = \frac{\text{Number of contracts won by named competitor}}{\text{Sum of rank order on tender list for your organisation across total number of tenders bid against the named contractor}}$$

A number of contractors operate sophisticated systems against these scores (see ... *On the Case 43*).[2]

A final figure of great value concerns working capital. Construction and consulting are essentially cash generators and survival depends upon cash flow. Expansion contains cash-flow risks. Efforts to expand sales must therefore be set against the capacity of the organisation.

$$\text{Working capital demands} = \frac{\text{Total working capital requirements}}{\text{Turnover generated}} \times 100\%$$

$$\text{Project risk concerning working capital} = \frac{\text{Project value}}{\text{Total working capital requirements}} \times 100\%$$

It has been emphasised that time series are important. Some of the

... On the case 43

A regional civil and building contractor that has an established market niche, which consistently yields above average margins, has developed a score sheet to monitor elements of its business against those of its competitors.

The data has been compared to a golf-style score sheet. In other words, the overall comparison against the competition is taken down to a detailed level.

The contractor has gone one step further. It is currently measuring its performance against performance achieved by leaders in other sectors, such as retailing and petrochemicals.

Consequently, performance marks have dropped. This is a reflection of the poor levels achieved in construction compared to other sectors, although the organisation claims it is still a leading performer amongst its peers.

outputs only become meaningful over a long period, especially beyond the need for tactical responses. Longer time horizons are used for marketing plan purposes.

How often should this type of data be collated? To some extent, the analysis of data will vary according to the market position of the organisation. There is little difficulty in collating the data with modern technology. The important issues are:

- Deciding to collect the information
- Having the time to analyse the data
- Acting upon the analysis.

Some data may be analysed monthly, particularly the more global data. Most should be analysed quarterly. All of the data should be reviewed in preparation for determining the annual business and marketing plans and setting new targets for the next year. It will also be used to refine strategy and tactics.

Segment sales evaluation

Many of the calculations undertaken above will need to be undertaken for each area of the business:

- Departments and divisions offering particular procurement routes (structural market segments)
- Procurement options, such as partnering or design and build
- Segments
- Niches
- Geographical areas of operation
- Core or key clients, such as those being sold services based upon relationship marketing.

This provides a more detailed level of market conditions in each market of operation.

The breakdown of these figures should be set alongside the score or priority given to each potential and existing client. Each priority was calculated for each client, giving a score as follows (see Chapter 2):

$$\text{Score} = \text{Weighting} \times \text{Rating}$$

It was advocated that time budgets and spend were allocated to clients, especially key or core clients. This should be monitored. Short-term fluctuations will occur, yet the trend over several months will be apparent. Adjustments between areas of under and overspend can be made.

Other qualitative factors must be weighed. The most important points concern answering the question, 'How close to targets is the organisation?' The closer, the more likelihood of a direct hit! This is especially crucial when using relationship marketing. For each client, monitoring must take place to establish:

- DMU mapping
  - Is every member of the DMU identified?
  - Has contact been established with each one?
  - Has ownership of the contact and reporting back been implemented?
- DMU profiling
  - Has the role and power base of every member of the DMU been identified?
  - Have their motivations been established?
  - Has their team characteristic been identified?

- Conditions of trust
  - What importance does the client place on conditions of trust?
  - What conditions of trust have been established?
  - What conditions need to be improved?
- What investment is the client making in the relationship?
  - What is the level of investment and commitment?
  - To what extent is investment with individuals?
  - To what extent is investment with the organisation?

It is possible to map some of these factors and rate them within a matrix (see Figure 14.2). The overall rating gives an indication of the value of the client and specifically the *relationship value*.

Care needs to be taken if this method is employed. It should not be used or confused with the client score, rating or weighting. What is rated is not the potential of the client to the organisation, but the success of the organisation in getting close to the client. What is very useful is the discipline of undertaking it. Once past the first couple of stages, there will almost certainly be a number of people involved and thus a rating will require bringing all

Client Name: ....

Priority Score: ....

	Door Opening	Courting	Pre-qualification	Tender	Contract
DMU Map	-2 to +2	etc			
DMU Profile	etc				
Conditions of Trust					
Relationship Investment					
Expectations					
Satisfaction					

Figure 14.2 Relationship rating matrix.

those involved together in order to rate the progress. This provides the forum for discussing the potential client fully and soliciting feedback on a specific and generic basis.

Some of the additional key measurements at the level of individual markets are:

$$\text{Organisation benefit of sale} = \text{Average size of sale} + (\text{Overall strike rate} \times \text{Total marketing and sales costs})$$

This is beneficial at a project level. Calculating the value of a client is also important. This is shown in Figure 14.3. Ensuring that the client value is worthwhile is important, especially if pursuing relationship marketing. What is more difficult to assess is the investment that the client is making into the relationship. However, there will be qualitative data about this factor. Investment by the client will also translate indirectly into the cost equation, for client management costs will go down and 'reduced costs' above the line will go up (see Figure 14.3). If these costs seem high to the

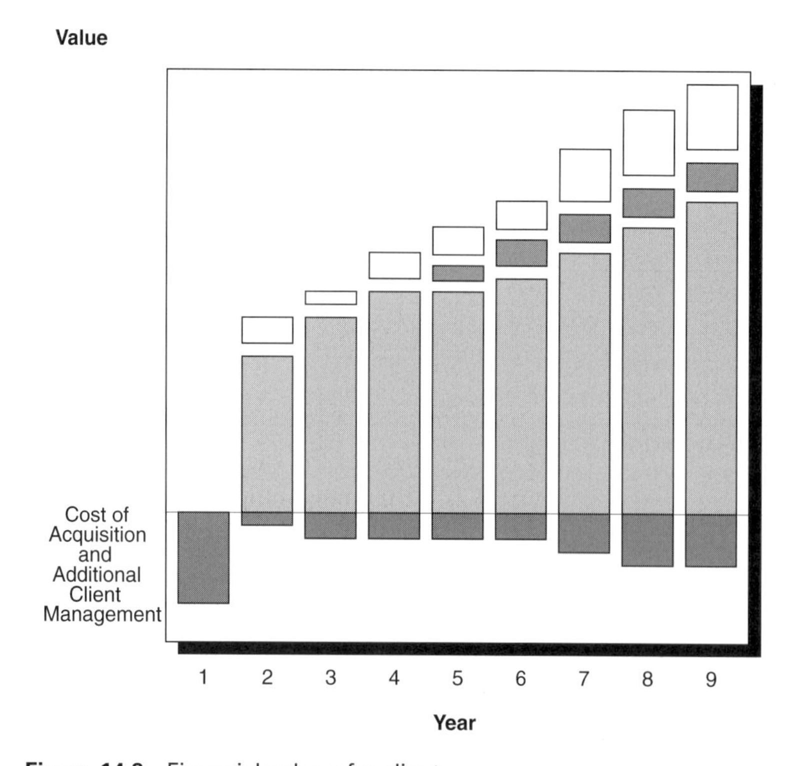

Figure 14.3 Financial value of a client.

organisation, then it may be the case that the client is not investing in the relationship. This is particularly important where there are formal partnering agreements (see Chapter 12).

This concludes the first part of monitoring of the second market view (see Figure 1.3), the second part being the marketing aspects concerned with delivery. Investment by the client is closely associated with this. The next level is about client handling.

Account monitoring

Monitoring how the handling of a potential or existing account is progressing is important in any situation where there is reliance upon a client. This can be for several reasons:

- A large project
- A series of concurrent projects
- A large influence in referral markets
- A large programme of work, with repeat business potential.

It is especially important to monitor account handling in the relationship marketing approach to selling. This is the case whether the importance of the relationship is informal or formalised through a partnering or similar arrangement.

What is monitored in these situations? In Chapter 4, several levels for monitoring were cited. These are:

- Strategic evaluation: a control function of monitoring targets
- Strategic actions: communication to solicit feedback
- Tactical processes: monitor targeting, empower sales, contract and site staff to create and deliver promises (cf. Table 4.1).

The management of the client using a framework is valuable (see Figure 6.2). At the beginning of the top line is *service quality*. The most 'easily identifiable' intangible of service quality is added value. The monitoring starts with the creation and promise of added value at the sales stage. The content of added value has to be identified at this stage in order to assess:

- Whether it matches client requirements
- Whether it is to be overtly sold or not
- How it can be delivered in practice.

Monitoring added value is therefore one of the most important facets of inputs. Added value is summed up as:

$$\text{Added value} = \text{Knowledge} + \text{Expertise} + \text{Relationship value}$$
$$= \text{Service quality}$$

How does one monitor this? This can be set out in a matrix format too (see Figure 14.4). How does the monitoring relate back to formulating the added value. The components, as listed above, will link through from strengths, identified in a SWOT analysis.

One advantage of this monitoring process is that an extended briefing document is prepared that contributes towards informing all parties across the functional roles of the organisation exactly what is trying to be achieved. This is important in all cases, but of particular importance in the relay team approach to account handling.

On the cost side of the management framework, from a sales point of view, it is the critical events or 'moments of truth', which are the easiest to measure and record (see Figure 6.2). Again, a simple matrix is helpful (see Figure 14.5; cf. Table 8.1 which covers critical events across a range of projects).

Client Name: ....

Project: ....

	Door Opening	Courting	Pre-qualification	Tender	Contract
Added Value 1	Description	Description	Description	Those Responsible	Delivery
Added Value 2	etc				
Added Value 3					
Added Value 4					
Added Value 5					
Added Value 6					

Figure 14.4 Record of added value.

Client Name: ....

Project: ....

	Door Opening	Courting	Pre-qualification	Tender	Contract
Critical Event 1	Description, Responsibility Reconciliation	etc			
Critical Event 2	etc				
Critical Event 3					
Critical Event 4					
Critical Event 5					
Critical Event 6					

Figure 14.5 Record of critical events.

Feedback of generic messages, as well as specific client and project actions, are critical to updating and refining the marketing plan as well as tactical client and project responses.

These are the sales monitoring procedures. All of this becomes part of the system for effective client management, as well as part of the sales management system (see Chapters 3, 5 and 6). However, the performance of the process in terms of service delivery must ultimately rest with the client.

Evaluating satisfaction

Whatever the subjective, and therefore sometimes unfair, evaluation of the contractor or consultant may be, it is nonetheless the basis on which decisions are made. Therefore, the client must be the 'best', if sometimes flawed, evaluator of service satisfaction. There need to be comprehensive performance measures, which indicate the level of satisfaction of the client.

Clients are motivated as follows:

$$\text{Motivation} = \text{Desire} + \text{Expectation}$$

Their commitment to the consultant or contractor depends upon how they perceive the overall outcome:

$$\text{Commitment} = \text{Implementation} + \text{Results}$$

It will be the levels of commitment, maintained by delivering the promises, which maximise the potential for repeat business. As was stressed in Chapter 6, this is not automatic, yet appropriate management helps to increase certainty (see Figure 6.2).

Marketing performance measures are still currently in their infancy,[3] but are evolving fast, some already incorporating a high degree of sophistication.[4] External consultants conduct a majority of periodic audits.[5] Of those establishing performance measures in-house, most have tended to focus upon outcomes alone (see ... *On the case 44*).[3,4] This approach is in line with expectations of industry,[6] but because the drive originates largely from industry, the marketing angle is overlooked or underplayed.[7]

In addition to outcomes, expectations require measurement. In a sense, this is the difference between a performance audit of site operations and an audit that embraces marketing.[4,7] Therefore, performance goes beyond the criteria of:

- Cost
- Programme
- Quality of work.

Performance measurement can be structured around the management framework (see Figure 6.2). There are some areas that are easier to act upon than others:

- Priority level 1
  - *Service quality* – the top line of the framework, which concerns commitment
  - *Perceived alternatives* – the top line of the framework, which concerns commitment
  - *Perceived sacrifice* – the bottom line of the framework, which concerns cost
  - *Critical events* – the bottom line of the framework, which concerns cost
- Priority level 2
  - *Client commitment* – the top line of the framework, which concerns commitment, especially changed attitudes

... On the case 44

An international contracting and aggregate conglomerate developed a performance audit system for the building side of its business. This commenced with the appointment of external consultants, who undertook a market research exercise, in order to:

- Establish the range of factors that influence the client perspective
- Establish how clients assess contractor performance
- Understand and quantify client expectations
- Benchmark current performance.

Several stages were involved in the initial market research:

1. A total of 16 face-to-face interviews
2. A total of 649 telephone interviews on a ratio of 1:5 existing clients to potential clients
3. Establishing 19 core measurement areas
4. Piloting the 19 core measurement areas
5. Refining the measurement areas
6. Providing a basis for setting up the internal performance measurement system (PMS).

This process took 18 months. The PMS audits are conducted as follows:

- Four weeks from start on-site
- Six to nine months into the contract
- Four weeks after handover.

The information is analysed centrally to measure the divisional performance. Remedial action is left to the discretion of the contract team.

- *Patronage concentration* – the top line of the framework, which concerns commitment, especially changed attitudes
- *Exit barriers* – the bottom line of the framework, which concerns cost and which might include structural change

- *Event configuration* – the bottom line of the framework, which concerns cost and which might include structural change

Whose satisfaction will be measured? This will vary from project to project. The client will remain constant. A more inclusive list could be drawn upon the following, depending upon the nature of the business:

- Instigator
  - Client
- Direct stakeholders
  - Design team
  - Funding institutions
  - Aid agencies
  - Central government
  - Municipality
  - Tenants and community groups
- Indirect stakeholders
  - Community groups and adjacent users
  - Training services and trainees
  - Schools education programme recipients.

Performance measurement is time consuming, so caution is needed as to how deep into the stakeholders it is taken.

It is important to compare expectations with the contract and post-contract stages. The first audit of performance establishes these expectations, although these may be moderated at times and recorded in subsequent audits. The audit stages could be formed as follows:

1. Four weeks prior to commencing on site or as soon as contract negotiations are completed
2. Four weeks after commencing on site
3. At six-to-nine-month intervals during the project
4. Four weeks after handover
5. One year after completion.

At each stage, clients are asked to score each factor on a scale of one to five, five scoring high, as to the importance of the factor to them – their expectations. This is the *weighting*. They also score each factor in terms of the experience being received from the contractor in that area. This is the *rating*. This is a now a familiar approach, used here for simple evaluation of client satisfaction.

A performance measure for each factor is produced as follows:

$$\text{Performance measurement} = \text{Weighting} \times \text{Rating}$$

In this way, measures can be produced for:

- Comparisons between client expectations and service delivery
- Comparisons of performance over contract duration.

What should be measured? Each organisation will wish to draw up its own criteria. Using the approach outlined in this book, it is suggested that performance measurement could be clustered around four broad areas:

1. Understanding client needs and motivations in the DMU
2. Conditions of trust
3. Client management framework and client handling model
4. Traditional areas: cost, programme and quality of work.

This covers both tangibles, see factor four above, and the intangibles. The criteria within each will relate to strengths and weaknesses of the organisation. Indeed, for core or key clients, specific factors may be added, derived from the added value matrix (see Figure 14.4).

The key to performance measurement is not, of course, carrying it out, but rather it is what is done with the information gathered in order to improve performance. Action needs to be taken as appropriate:

- Adjust service against specific factors – project management
- Improve client service overall – client management
- Apply general lessons across all clients – enhance strengths and overcome weaknesses
- Analyse for areas of potential competitive service advantage.[7]

This action review harks back to and invokes many of the issues covered in previous sections of the book – especially Chapters 1–3, 8–10 and Chapter 12. Actions flowing from the data relies upon having adequate systems in place (see Chapter 3 especially).

Collating information

A collection and collating system provides an important resource. It becomes sub-system within the overall sales system. A database provides up-to-date information, providing procedures are followed (cf. the Paradox of the *Little Black Book Syndrome*, Appendix A – see also Figure 3.2). There are complex interactive systems available today, including relational databases and ones that link through to accounting and other functions (see ... *On the Case 45*).[8]

These specialist software products have been tailored to the project market and based around business-to-business relationships. Yet their primary limitations are that they are developed from products or out of an understanding that is derived from the more traditional sales environment, that is consumer marketing and standardised service packages. The specific demands of contracting and consulting have yet to be fully integrated. The approach below tries to encapsulate some of the ideas raised in this context

... On the case 45

An international contractor uses a specialist database system established for the global market. It enables recording of:

- Contacts
- Leads
- Tenders
- Contracts
- All meetings
- Correspondence
- Diary activity.

Access is available to anyone in the organisation, stores project history and can be linked into purchasing. It is therefore a relational database.

A national contractor, comprising a series of regional and specialist regional contractors has established its marketing database from a propriatary system to track 'client for life relationships'. Its features include:

- Contact register and diary function
- Automatic appointment confirmation
- Action prompts
- Lead tracking and project-to-client event tracking
- Mailshots logged against individual names.

during the book, including the first part of the chapter, in order to provide a skeleton from which a more comprehensive approach can be adopted on an organisation-by-organisation basis. The usual caveat prevails; that is, precise implementation must take into specific market and organisation factors, the success of doing so being part of the competitive edge of the organisation. The framework is provided in Figure 14.6.

It is only a framework. Many of the 'fields' could be broken down into a series of fields. Against projects, under 'details' there are a number of additional fields that could be developed including:

- ❑ Project type, location
- ❑ Segment and niche
- ❑ Extent of competition
- ❑ Project strike rate.

There are also opportunities to create links, for example with added value and critical event matrices (see Figures 14.4 and 14.5). Having a comprehensive system is beneficial, yet this has to be balanced against a realistic assessment as to whether staff have the time and inclination to keep databases up to date. Inclination tends to be lacking, but there is a need to change this situation so that the personality culture is overturned in favour of contractors owning the information through the systems. It is, however, important not to upgrade or impose systems that are too daunting. It is vital to integrate marketing and project management. This is necessary to monitor satisfaction levels. To confirm this need, even under partnering arrangements, it was said by clients in the USA, 40% of all problems concern job management.[9] Having a database that is reasonably comprehensive, designed with an eye to upgrading is necessary. As has been stated in Chapter 3, accounting requires careful procedures and sales do too.

Having a good database is not a system in itself. People run systems. An effective database certainly makes a useful contribution, yet it is also necessary for people to talk to each other. Informal discussions are useful, but cannot be relied upon alone. Formal meetings are needed.

Progress reports

There has been minimal research on this. A pilot study found from seven case studies that there was no overall pattern emerging of

Client 1:	Organisation Name	Contact Address & Numbers			Organisation Details:
Future Projects:	Size and Duration of Programme				Priority Score
	Project 1	Details	Size	Timing	Client Sales Target
	Project 2	Details	Size	Timing	Sales Budget
	Project ...n	Details	Size	Timing	Time Budget
Current Projects:	Project 1	Details	Turnover	Margin	Repeat Business
	Project 2	Details	Turnover	Margin	Value of Client
	Project ...n	Details	Turnover	Margin	
Completed Projects:	Project 1	Details	Value	Profit	Log of Past Contacts: Date, Stage of Relationship, Initiator, Objectives & Outcomes
	Project 2	Details	Value	Profit	
	Project ...n	Details	Value	Profit	
DMU:	Contact Name 1	Position, Contact Address & Numbers			Account Handler
	Power Position in DMU	Motivation			First Contact
	Team Characteristics	Personal Interests, etc			Last Contact
	Line Manager	Head of Function			Interim Action
	Projects involved with and function				Next Contact
	Contact Name 2	Contact Address & Numbers			Owner 2
	Power Position in DMU	Motivation			First Contact
	Team Characteristics	Personal Interests, etc			Last Contact
	Line Manager	Head of Function			Interim Action
	Projects involved with and function				Next Contact
	Contact Name ...n	Contact Address & Numbers			Owner ...n
	Etc.	Etc.			Etc.
Others:	Contact Name 1	Contact Address & Numbers			Owner ...n
	Relationship to DMU	Relationship to Projects			First Contact
	Contact Name ...n	Contact Address & Numbers			Owner ...n
	Etc.	Etc.			Etc.

Figure 14.6 Sales database framework.

the effectiveness and regularity of progress meetings. There was also no discernible pattern emerging among the two main sub-groups – consultants and contractors. There appeared to be a great variety in the amount of reporting, it being minimal in one case study and very comprehensive for another.[10] All the organisations were successful, so they were considered to be doing something 'right'. The pilot study showed that organisations with 'D' type market positions of an *analytical approach* appeared to require weekly meetings. Organisations occupying other market positions tended to prefer monthly meetings. This, when coupled with the qualitative data, would suggest that strong management oversight is required. It also accords with the findings of others,[11] which suggest that the *routinised* organisations rely on standard procedures and systems and the *innovative approach* is conducive to a more autocratic management style whereby the senior figures hold the reins throughout.

The *value orientation* was important. Market positions 'A', 'C' and 'E' all have a *paternal/practice value orientation* and these had an emphasis upon monthly meetings at director level. Positions B, D and F had a *corporate orientation*. There was no overall picture; however, market position 'D', that is an *analytical approach* coupled with a *paternal/practice orientation*, appeared to exhibit a very intensive requirement for reporting. Operational reporting at project or sales levels seemed to be mandatory and director level reporting appeared to be a strong additional requirement. Reporting was weekly.[10] While it is impossible to make generalisations, it is possible to state that organisations would benefit from reviewing their reporting procedures on marketing and sales matters.

Rewarding staff

Sales monitoring has an impact upon staff. While the organisation is being evaluated, so are all those involved in selling services. Gauging the effort of all sales staff is complex.

The database provides raw material for evaluating the level and type of contacts (see Figure 14.6). A calculation can be carried out:

$$\text{Number of contacts} = \text{Number of target clients} \times \text{Number of contacts required per client}$$

Evaluating whether the workload is realistic is aided by the following calculation:

$$\text{Size of sales effort in terms of full time equivalent sales staff} = \text{Number of target clients} \times \text{Number of contacts required per client}$$

This calculation can be carried out for each type of client. It can also be carried out for each market segment. Having carried out this analysis, it is possible to build up the total costs of sales:

$$\text{Average cost per contact} = \frac{\text{Number of contacts}}{\text{Full-time sellers (salary + overheads + expenses)} + \text{Proportional part-time sellers (salary + overheads + expenses)}}$$

$$\text{Total contact costs} = \text{Average cost per contact} \times \text{number of contacts}$$

The measures above focus upon required output. It is frequently assumed that providing financial incentives will motivate staff. Staff motivations and incentives can be confused:

- ❑ Motivation is the drive people have to do things
- ❑ Incentives work on the motivations as an encouragement.

It has been shown that most financial incentives are ineffective if carried out on sales-achieved basis. This sort of sales incentive payment encourages secrecy and thus the *little black book syndrome* (see Paradox 9, Appendix A). Evaluating the quality of contacts is the most appropriate way. Any financial rewards should embrace this too, bearing in mind that:

1. Job satisfaction includes job security, the working conditions, remuneration and the nature of the work
2. Regular incentives come to be seen as part of the standard remuneration package, thus eroding their real effect
3. Incentives can only work on the current level of people's motivation
4. Individual aspirations do not necessarily correspond to company mission statements, aspirations and culture
5. Recognition and affiliation can be more powerful reinforcers of motivational drives than money
6. Incentives must be realistic in assessment criteria and pitched at an appropriate level; that is, worth having without demotivating further down the line.

A contacts or relationship marketing perspective for staff evaluation will take into account qualitative data:

- ❑ Depth of knowledge about and secured from potential client
- ❑ Closeness and extent of communication with the potential client.

The aggregate evaluation of the sales performance of a person is therefore a picture of the depth of the relationships achieved across all their targets as well as sales performance. This is absolutely essential otherwise sales efforts will always be geared to short-term turnover rather than to long-term repeat business, where turnover is potentially much larger with lower marketing costs.

How can an organisation move to an approach for evaluation based upon contacts and relationship marketing sales? One way is to estimate current sales and then target sales. Qualitative results can then be weighted alongside sales targets. The sales effort and evaluation of that effort for the current year is set according to where the organisation wants to be in 5 years time. This may have a short-term detrimental effect on sales unless resources are increased because there will be some sacrifice of opportunistic sales. However, it is precisely these sales that are at the expense of the long-term company future. One way of increasing the sales resources without increasing overheads is to begin the transition of all staff towards the part-time seller or marketing roles.

How does this work for the sales staff? In one year, if a sales person reaches their new sales target, it counts towards, say, 50 points for their evaluation. If a sales person secures their repeat business target, they get 20 points towards their evaluation. If they identify the key qualitative data, for example the DMU map of the members, team characteristics and motivations, their ranking in the power structure, they receive five points towards their evaluation, and so on.

This is carried out in a sales policy context of:

- ❑ Highly selective targeting of clients into
  - ■ Core clients with whom close relationships are being sought for high levels of repeat business
  - ■ Experimentation
  - ■ Other targets where leads are pursued for new sales
  - ■ Opportunistic response to enquiries and request for proposals.
- ❑ Setting time budgets for
  - ■ Repeat business marketing (40%)

- Experimentation (10%)
- New sales (30%)
- Other targets (10%)
- Opportunistic responses (10%).

The effects can be illustrated in a histogram, with adjustments being made in subsequent years as to the points scored in sales staff evaluation. The criteria used in this illustration are shown in Table 14.1:

Figure 14.7 shows in parallel the effort *vis-à-vis* sales. It can be seen that there is a time lag between effort and sales coming in, especially repeat business, as the balance between new and repeat business shifts in favour of the latter. At first, there is even a short-term decline in overall sales. This is the lag between the effort and the trade-off concerning lost opportunistic new business. This method of staff evaluation and appraisal feeds into the data requirements of the database. It provides an encouragement for sales staff to manage the process. Indeed, management should not reward staff who verbally demonstrate they have the information, but only reward staff for information recorded. This moves the management of the sales process away from an approach based on a *personality culture* to one based upon a management system. An organisation now owns the sales effort and it does not 'walk out of the door' when a sales person leaves. In return, the sales person becomes recognised for what they do, rewarded accordingly and the beginnings of a genuine career structure can be put in place.

The company has the data in order to amend and adjust its offers to individual clients, amend its strategy derived from gen-

Table 14.1 Staff evaluation of selling progress

Figure 14.7 label	Evaluation criteria		Maximum points
–	New sales		50
–	Core clients:		
	Repeat business	20	30
	Loyalty	10	
A	DMU motivations		5
B	DMU team roles		5
C	DMU profile map		2
D	Identifying corporate needs		2
E	Development programme		2
F	Project needs		2
G	Background information		2

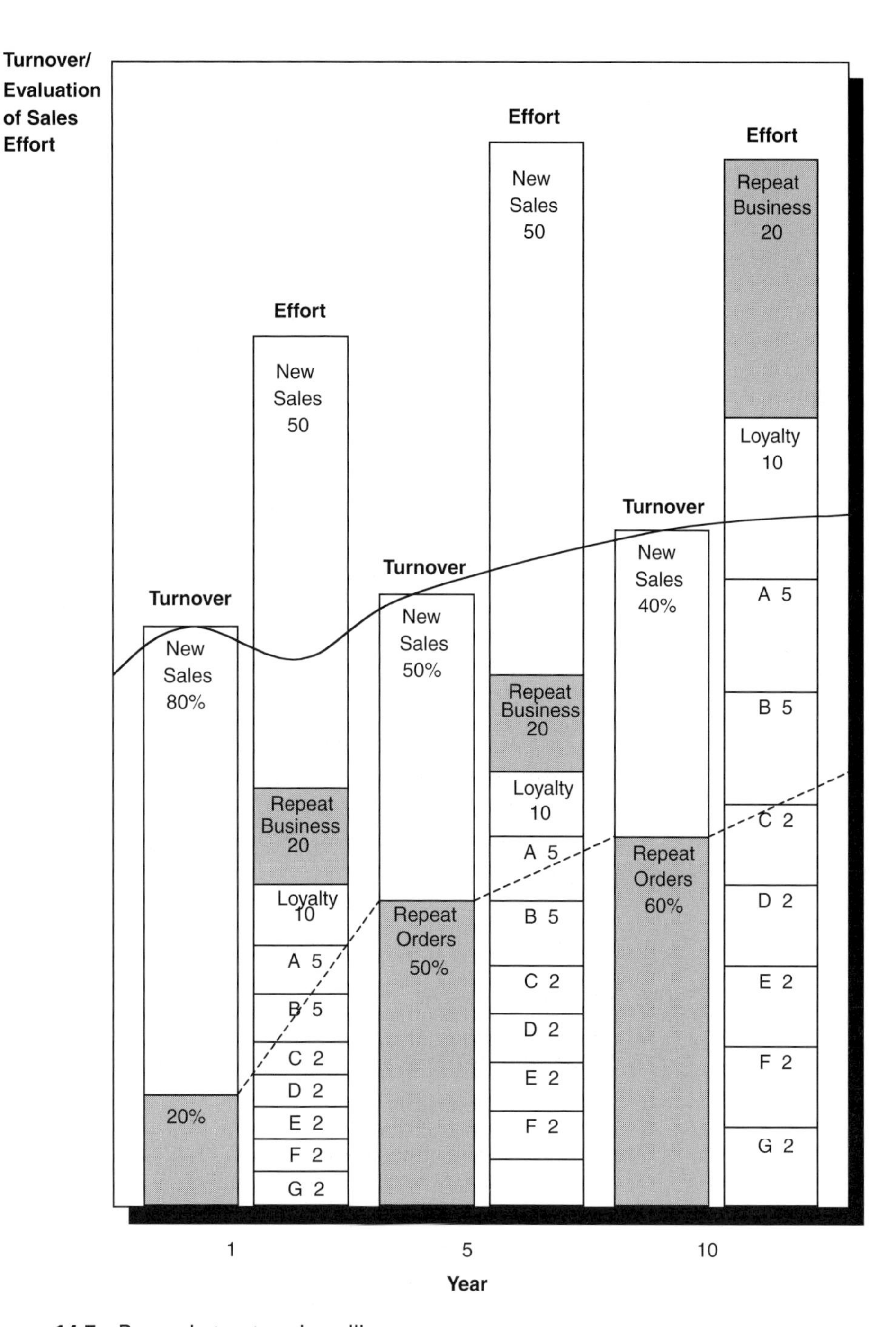

Figure 14.7 Reward structure in selling.

eric lessons from the data feedback, thus having the potential to become a learning organisation in marketing and sales terms.

Staff can also be evaluated according to their sales competencies, as described in Chapter 13. The competencies operate at several levels. The first level looks at moving towards closing a sale:

- Opening doors
- Describe the features, yet sell the benefits
- Overcome objections
- Acknowledge problems, sell solutions
- Turn the person to whom you are selling into an advocate, so they do the selling in the DMU
- Ask for regular update on how you are doing and how long before the sale is closed.

Competencies for developing relationships embrace:

- Focusing upon the target audience
- Being certain that the targets are interested in your services
- Understanding the benefits they need
- Using testimonies
- Gaining their attention
- Motivating the targets to respond or act
- Assuring the targets that you deliver your promises – assured satisfaction!

The third level concerns the competencies, which are operating within the previous two levels:

- Gain attention – decide what to say
  - Make an appeal
  - Develop a theme
  - Present an idea
- Hold interest – decide how to say it logically
  - Appeal to motivations of DMU
  - Appeal to quality, economy and value
- Arouse desire – say it symbolically
  - Performance satisfaction
  - Appearance satisfaction
  - Overcoming shame
  - Avoiding blame
  - Creating joy
- Elicit action – decide who should respond in the DMU.[12]

These are qualitative assessments and are not always easy to gauge. Sales people will still work on their own some of the time; however, the approach set out in this book is more facilitative than controlling and therefore is even more likely to involve other people from time to time.

Conclusion

This chapter has drawn together many of the themes raised in previous chapters and sections. The endeavour has been to address *how* to monitor the sales process. A number of options have been set out and organisations will be selective according to their approach to business and the market. The promise is that there are many actions that can be taken, striking a balance between the generally poor level and methodology of monitoring *and* the requirements to 'get on with the job' in a reasonably unfettered way in a stressful working environment. While a balance must be found, it has been argued that greater sophistication is needed and will in the longer run help to relieve some of the rapid response stresses of opportunistic selling.

Summary

1. The purpose of this chapter has been to demonstrate that selling requires adequate levels of monitoring.
2. Particular consideration has been given to:
   - The more complex the selling process, the greater the levels of monitoring
   - The need to develop different methods of monitoring to oversee and facilitate the sales process.

References and notes

1. Maister, D. (1989) Marketing to existing clients, *Journal of Management Consultancy*, **5**, 25–32.
2. Chevin, D. (1998) The good knight, *Building*, 4 September.
3. Bean, M. (1997) Developing and supporting a trial performance measurement system, *Proceedings of the 2nd National Construction Marketing Conference*, 3 July, Centre for Construction Marketing in association with CIMCIG, Oxford Brookes University, Oxford.

4. Smyth, H.J. (1999) Performance audits and client satisfaction, *Proceedings of the CIB Symposium on Customer Satisfaction*, September, Cape Town.
5. See, for example, Leading Edge (1994) *Capturing Clients in the 90's: A Benchmark Study of Client Preferences and Procurement Routes*, Leading Edge Management Consultancy, Welwyn.
   Buttle, F. (1996) Service quality in the construction industry, *Proceedings of the 1st National Construction Marketing Conference*, 4 July, The Centre for Construction Marketing in association with CIMCIG, Oxford Brookes University, Oxford.
   Preece, C.N., Putsman, A. and Walker, K. (1996) Satisfying the client through a more effective marketing approach in contracting – a case study, *Proceedings of the 1st National Construction Marketing Conference*, 4 July, The Centre for Construction Marketing in association with CIMCIG, Oxford Brookes University, Oxford.
   Pratt, J. (1997) Audit of customer expectations, *Proceedings of the 2nd National Construction Marketing Conference*, 3 July, Centre for Construction Marketing in association with CIMCIG, Oxford Brookes University, Oxford.
6. For example, see Egan Report (1998) *Rethinking Construction*, The Report of the Construction Task Force, Department of the Environment, Transport and the Regions, London.
7. Smyth, H.J. and Thompson, N. (1998) A client orientated service for partnering, unpublished paper, Centre for Construction Marketing, Oxford Brookes University, Oxford.
8. Nawacki, N. (1997) Building an edge, *Construction Computing*, December 1997/January 1998.
   McCormack, S. (1997) Staying ahead of the competition, *Construction Computing*, December 1997/January 1998.
9. Stephenson, R.J. (1996) *Project Partnering for the Design and Construction Industry*, John Wiley, New York.
10. Smyth, H.J. (1998) The internal market monitoring systems in contracting and consulting firms, *Proceedings of the 3rd National Construction Marketing Conference*, 9 July, The Centre for Construction Marketing in association with CIMCIG, Oxford Brookes University, Oxford.
11. Coxe, W., Harting, N.F., Hochberg, H. *et al.* (1987) *Success Strategies for Design Professionals: Superpositioning for Architecture and Engineering Firms*, McGraw-Hill, New York.
12. Smyth, H.J., Branch, R.F. and McIlveen, A. (1995) *Developing Unique Selling Propositions in Fragile Construction Markets*, Centre for Construction Marketing, Oxford Brookes University, Oxford.

Chapter 15

Sales as Market Research

Themes

1. The *aim* of this chapter is to look at the ways and means of conducting market research in construction and consultant organisations.
2. The *objectives* of this chapter are to consider:
   - ❑ Formal ways of obtaining market data
   - ❑ Informal ways of obtaining market data
   - ❑ Integration of the two ways.
3. The primary *outcome* of this chapter is to gain an understanding of the specific ways in which market research can be developed in imaginative and systematic ways appropriate to organisations within the sector.

Keywords

Market research, Four market views, Feedback

Introduction

Market research is a neglected area in construction. Contractors and consultants tend to carry out little market research, at least judged by the conventions of other industries. The reason for this is twofold. One is to do with attitude and approach to business in the sector. The second gives understandable reasons for the reluctance to engage with market research. Contracting and consulting organisations operate at the top end of the supply chain and sector specifics render it difficult to transfer the processes used elsewhere. Yet good market information is important in being effective in selling construction or design services. It is part of the systematic approach to selling advocated throughout the book. It is also a raw material for developing and evolving marketing plans.

Contractors and consultants do, of course, obtain market information and many analyse the findings:[1]

> *In practice, most contracting companies carry out market analysis in some form. The details of the approach will vary from company to company, but in essence three stages are clearly definable:* ***identifying, promoting and satisfying*** *the customer needs at a profit.*

As can be seen from the above quotation from Harris and McCaffer, the main view is that market research and selling are somehow synonymous or at the very least closely linked. Market analysis uses the raw data from research, gleaned in this instance by identifying needs. Market analysis is undertaken through promotion and selling. This is a type of *informal market research*. However, Gummesson[2] states:

> *The knowledge about customers is often based on market research, with a preference for statistical studies as allegedly being more scientific. Yet we know that asking the customer, particularly through structured questionnaires, only reveals a superficial layer of attitudes and behaviour, not the roots. We accept a customer's statement as a phenomenological fact, but the interpretation is loaded with uncertainties.*

We need to be cautious about market research that is quantitative in business-to-business relationships. Sales, therefore, provides excellent market intelligence in the way it has been portrayed in this book. It is part of the research process. There is also scope for more formal research that asks qualitative questions.

Formal market research is usually associated with audits and surveys, conducted mainly in mass consumer markets and by political parties. Indeed, the unrepresentative industry sample that has been referred to in previous chapters adds weight to this: only five people from a response, which represented 133 proposed actions for improving marketing, indicated market research as something to change or improve.[3] Market research is a low priority.

Scope for market research

There are a number of avenues for market research in construction. The largest practices can emulate their contractor counterparts, using a similar, yet less intensive, approach. One of the main reasons for placing market research towards the end of the book is to recognise that construction is located towards the upper end of the supply chain, so selling itself becomes an important part of market research. Another reason is the historical reluctance to embrace the activity and finally, there is a close link between research and monitoring the sales effort in construction.

In Figure I.1 the changing trajectory to marketing and sales is illustrated. The emphasis is towards more strategic support, characterised by higher levels of commitment and investment across the organisation to marketing. One aspect of that effort is market research. Figure 1.3 demonstrated that there were four main views of the marketplace. To be effective in the market requires having the raw material from each of those views in order to make informed decisions and take decisive action. The necessary information from those four views is the data from *market research*. In summary, the four views are:

- Senior management view
- Sales and contract staff
- Client audits
- Formal market research.

In a broad sense, all these views can be seen as the overall market research exercise. The first two views come indirectly into the market research process. That is market data that has been absorbed and interpreted by others in the organisation. Client audits come from the external environment. Audits at the head of the market planning process can be seen as formal market research, but a great deal will be performance measurement, as described in the previous chapter. Formal market research includes specific auditing and survey work, unsolicited data that comes from clients, if it is systematically collected and collated. Figures 1.4 and 2.1 show how the feedback from these processes feed back into market planning. Figure 3.4 categorises market research requirements under three headings:

1. Overall market – for example market surveys
2. Segments – for example focus groups
3. Target clients – for example (potential) client audits.

It can be seen from this diagram that some market research is conducted in a *top-down* fashion, while some is fed back in a *bottom-up* fashion. The *top-down* processes are:

- Business plan objectives for formulating formal market research into the *overall market*
- Business plan objectives for formulating formal market research into *market segments*
- Client audits for formulating formal market research into the *overall market* and *market segments*.

The *bottom-up* processes are:

- Market information from targeted selling, mediated through database and reporting structures, on *market segments* and *market targets*
- Market information from experimental selling, mediated through database and reporting structures, on potential *market segments* and potential *market targets*
- Market information from performance measures on *market segments* and *market targets*.

Whether the process is top down, bottom up or both, it is senior management that will make the final decisions based upon the information. The key to success in this area of market research is therefore to analyse the data. Senior management in particular need market research data because of their partial view of the market without the data and analysis (see Figure 1.3).

Management may concede that they may not always know when a project is going to be initiated, nor the details of the requirements for that client – that is why they employ sales people. They may concede that some of the feedback is useful in refining the plan. However, they may assert that they have a better and more rounded view of the marketplace and an excellent overall perception of what clients want. Yet research does not support this view. While they may have a particular view of the market, it is partial. Investigating maintenance, product repair, securities brokerage, credit card and retail banking companies, Parasuraman *et al.*[4] found that not to be the case, which has been supported in construction by other work.[5] Parasuraman *et al.* found discrepancies regarding executive perceptions of service quality and consumer perception of service quality. These discrepancies can be barriers to delivering a service that clients would perceive as being of high quality.[4]

In summary, the aim is to build bridges between the current situation and the preferred ones for effective selling, using market research. This is illustrated in Figure 15.1.

Methods and practice

What is presented above is a formal approach and system, one that fits into a systematic approach to selling (see Figure 3.2). The same caveat is necessary in a market research context, as has been stated

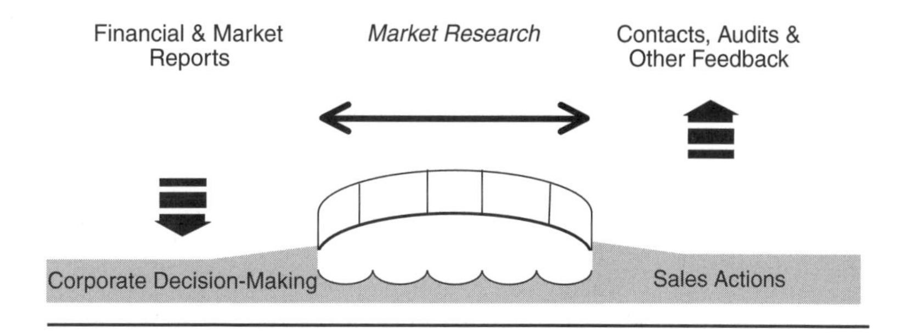

Figure 15.1 The market research as a bridge to improve sales.

for sales systems. If existing or informal ones work, then do not change them without consideration. If they work, then the issue is whether they are sufficiently good to yield competitive advantage. If they fail on this test, then they need upgrading. Upgrading can lead to ineffectual functioning or breakdown, especially if the systems are too onerous.

There are cultural differences as to what constitutes market research. The perception of many in the West is that it concerns consumer and political polls. On the other hand, the Japanese tend to focus upon highly specific matters for market research.[6] The detailed feedback is very valuable, the downside being the danger of missing the overview:[7]

> *Japanese marketing fosters an inductive, bottom-up approach to business planning and problem solving, whereas US and European management favour more deductive, top-down planning methods.*

A combination of top-down and bottom-up approaches is clearly an optimal situation, all four views of the market having the capacity for integration. It is instructive to use a computer organisation to illustrate a more integrated approach. The world's largest software company gears product development to specific targets using survey feedback as to why people purchase packages and the levels of satisfaction experienced. In developing a new package, they tested it with users of a rival across the USA, from which they generated an advertising campaign to say that eight out of ten users of the rival package preferred their new product. This was linked with direct selling and direct mail, with the consequence that there was a massive switch to the new product at the expense of the competition.

Conclusion

The aim of market research is not simply to feed into the development of market research. The main and continuous function is to use the data as the raw material for refining the existing marketing plan and sales process. This keeps the marketing plan dynamic and the sales process relevant in the market and to the needs of the organisation.

This has been a succinct chapter, but it puts in place the missing piece in the jigsaw. In Figures 1.4 and 2.1, two routes to developing and implementing marketing plans were set out. Whichever route is pursued, the end result is a continuous *monitoring* process. A feedback loop is then required to feed in at the top of the cycle. Market research is one important part of that link. It feeds decisions made about marketing and selling by marketing managers and senior management within the organisation.

Market research therefore makes marketing an iterative process. Together with sales monitoring, as set out in Chapter 14, the loop is completed. A break in the chain is induced by one of two factors:

- ❑ Poor implementation
- ❑ Restructuring of the organisation:
  - ■ Marketing level
  - ■ Multi-level.

Summary

1. The purpose of this chapter has been to assert that market research has a key role for consulting and contracting organisations:
   - ■ Completing the feedback loops from the marketplace to strategic decision making
   - ■ Inducing a dynamic marketing and sales process.
2. This chapter has:
   - ■ Considered the sources and components of market research
   - ■ Created a context for improving market research in consulting and contracting organisations.

References and notes

1. Harris, F. and McCaffer, R. (1995) *Modern Construction Management*, 4th edn, Blackwell Science, Oxford, p. 191.
2. Gummesson, E. (1994) Service management: an evaluation and the future, *International Journal of Service Industry Management*, **5**(1), 77–96.
3. The sample is derived from questionnaire responses among delegates attending courses run by the Centre for Construction Marketing, Oxford Brookes University, Oxford, which runs short courses and training events, conferences and publishes research.
4. Parasuraman, A., Zeithaml, V.A. and Berry, L.L. (1985) A conceptual model of service quality and its implications for future research, *Journal of Marketing Management*, **49**, Fall, 41–50.
5. In construction, see Buttle, F. (1996) Service quality in the construction industry, *Proceedings of the 1st National Construction Marketing Conference*, 4 July, The Centre for Construction Marketing in association with CIMCIG, Oxford Brookes University, Oxford.
   In the building materials sector, see Pratt, J. (1997) Audit of customer expectations, *Proceedings of the 2nd National Construction Marketing Conference*, 3 July, Centre for Construction Marketing in association with CIMCIG, Oxford Brookes University, Oxford.
6. Johansson, J.K. and Nonaka, I. (1987) Market research the Japanese way, *Harvard Business Review*, May–June, 16–22.
7. Okamoto, T. and Preece, C.N. (1998) Japanese contractors' overseas job opportunities and the development of more effective international marketing strategies, *Proceedings of the International Construction Marketing Conference*, 26–27 August, University of Leeds, Leeds, p. 272.

Conclusion

This book is a beginning, not an end. It has provided practical, 'hands-on' information on *what* can be done and *how* to do it through providing guidance and illustrating concepts and approaches with a great many case studies.

It has adopted a strategic approach to the investigation of marketing and selling. In this sense, it is a text to stimulate and encourage pioneers. It is hoped that it has stimulated you to want to review all or a range of marketing and sales activities. It is also anticipated that specific improvements can now be made. Yet the main benefit is to absorb the potential for re-evaluating and refining activities over a longer term to render changes in:

- ❑ Attitudes
- ❑ Structures
- ❑ Process.

Indeed, many of the ideas will challenge the way in which organisations are run across the board.

In some organisations a few changes may emanate. For others, a radical change of perspective may result. Yet the opportunities for change are not totally within your grasp, either because you have still to enter industry or are not in a position of power to effect change. A long-term view is needed. The desired outcome is a fundamental shift in the way marketing is conducted across the sector. Those that take forward some of the ideas have the possibility to create competitive advantage. The essence of doing so is to induce:

- ❑ Higher turnover and profit levels
- ❑ Concentration of ownership of the market into fewer hands
- ❑ Erection of barriers to increased competition and sector fragmentation
- ❑ Improved ability to intervene in and manage the market.

All of this may seem to be flying in the face of the market. However, care has been taken throughout the book to consider the processes leading to the desired objectives as operating *within* the market. In most countries, clients are not receiving what they want. So industry has yet to serve clients in a way that goes beyond current expectations. Change incurs costs, yet there are theoretically benefits that flow too. The practical potential and implications have been spelt out. The onus is on each person to take what is perceived as useful to create competitive advantage. The net result should be an improvement in the image of the sector, especially for contractors. The substance may improve, but so too must the sector, in order to encourage increased spend on construction. An improved image induces clients to invest in projects rather than other expenditure with more traditional procurement criteria. In short, the investment in construction must yield higher benefits than other types of expenditure and must be perceived to do this. This hurdle is as much to do with the poor sector image and uncertainty associated with that image.

Two theoretical approaches have been explored in the book. The *marketing mix*, it has been demonstrated has considerable scope for extension and development in contracting and to some extent amongst consultants too. The *relationship marketing* approach is the prime alternative. The emphasis has been upon the *relationship marketing* approach, it being argued that this is the more natural body of theory for business-to-business transactions in the service sector.

The content of the book implies additional investment. The degree of additional investment varies according to the theoretical concepts adopted and the level of commitment being adopted. The investment is short term, until the receipts and profit come through. Without doubt, there are considerable tensions between the short and long term. Expenditure has to be balanced against other requirements. However, the accountancy mentality is that marketing and selling are costs, even though they supply the raw materials for the construction and design team operations. A balance needs to be struck. In a broader sense, this argument, although couched in terms of survival against the competition and thus essential for success, flies in the face of other trends. There is a current emphasis on:

- Decentralised profit centres
- Focusing on core activities
- Delayering management
- Management focusing on change so that they will *not* be accountable
- Individualist performance measures for others being the sole success criteria.

All of these tend towards people and resources being stretched or overstretched, rather than empowered or conserved. The consequences are that people do not have the time to invest in marketing, concentrating upon short-term issues against which they will be evaluated and rewarded and organisations having inadequate resources to support the efforts. The consequence of such outcomes for selling is that it will become more price based and competition even less differentiated. The industry will be more fragmented.

Overall, industry is in danger of creating the anorexic organisation, incapable of reacting to change. Those working within industry will be without character, playing mere roles. The corrosion of the ability to learn, mature and inject substance into a service will make working in the service industry on a par with flow-line production employment. Marketing, and *relationship marketing* in particular, injects the reverse. The marketing investment potentially builds substance into the service. It builds job satisfaction and character into the individual and the organisation. There is a specific need for trust to be one of the things that grows out of such investment. There is a deep-seated need for trust to be established between parties, trust that goes beyond measures spoken of in partnering agreements. Trust is the oil in the wheels of relationship marketing. In so doing, an ethical basis is established in the construction business.

Marketing and selling construction services is about making a difference in theory and practice. The signposts are favourable and it is hoped that they will act as a counterforce to the trends of corporate anorexia and character corrosion that is sweeping across industry. The time for sitting on the fence is over! It is a time for commitment! It is time for change today and preparation for long-term change. It is no longer viable for the long term in construction to be defined as, 'that which never comes'.

Appendix A Paradoxes of Marketing and Sales

The paradoxes of marketing and sales have been pulled out and arranged into an appendix because they highlight problems that managers face and need to address. Addressing paradoxes is uncomfortable. Some can be resolved, others require coping strategies. Facing paradoxes head on is a sign of successful management. They also provide a useful learning, training and equipping medium for using with others.

The paradoxes are grouped in order by chapter, according to when they first appear.

Purpose of the book

1. The paradox of the street-fighting man

The *street-fighting man* style of management – banter and joking at the expense of others, verbal abuse and shouting as the weapons to instil fear and maintain power in the office corridor or in the site hut have an internal focus founded on *fear*. The paradoxical solution is focusing outside oneself, as a focus outside oneself encourages helping others: the client and the staff. A *client orientation* requires understanding and responding to client needs: delivering the goods. The staff is the second category. Help them to overcome their street-fighting mentality both through direct help, such as training, and indirectly through demonstration, so that they too become more eager to serve. Contracting and consultant practice concerns working *with* people, not against them. Marketing has relationship building rather than destruction at its core.

2. The paradox of a client orientation

Most staff have an operational orientation, derived from their decision to work in the sector, reinforced by their training. A *client orientation* has an external focus. It requires an attitude shift before it begins to impact upon behaviour, regardless of the processes and structures that senior management may put in place. Indeed, one of the paradoxes of marketing is that the closer one looks at specific actions geared to help the client, the more it tends to end up with an internal focus of shifting policy, process, structure and attitudes that others hold onto as their security. It reverses the paradox of the *street-fighting man* as the focus moves from the sum of individual staff development to the company development through a *client orientation*.

3. The paradox of the schizophrenic marketing and sales function

Marketing as a function requires a predominantly *cerebral* approach. Selling requires a predominantly *intuitive* approach. In contracting and professional organisations the same person is frequently responsible for co-ordinating marketing strategy and selling. This can lead to a kind of *schizophrenia*. The trend, however, is towards an active integration of marketing and selling functions across a number of sectors. This *schizophrenia* is therefore to be fostered, the *court jester* being a good analogy, which combines strategic intervention and invention *with* tactical and intuitive responses. Integrating marketing and sales has effective management at its heart.

Chapter 1. The business plan and marketing strategy

4. The paradox of the zero-sum game of marketing

Identifying the growth market in the sector may not be the same as identifying the growth market for the organisation. Growth markets in the sector tend to attract all comers and therefore result in high levels of competition, except under boom conditions. This is the *zero-sum game*. It may be more advantageous to stay in markets that match your strength and profile, that are stable, and even, on occasions, that are experiencing decline. The more fragmented the market, the more possible it is to sustain or grow market share in an area of decline.

5. The paradox of risk certainty

Managing the market is designed to create more certainty in an uncertain environment. Service differentiation and diversification, the means by which growth in existing and new markets is achieved, carry risks because both differentiation and diversification themselves create uncertainty and in turn need management.

Chapter 2. From strategic market positioning to selling

6. The paradox of getting the job done

Contractor culture is action. A thoughtful attention to detailed analysis of market specifics, namely market position, segmentation, niches and targets, can be received with impatience or perceived irrelevance to the job of getting work in and completing the projects. Paradoxically, thoughtful attention is essential for getting the jobs that can be done best, to the highest levels of profitability and with client satisfaction.

7. The paradox of international presence

To be a large and successful contractor within one's own country demands being an international contractor and will increasingly demand being a global player.

8. The paradox of accountability bringing freedom

Accountability, as opposed to fearful control, yields greater freedom to sell in a facilitative way. It makes it legitimate for those selling to be imaginative and for this to be monitored. The emphasis has shifted to process rather than merely sales results.

9. The paradox of the 'little black book' syndrome

Although contractors value sales contacts, they hire and fire on the basis of individuals having these, yet attempts for the employer to own this information about contacts in a systematic way is weak – see also Paradox 8.

10. The paradox of the corporate servant

What looks good to the employer may be good enough, yet it may not be the best and so the employer is inadvertently being short-changed. In a sales context, short-term turnover may not yield the greatest long-term turnover and profit. In other words, they tend to look inwards towards 'pleasing' the employers, rather than serving.

Section II. Selling

11. The lifeblood paradox

Sales occupies a low status within the consultant and contracting organisation in both operational and career terms, yet it is the lifeblood of the organisation, feeding work to both the organisation and employees. The consequence of this paradox is that is tends to be inadequately or inappropriately managed.

Chapter 5. Approaches to sales promotion

12. The sales paradox

The sales effort must be greatest during the upturn of the economic cycle. Because of the time-lag in courting and contracts coming through, it is this effort that will reap rewards during the downturn, aid survival compared to the competition and help to build market share for the next upturn. The paradox is that the organisation is putting the greatest time commitment to sales when it is at its busiest looking after existing clients.

13. The paradox of the lighthouse keeper

The *lighthouse keeper* is beaming out their communication warning to ships so they may steer the correct course. Clients want more direct person–person communication today – a hands-on sales approach where needs and issues can be discussed. The paradox is that most people went into contracting and especially the professions because they were attracted to the activity more than to relationship.

Appendix B

Suggested Further Reading

Purpose of the book

Coxe, W. (1983) *Marketing Architectural and Engineering Services*, 2nd edition, van Nostrand Reinhold, New York.

Levitt, T. (1983) *The Marketing Imagination*, The Free Press, Macmillan, New York.

Articles in the *International Journal for Construction Marketing*, http://www.brookes.ac.uk/other/conmark/IJCM/

Proceedings of the *Annual National Construction Marketing Conference* (1996 onwards), Centre for Construction Marketing in association with CIM-CIG, Oxford Brookes University, Oxford.

Proceedings of the *International Construction Marketing Conference*, 26–27 August, University of Leeds, Leeds.

Section I. What is marketing?

Coxe, W., Harting, N.F., Hochberg, H. *et al.* (1987) *Success Strategies for Design Professionals: Superpositioning for Architecture and Engineering Firms*, McGraw-Hill, New York.

Joby, J.S. (1983) *The Railway Builders: Lives and Works of the Victorian Railway Contractors*, David & Charles, London.

Kotler, P., Armstrong, G., Saunders, J. and Wong, V. (1996) *Principles of Marketing*, Prentice Hall, London.

Linder, M. (1994) *Projecting Capitalism: A History of the Internationalization of the Construction Industry*, Greenwood Press, London.

MacDonald, M. (1995) *Marketing Plans*, Butterworth-Heinemann, Oxford.

Smyth, H.J. (1985) *Property Companies and the Construction Industry in Britain*, Cambridge University Press, Cambridge.

Chapter 1. The business plan and marketing strategy

Coxe, W., Harting, N.F., Hochberg, H. *et al.* (1987) *Success Strategies for Design Professionals: Superpositioning for Architecture and Engineering Firms*, McGraw-Hill, New York.

Kotler, P. (1991) *Marketing Management*, Prentice Hall, New York.

Kotler, P., Armstrong, G., Saunders, J. and Wong, V. (1996) *Principles of Marketing*, Prentice Hall, London.

MacDonald, M. (1995) *Marketing Plans*, Butterworth-Heinemann, Oxford.

Simkin, L. (1996) People and processes in marketing planning: the benefits of controlling implementation, *Journal of Marketing Management*, **12**, 375–390.

Smyth, H.J. (1998) The competitive stakes and mistakes: the position of British contractors in Europe, *Proceedings of the 3rd National Construction Marketing Conference*, 9 July, The Centre for Construction Marketing in association with CIMCIG, Oxford Brookes University, Oxford.

Smyth, H.J. and Stockerl, K. (1998) Strategic marketing planning by UK contractors, *Proceedings of the International Construction Marketing Conference*, 26–27 August, Leeds University, Leeds.

Stockerl, K.C. (1997) The importance of strategic marketing planning for the UK construction industry in a changing European business environment, *Proceedings of the 2nd National Construction Marketing Conference*, 3 July, Oxford Brookes University, Oxford.

Valikangas, L. and Lehtinen, U. (1994) Strategic types of services and international markets, *International Journal of Service Industry Management*, **5**, 72–84.

Chapter 2. From strategic market positioning to selling

Bean, M. (1997) Developing and supporting a trial performance measurement system, *Proceedings of the 2nd National Construction Marketing Conference*, 3 July, Centre for Construction Marketing in association with CIMCIG, Oxford Brookes University, Oxford.

Cova, B. (1996) Construction marketing in France: from reaction to anticipation, *Proceedings of the 1st National Construction Marketing Conference*, 4 July, The Centre for Construction Marketing in association with CIMCIG, Oxford Brookes University, Oxford.

Doyle, P., Saunders, J. and Wong, V. (1992) Competition in global markets: a case study of American and Japanese competition in the British market, *Journal of International Business Studies*, **23**, 419–442.

Fellows, R. and Langford, D. (1993) *Marketing and The Construction Client*, CIOB.

Hand, P.W. (1998) Cast the net wide – but use a wide mesh, *Opportunities and Strategies in the Global Marketplace – Proceedings of the 1st International Construction Marketing Conference*, 27–28 August, University of Leeds, Leeds.

Langford, D.A. and Rowland, V.R. (1995) *Managing Overseas Construction Contracting*, Thomas Telford, London.

Ostler, C.H. (1998) Country analysis: its role in the international construction industry's strategic planning procedure, *Opportunities and Strategies in the Global Marketplace – Proceedings of the 1st International Construction Marketing Conference*, 27–28 August, University of Leeds, Leeds.

Payne, A. (1991) *Relationship Marketing: The Six Markets Framework*, Working Paper, Cranfield School of Management, Cranfield.

Porter, M. (1986) *Competition in Global Industries*, Harvard Business School.

Pratt, J. (1998) Re-segmentation – a new route to better margins, *Proceedings of the 3rd National Construction Marketing Conference*, 9 July, The

Centre for Construction Marketing in association with CIMCIG, Oxford Brookes University, Oxford.

Siehler, B. (1998) Different approaches of European contractors to be a global player, *Proceedings of the 3rd National Construction Marketing Conference*, 9 July, The Centre for Construction Marketing in association with CIMCIG, Oxford Brookes University, Oxford.

Smyth, H.J. (1998) *Innovative Ways of Segmenting the Market: Practice Guide No. 1*, Centre for Construction Marketing, Oxford Brookes University, Oxford.

Smyth, H.J. (1998) The competitive stakes and mistakes: the position of British contractors in Europe, *Proceedings of the 3rd National Construction Marketing Conference*, 9 July, The Centre for Construction Marketing in association with CIMCIG, Oxford Brookes University, Oxford.

Smyth, H.J. and Stockerl, K. (1998) Strategic marketing planning by UK contractors in an international business environment, *Proceedings of the International Construction Marketing Conference*, 26–27 August, University of Leeds, Leeds.

Vincent, S. (1998) Selling in Specialist Markets, *Proceedings of the 3rd National Construction Marketing Conference*, 9 July, The Centre for Construction Marketing in association with CIMCIG, Oxford Brookes University, Oxford.

Chapter 3. Sales systems

Ballantyne, D. (1997) Internal networks for internal marketing, *Journal of Marketing Management*, **13**, 343–366.

Ballantyne, D., Christopher, M. and Payne, A. (1995) Improving the quality of services marketing: service (re)design is the critical link, *Journal of Marketing Management*, **2**, 25–28.

Maister, D. (1989) Marketing to existing clients, *Journal of Management Consultancy*, **5**, 25–32.

Markham, C. (1987) *Practical Consulting*, Institute of Chartered Accountants in England and Wales, London, p. 79.

Smyth, H.J. (1985) *Property Companies and the Construction Industry in Britain*, Cambridge University Press, Cambridge.

Chapter 4. Market vehicles

Doyle, P., Saunders, J. and Wong, V. (1992) Competition in global markets: a case study of American and Japanese competition in the British market, *Journal of International Business Studies*, **23**, 419–442.

Langford, D.A. and Rowland, V.R. (1995) *Managing Overseas Construction Contracting*, Thomas Telford, London.

Madsen, T.K. (1989) Successful export marketing management: some empirical evidence, *International Marketing Review*, **6**, 41–57.

Smyth, H.J. (1998) The competitive stakes and mistakes: the position of British Contractors in Europe, *Proceedings of the 3rd National Construction Marketing Conference*, 9 July, The Centre for Construction Marketing in association with CIMCIG, Oxford Brookes University, Oxford.

Stockerl, K.C. (1997) The importance of strategic marketing planning for the UK construction industry in a changing European business environment, *Proceedings of the 2nd National Construction Marketing Conference*, 3 July, Oxford Brookes University, Oxford.

Section II. Selling

Dibb, S. and Wensley, R. (1987) Energy efficient house design: the analysis of customer choice, *Reviewing Effective Research and Good Practice in Marketing* (ed. R. Wensley), Marketing Education Group, Warwick University.

Gummesson, E. (1991) Marketing-orientation revisited: the crucial role of the part-time marketer, *European Journal of Marketing*, **25**(2), 60–75.

Knutt, E. (1997) The scouts, *Building*, 21 February.

Murdoch, A. (1998) Keep your sales team sweet, *Management Today*, January.

Schneider, B. (1994) HRM – a service perspective: towards a customer-focused HRM, *International Journal of Service Industry Management*, **5**(1), 64–76.

Smyth, H.J., Stallwood, P. and Thompson, N. (1997) *The Market for Energy Efficient Homes: Research Report No. 1*, Centre for Construction Marketing, Oxford Brookes University, Oxford.

Chapter 5. Marketing mix and sales promotion

CIOB (1993) *Marketing and the Construction Client*, CIOB, Englemere.

Morgan, R.E. and Morgan, N.A. (1991) An exploratory study of market orientation in the UK consulting engineering profession, *International Journal of Advertising*, **10**, 333–347.

Kotler, P., Armstrong, G., Saunders, J. and Wong, V. (1996) *Principles of Marketing:The European Edition*, Prentice Hall, London.

Cummins, J. (1989) *Sales Promotion: How to Create and Implement Campaigns that Really Work*, Kogan Page, London.

Eldridge, N. and Carvell, P. (1986) *Promoting the Professions: Which Way Do We Go?* Surveyor's Publications, London.

Preece, C.N., Moodley, K. and Smith, A.M. (1998) *Corporate Communications in Construction: Public Relations Strategies for Successful Business and Projects*, Blackwell Science, Oxford.

Wilson, A. (1991) *New Directions in Marketing: Business-to-Business Strategies for the 1990s*, Kogan Page, London.

Chapter 6. Relationships and sales promotion

Gummesson, E. (1990) Making relationship marketing operational, *International Journal of Service Industry Management*, **5**(5), 5–20.

Preece, C.N., Moodley, K. and Smith, A.M. (1998) *Corporate Communications in Construction: Public Relations Strategies for Successful Business and Projects*, Blackwell Science, Oxford.

Storbacka, K., Strandvik, T. and Gronroos, C. (1994) Managing customer relationships for profit: the dynamics of relationship quality, *International Journal of Service Industry Management*, **5**, 21–38.

Chapter 7. Selling the service and product

Akintoye, A. (1994) Design and build: a survey of construction contractor's views, *Construction Management and Economics*, **12**, 155–163.

CIOB (1993) *Marketing and the Construction Client*, CIOB, Englemere.

Connaughton, J.N. (1994) *Value by Competition: A Guide to the Competitive Procurement of Consultancy Services for Construction*, CIRIA, London.

Construction Manager (1997) Why are design & build clients unhappy? *September*, **3**(7), 24–25.

Levitt, T. (1983) After the sale is over, *Havard Business Review*, September–October, 87–93.

Cravens, D.W. and Piercy, N.F. (1994) Relationship marketing and collaborative networks in service organizations, *International Journal of Service Industry Management*, **5**(5), 39–53.

Knutt, E. (1997) The scouts, *Building*, 21 February.

Pearce, P. (1992) *Construction Marketing: A Professional Approach*, Thomas Telford, London.

Preece, C.N., Putsman, A. and Walker, K. (1996) Satisfying the client through a more effective marketing approach in contracting – a case study, *Proceedings of the 1st National Construction Marketing Conference*, 4 July, The Centre for Construction Marketing in association with CIM-CIG, Oxford Brookes University, Oxford.

Smith, P.R. (1993) *Marketing Communications: An Integrated Approach*, Kogan Page, London.

Valikangas, L. and Lehtinen, U. (1994) Strategic types of services and international markets, *International Journal of Service Industry Management*, **5**, 72–84.

Chapter 8. Selling through relationships

Belbin, R.M. (1993) *Team Roles at Work*, Butterworth-Heinemann, Oxford.

Desouza, G. (1992) Designing a customer retention plan, *The Journal of Business Strategy*, **13**(2), 24–28.

Donay, P.M. and Cannon, J.P. (1997) An examination of the nature of trust in buyer–seller relationships, *Journal of Marketing*, **61**, 35–51.

Gronroos, C. (1990) *Service Management and Marketing: Managing the Moments of Truth in Service Competition*, Free Press/Lexington Books, New York.

Gummesson, E. (1994) Service management: an evaluation and the future, *International Journal of Service Industry Management*, **5**(1), 77–96.

Levitt, T. (1983) After the sale is over, *Havard Business Review*, September–October, 87–93.

Levitt, T. (1983) *The Marketing Imagination*, The Free Press, Macmillan, New York.

Madsen, T.K. (1989) Successful export marketing management: some empirical evidence, *International Marketing Review*, **6**, 41–57.

Maister, D. (1989) Marketing to existing clients, *Journal of Management Consultancy*, **5**, 25–32.

Management Today (1998) Keep your sales team sweet, *January*.

Page, M., Pitt, L., Berthon, P. and Money, A. (1996) Analysing customer defections and their effects on corporate performance: the case of IndCo, *Journal of Marketing Management*, **12**, 617–627.

Schurr, P.H. and Ozanne, J.L. (1985) Influences on the exchange process: buyer's preconceptions of a seller's trustworthiness and bargaining toughness, *Journal of Consumer Research*, **11**(4), 939–953.

Senge, P.M. (1992) *The Fifth Discipline: The Art and Practice of the Learning Organisation*, Century Business, London.

Smyth, H.J. (1998) The competitive stakes and mistakes: the position of British contractors in Europe, *Proceedings of the 3rd National Construction Marketing Conference*, 9 July, The Centre for Construction Marketing in association with CIMCIG, Oxford Brookes University, Oxford.

Building (1994) 2 December; (1997) December.

Smyth, H.J. and Stockerl, K. (1998) Strategic marketing planning by UK contractors in an international business environment, *Proceedings of the International Construction Marketing Conference*, 26–27 August, University of Leeds, Leeds.

Stockerl, K.C. (1997) The importance of strategic marketing planning for the UK construction industry in a changing European business environment, *Proceedings of the 2nd National Construction Marketing Conference*, 3 July, The Centre for Construction Marketing in association with CIMCIG, Oxford Brookes University, Oxford.

Storbacka, K., Strandvik, T. and Gronroos, C. (1994) Managing customer relationships for profit: the dynamics of relationship quality, *International Journal of Service Industry Management*, **5**, 21–38.

Williamson, O.E. (1981) Contract analysis: the transaction cost approach, *The Economic Approach to Law* (eds P. Burrows and C.G. Veljanovski), Butterworths, London.

Williamson, O.E. (1985) *The Economic Institutions of Capitalism*, Free Press, New York.

Chapter 9. Selling added value and the product

Akintoye, A. (1994) Design and build: a survey of construction contractor's views, *Construction Management and Economics*, **12**, 155–163.

Blois, K. (1997) Are business-to-business relationships inherently unstable? *Journal of Marketing Management*, **13**, 367–382.

Reichheld, F.A. (1994) Loyalty and the renaissance of marketing, *Marketing Management*, **2**(4), 10–21.

Smyth, H.J. (1996) Design and build marketing: issues and criteria for architecture selection, *Proceedings of the 1st National Construction Marketing Conference*, 4 July, The Centre for Construction Marketing in association with CIMCIG, Oxford Brookes University, Oxford.

Smyth, H.J. and Stockerl, K. (1998) Strategic marketing planning by UK contractors in an international business environment, *Proceedings of the International Construction Marketing Conference*, 26–27 August, University of Leeds, Leeds.

Stockerl, K.C. (1997) The importance of strategic marketing planning for the UK construction industry in a changing European business environment, *Proceedings of the 2nd National Construction Marketing Conference*, 3 July, Oxford Brookes University, Oxford.

Chapter 10. Selling added value and the service

Addis, W. and Al-Ghamdi, M. (1998) The international competitiveness of consulting engineers: the role to be played by technical marketing, *Opportunities and Strategies in the Global Marketplace – Proceedings of the 1st International Construction Marketing Conference*, 27–28 August, University of Leeds, Leeds.

Association of Consulting Engineers (1994) *Client Perception Study*, The Association of Consulting Engineers.

Blois, K. (1997) Are business-to-business relationships inherently unstable? *Journal of Marketing Management*, **13**, 367–382.

Donay, P.M. and Cannon, J.P. (1997) An examination of the nature of trust in buyer–seller relationships, *Journal of Marketing*, **61**, 35–51.

Hall, M.A., Melaine, Y. and Sheath, D.M. (1998) Operations by British contractors during the procurement process in a global and multicultural environment: some recent experiences, *Opportunities and Strategies in the Global Marketplace – Proceedings of the 1st International Construction Marketing Conference*, 27–28 August, University of Leeds, Leeds.

Institute of Civil Engineers (1995) *Wither Civil Engineering*, Institute of Civil Engineers, London.

Leveson, R. and Pickrell, S. (1998) Partner or competitor? Re-framing relationships in construction, *Proceedings of the 3rd National Construction Marketing Conference*, 9 July, The Centre for Construction Marketing in association with CIMCIG, Oxford Brookes University, Oxford.

Smith, P.B. (1998) CARE^{3}S in action: a practical philosophy for customer and client care in the construction and property industry, *Opportunities and Strategies in the Global Marketplace – Proceedings of the 1st International Construction Marketing Conference*, 27–28 August, University of Leeds, Leeds.

Smyth, H.J. and Thompson, N. (1999) Partnering and trust, *Proceedings of the CIB Symposium on Customer Satisfaction*, September, Cape Town.

Storbacka, K., Strandvik, T. and Gronroos, C. (1994) Managing customer relationships for profit: the dynamics of relationship quality, *International Journal of Service Industry Management*, **5**, 21–38.

Thompson, N.J. (1997) Evidence on evidence of trust, *Proceedings of the 2nd National Construction Marketing Conference*, 3 July, The Centre for Construction Marketing in association with CIMCIG, Oxford Brookes University, Oxford.

Thompson, N. (1998) Can clients trust contractors? Conditional, attitudinal and normative influences on client's behaviour, *Proceedings of the 3rd National Construction Marketing Conference*, 9 July, The Centre for Construction Marketing in association with CIMCIG, Oxford Brookes University, Oxford.

Wood, B.R. and Smyth, H.J. (1996) Construction market entry and development: the case of just in time maintenance, *Proceedings of the 1st National Construction Marketing Conference*, 4 July, The Centre for Construction Marketing in association with CIMCIG, Oxford Brookes University, Oxford.

Chapter 11. Selling and the construction project team

Ahmad, I.U. and Sein, M.K. (1997) Construction project teams for TQM: a factor-element impact model, *Construction Management and Economics*, **15**, 457–467.

Buttle, F. (1996) Service quality in the construction industry, *Proceedings of the 1st National Construction Marketing Conference*, 4 July, The Centre for Construction Marketing in association with CIMCIG, Oxford Brookes University, Oxford.

Cherns, A.B. and Bryant, D.T. (1983) Studying the client's role in construction management, *Construction Management and Economics*, **1**, 177–184.

Mohr, J.J., Fisher, R.J. and Nevin, J.R. (1996) Collaborative communication in interfirm relationships: moderating effects of integration and control, *Journal of Marketing*, **60**, 103–115.

Mulder, L. (1997) The importance of a common project management method in the corporate environment, *R&D Management*, 27(3), 189–196.

Smith, J.B. and Barclay, D.W. (1997) The effects of organisational differences and trust on the effectiveness of selling partner relationships, *Journal of Marketing*, **61**, 3–21.

Chapter 12. The client perspective

Bennett, J. and Jayes, S. (1995) *Trusting the Team: The Best Practice Guide to Partnering in Construction*, Centre for Strategic Studies in Construction, University of Reading.

Blois, K. (1997) Are business-to-business relationships inherently unstable? *Journal of Marketing Management*, **13**, 367–382.

Campbell, N. (1995) An interaction approach to organisational buying behaviour, *Relationship Marketing for Competitive Advantage* (eds A. Payne, M. Christopher, M. Clark and H. Peck), Butterworth Heinemann, Oxford.

Donaldson, W. (1996) Industrial marketing relationships and open-to-tender contracts: co-operation or competition? *Journal of Marketing Practice and Applied Marketing Science*, **2**(2), 23–34.

Egan Report (1998) *Rethinking Construction*, The Report of the Construction Task Force, Department of the Environment, Transport and the Regions, London.

Jackson, B.B. (1985) Build customer relationships that last, *Havard Business Review*, November/December, 120–128.

Latham Report (1994) *Constructing the Team*, HMSO, London.

Leading Edge (1994) *Capturing Clients in the 90's: a Benchmark Study of Client Preferences and Procurement Routes*, Leading Edge Management Consultancy, Welwyn.

Reichheld, F.F. (1994) Loyalty and the Renaissance of Marketing, *Marketing Management*, **2**, 10–21.

Smyth, H.J. (1997) Partnering and the problems of low client loyalty incentives, *Proceedings of the 2nd National Construction Marketing Conference*, 3 July, Centre for Construction Marketing in association with CIMCIG, Oxford Brookes University.

Smyth, H.J. and Thompson, N. (1999) Partnering and trust, *Proceedings of the CIB Symposium on Customer Satisfaction*, September, Cape Town.

Stephenson, R.J. (1996) *Project Partnering for the Design and Construction Industry*, John Wiley, New York.

Williamson, O.E. (1981) Contract analysis: the transaction cost approach, *The Economic Approach to Law* (eds P. Burrows and C.G. Veljanovski), Butterworths, London.

Williamson, O.E. (1985) *The Economic Institutions of Capitalism*, Free Press, New York.

Section III : Which sales and evaluation techniques?

Cravens, D.W. and Piercy, N.F. (1994) Relationship marketing and collaborative networks in service organizations, *International Journal of Service Industry Management*, **5**(5), 39–53.

Chapter 13. Sales messages

Blois, K. (1997) Are business-to-business relationships inherently unstable? *Journal of Marketing Management*, **13**, 367–382.
Boisot, M. (1986) Markets and hierarchies in a cultural perspective, *Organisation Studies*, **7**, 135–158.
Morgan, G. (1986) *Images of Organization*, Sage, London.
Preece, C.N., Moodley, K. and Humphrey, J. (1997) Effective media relations strategies for civil engineering and building contractors, *Proceedings of the 2nd National Construction Marketing* Conference, July 3, Centre for Construction Marketing in association with the CIMCIG, Oxford Brookes University, Oxford.
Preece, C.N., Moodley, K. and Smith, A.M. (1998) *Corporate Communications in Construction: Public Relations Strategies for Successful Business and Projects*, Blackwell Science, Oxford.
Preece, C.N., Putsman, A. and Walker, K. (1996) Satisfying the client through a more effective marketing approach in contracting – a case study, *Proceedings of the 1st National Construction Marketing Conference*, 4 July, The Centre for Construction Marketing in association with CIMCIG, Oxford Brookes University, Oxford.
Yung, F. and Fellows, R.F. (1996) Construction firms' and clients' attitudes towards advertising in the Hong Kong construction industry, *Asia Pacific Building and Construction Management Journal*, **2**(1), 25–32.

Chapter 14. Sales monitoring

Bean, M. (1997) Developing and supporting a trial performance measurement system, *Proceedings of the 2nd National Construction Marketing Conference*, 3 July, Centre for Construction Marketing in association with CIMCIG, Oxford Brookes University, Oxford.
Buttle, F. (1996) Service quality in the construction industry, *Proceedings of the 1st National Construction Marketing Conference*, 4 July, The Centre for Construction Marketing in association with CIMCIG, Oxford Brookes University, Oxford.
Leading Edge (1994) *Capturing Clients in the 90's: a Benchmark Study of Client Preferences and Procurement Routes*, Leading Edge Management Consultancy, Welwyn.
Pratt, J. (1997) Audit of customer expectations, *Proceedings of the 2nd National Construction Marketing Conference*, 3 July, Centre for Construction Marketing in association with CIMCIG, Oxford Brookes University, Oxford.
Preece, C.N., Putsman, A. and Walker, K. (1996) Satisfying the client through a more effective marketing approach in contracting – a case study, *Proceedings of the 1st National Construction Marketing Conference*, 4 July, The Centre for Construction Marketing in association with CIMCIG, Oxford Brookes University, Oxford.

Smyth, H.J. (1998) The internal market monitoring systems in contracting and consulting firms, *Proceedings of the 3rd National Construction Marketing Conference*, 9 July, The Centre for Construction Marketing in association with CIMCIG, Oxford Brookes University, Oxford.

Smyth, H.J. (1999) Performance audits and client satisfaction, *Proceedings of the CIB Symposium on Customer Satisfaction*, September, Cape Town.

Chapter 15. Sales as market research

Gummesson, E. (1994) Service management: an evaluation and the future, *International Journal of Service Industry Management*, **5**(1), 77–96.

Johansson, J.K. and Nonaka, I. (1987) Market research the Japanese way, *Harvard Business Review*, May–June, 16–22.

Okamoto, T. and Preece, C.N. (1998) Japanese contractors' overseas job opportunities and the development of more effective international marketing strategies, *Proceedings of the International Construction Marketing Conference*, 26–27 August, University of Leeds, Leeds, p. 272.

Index